BASIC MATRIX
ALGEBRA
WITH
ALGORITHMS AND
APPLICATIONS

CHAPMAN & HALL/CRC MATHEMATICS

BASIC MATRIX ALGEBRA WITH ALGORITHMS AND APPLICATIONS

ROBERT A. LIEBLER

CHAPMAN & HALL/CRC

A CRC Press Company

Boca Raton   London   New York   Washington, D.C.

Library of Congress Cataloging-in-Publication Data

Liebler, Robert.
    Basic matrix algebra with algorithms and applications / Robert Liebler.
        p. cm. — (Chapman & Hall/CRC mathematics)
    Includes index.
    ISBN 1-58488-333-2 (alk. paper)
    1. Matrices. I. Title. II. Series.

QA188 .L49 2002
512.9′434—dc21   2002034768

Visit the CRC Press Web site at www.crcpress.com

© 2003 by Chapman & Hall/CRC

No claim to original U.S. Government works
International Standard Book Number 1-58488-333-2
Library of Congress Card Number 2002034768
Printed in the United States of America   3 4 5 6 7 8 9 0
Printed on acid-free paper

To Roger Soucie
and my many other teachers

Contents

Preface

Linear algebra is recognized to be of central importance in mathematics for both pure mathematical reasons and practical application. The mathematics major must master linear algebra because it stands with Euclidean geometry and calculus and number theory as one of the cornerstones of modern mathematics. In many cases, a technical subject can be regarded as a generalization and/or amplification of some aspects of some of these fundamental subjects.

Computers have allowed researchers in all disciplines to seek new ways to better understand their own subjects. Because matrix methods are very well-suited to automatic computation, and because of its fundamental position in modern mathematics, a wide variety of linear models are now in use by scientists, engineers and bureaucrats around the world.

Yet matrices and linear algebra have not taken a central position in the university mathematics curriculum. This subject is often postponed by majors until their last year, while at the same time portions of the subject appear in courses all across the university, with each presenting just enough to allow their students to work with the linear models currently used in that discipline.

This text is an attempt to begin to rectify this situation. In short, this book aims to be a good beginning. It develops computational and formal facility while presenting a wide variety of applications and developing a solid algorithmic foundation. Abstraction itself is minimized, but an appetite for the abstraction that a proper sequel linear algebra course possesses is encouraged through many specific objectives.

This book is written with two audiences in mind. One audience is the freshman major in a mathematical subject (e.g., mathematics, physics or computer science). Years of experience have taught me that these students need to understand some meaty examples before they can appreciate abstract concepts or theoretical arguments. This book offers such students basic matrix arithmetic, a few nontrivial mathematical algorithms and supporting geometric interpretations in $\mathbb{R}^2$ and $\mathbb{R}^3$, together with several forward references to more advanced mathematical topics both pure and applied.

The second audience is rather more mature, but less technical. These are upper divisional students in, for example, Natural Resources, Social Sciences or Business. They have had a brush with a linear model or two but seek a broader horizon so that they might actually be able to use these methods in new situations. This book offers such students a solid algorithmic foundation upon which they may build improved understanding of the linear models in their fields of study, and a few hints about where to find related mathematics.

Although originally conceived for a course with 24 classes, this book contains enough material for at least 36 class periods. This leaves the instructor with substantial flexibility, but care is required because the book is purposefully small and tightly organized. There is a companion **Instructor's Manual** containing possible course suggestions and other supporting material.

This book is organized into "bite-sized" objectives (appearing as subsections in the Table of Contents), with interrelated examples and "major results/proofs" (each listed in the front matter). It presents the subject in both a formal and a visual way, with numerous figures and diagrams. Work with both a graphing calculator (TI-83) and complex numbers is an integral part of the presentation. All sections end with summaries and many also have brief "looking further" remarks about a broader context with references.

The writing style is informal — touching on history, linguistics, anthropology and even philosophy. Some effort has been made to entice the student into reading more carefully than just the examples, as suffices in most books at this level. The demands on the reader slowly escalate, and formal proofs are more evident in the later chapters. Many of the most difficult concepts are avoided by the emphasis on algorithms and most uniqueness questions come up only in exercises.

An exceptionally wide variety of applications, both "pure" and "applied," are included. These include traditional topics like least squares, changing plane coordinates, Markov chains and Leslie population models as well as some that are relatively unique at this level such as the fast Fourier transform, transfer matrices, principal component matrix compression and integer lattices.

Although it shares many of the philosophical premises of the Linear Algebra Curriculum Study Group Recommendations for the first course in linear algebra, this book is very much an independent project. Major differences are that we

- Emphasize algorithms and recursion at the expense of the basic theory of abstract vector spaces — which does appear in an appendix.
- Use interplay between polynomials and matrices to obtain all of the familiar eigenvalue results without Gram-Schmidt orthogonalization or determinants.
- Integrate complex numbers throughout the text and work with a variety of "coefficient rings" — including the integers.

The emphasis on minimal polynomials and their computation by means of the Krylov algorithm breaks new pedagogical ground. (Determinants are presented principally in their geometric context and as a vehicle for a grand review of the core course.)

This project has been ongoing for more than 15 years. Any success that this project might enjoy must be shared by many people. Special thanks are due to Colorado State University Mathematics Department Heads, R. E. Gaines and H.P. Miranda and to my many thoughtful colleagues, especially K. Klopfenstein. Graduate assistants who have helped sustain and improve this long project include Mark Vincent, Matt Gibbs, Ann Cushman and Elizabeth Scott-Janda. It is a joy to also acknowledge the many students who, through their questions, continue teaching me much needed communication skills. Finally, it is a special pleasure to acknowledge the invaluable encouragement and support of Linda Hamilton.

Robert A. Liebler

Examples

Major results/proofs

Chapter 1

Systems of linear equations and their solution

Mathematics, as you know, is full of symbols, and much of this notation does not appear in other subjects. In this course you will see a lot of new notation. This notation is really a carefully constructed language.

You may not be aware of the importance of notation in mathematics. Good notation is designed to accentuate the significant aspects of the problem being studied and to suppress its insignificant aspects. Because what is significant is sometimes a matter of judgment, our notation (and language) influences our perception.

For example, situations need to be discussed in which a great many variables are present. The exact number of variables n is really not important. What should the variables be named? The usual choice of letters of the alphabet starting with x has the disadvantage that there may not be enough letters. A more serious disadvantage is that you can't even write them all down without implicitly specifying their number. This dilemma is solved with new notation. Then we choose a compound notation and write the variables $x_1, x_2, \ldots, x_n$. The x part of the notation tells you it is (most likely) a variable and the subscript distinguishes the different variables without determining n.

Notice also the ellipsis "$\ldots$". In this course, you will always be safe if you think of "$\ldots$" as shorthand for "and some omitted terms." A consequence of the power of this notion is ambiguity. The exact nature of these terms is not specified and must be determined by context.

This chapter builds on **your** familiarity with elementary functions, analytic geometry, and basic trigonometry to present matrices in a variety of algebraic and geometric settings. The central objects of study are systems of linear equations. After developing a method for solving any "linear system" we develop a geometric understanding of their solution by means of transform plots of linear functions.

Because this subject is very unforgiving of arithmetic errors, we use calculators or computer software. This course requires a graphing calculator that deals with complex numbers and matrices. Occasionally the text includes explicit instructions for the TI-83 calculator but **there are many other brands and models that could be used equally well.**

1.1   Recognizing linear systems and solutions

Objective 1

Given a system of equations and a collection of variables, determine whether the system is linear in the given variables.

An equation is ***linear in the variables*** $x_1, x_2, \ldots, x_n$ if it can be simplified into the ***standard form***:

$$a_1 x_1 + a_2 x_2 + \cdots + a_n x_n = b,$$

where the ***constant term*** b and the ***coefficients*** a_i are numerical constants (or at least do not depend on any x_i). The most general equation that is linear in x, y has the form:

$$ax + by = c$$

where a, b and c DO NOT depend on the variables x, y.

Notice: Any equation is automatically linear in any variable that does not appear explicitly in the equation because zero is an acceptable coefficient.

An equation is called a ***linear equation*** (with no variables specified) if it is linear in the collection of ALL its variables. A ***system of linear equations***, or ***linear system***, in the variables $x_1, x_2, \ldots, x_n$ is a collection of linear equations in the variables $x_1, x_2, \ldots, x_n$.

Example 1.1 (Linear equations and linear systems.)

     a. The equation $-5(3 + 2x - y) + 7x + 3y = 3(x - 2y) + 4$

        is a **linear equation in** x **and** y because it can be algebraically simplified into:

$$-6x + 14y = 19$$

        Also, we can simply say that this is a **linear equation** (and not list any variables at all!) because it is linear in all of its variables.

     b. The equation $3x + 2(y + 4) - y^2 = 5$

        can be algebraically manipulated into:

$$3x + 2y - y^2 = -3.$$

        This equation is linear in x, but not linear in y (or in x, y,) because of the $-y^2$.

     c. The equation $xy = 1$

        is not linear. BUT, it is linear in the variable y alone because each symbol, for example x, is presumed to denote a constant unless it is explicitly identified as a variable. This shows that the variable list is crucial to deciding linearity.

It also shows that an equation may be linear in each of the variables x and y separately but not linear in x, y.

d. The system
$$3x + 2y + 7z = 3$$
$$2x - 4y \qquad = 0$$

is a linear system of two equations in the three variables x, y and z. The second equation can be rewritten in standard form:

$$2x + (-4)y + 0z = 0$$

e. The system
$$3a \qquad + 2cx = \sin \pi/6$$
$$ax^2 + bx + \quad c = \quad y$$

is not linear in the variable x because of the term ax^2. However, this system is linear in the variables a, b, c and y. The list of variables must be known before you can decide a system's linearity. The term "$\sin \pi/6$" is the value of the trigonometric function $\sin \theta$ at $\theta = \pi/6$ and $\sin \pi/6 = 1/2$.

f. The equation
$$\sin^2 x + \cos^2 x = 3$$

is a little tricky. Since $\sin^2 x + \cos^2 x = 1$ is a trigonometric identity, this equation simplifies to $1 = 3$, which is linear but has no solution. □

Linear systems are an important part of mathematics and form the basis of many mathematical models. You have already studied linear systems of two equations in two variables when, in other courses, you computed the intersection of two straight lines in the Cartesian plane (see Example 1.3 on page 5). In this course, larger systems are considered so we cannot rely solely on such geometric interpretation (it would lead directly to higher dimensional geometry for which we have no developed intuition).

Objective 2
Given a system of linear equations in n variables, determine whether a given n-tuple is a solution.

An equation is an *identity* if it can be simplified to the form:

$$0 = 0.$$

The process of simplification is usually one of algebraic manipulation but could also involve the use of other identities as in part f of Example 1.1. An n-tuple $(s_1, s_2, ..., s_n)$ of numbers is a *solution* to a linear equation $a_1 x_1 + a_2 x_2 + \cdots + a_n x_n = b$, if the substitutions s_1 for x_1 , s_2 for x_2, ..., s_n for x_n in the equation result in an identity. An n-tuple is a solution to a linear system if it is simultaneously a solution to each of the system's equations. In this case we also say that

$(s_1, s_2, ..., s_n)$ *satisfies the system*. The number s_i is called the **i-th coordinate** or **i-th entry** of $(s_1, s_2, ..., s_n)$.

Warning: Occasionally the simplification process may involve multiplication by an expression that is zero for certain values of the variables. When this happens, the resulting system can have **extraneous solutions** that are not solutions to the original system.

Example 1.2 (Solutions to linear systems.)

a. The equations
$$50c_1 + c_2 + 7c_3 + c_4 = 0$$
$$40c_1 + 6c_2 + 2c_3 + c_4 = 0$$
$$52c_1 + 4c_2 + 6c_3 + c_4 = 0$$

form a system of 3 linear equations in the 4 unknowns (variables) $c_1, ..., c_4$. The 4-tuple $c_1 = 1, c_2 = -2, c_3 = -4, c_4 = -20$ is a solution to this system because
$$50(1) + (-2) + 7(-4) + (-20) = 0$$
$$40(1) + 6(-2) + 2(-4) + (-20) = 0$$
$$52(1) + 4(-2) + 6(-4) + (-20) = 0.$$

We can express this in several more compact ways:

$(c_1, c_2, c_3, c_4) = (1, -2, -4, -20)$ or
$(c_1 \ c_2 \ c_3 \ c_4) = (1 \ -2 \ -4 \ -20)$ or

$$\begin{pmatrix} c_1 \\ c_2 \\ c_3 \\ c_4 \end{pmatrix} = \begin{pmatrix} 1 \\ -2 \\ -4 \\ -20 \end{pmatrix}.$$

b. Consider
$$3x + 2y + 7z = 3$$
$$2x - 4y = 0.$$

It is a linear system. The indicated 3-tuple is not a solution because the first equation fails $(3(2)+2(1)+7(3) \neq 3)$ even though the second equation holds $(2(2) - 4(1) = 0)$.

$$\begin{pmatrix} x \\ y \\ z \end{pmatrix} = \begin{pmatrix} 2 \\ 1 \\ 3 \end{pmatrix}.$$

This linear system has $x = -15, y = -7.5, z = 9$ as a solution because
$$3(-15) + 2(-7.5) + 7(9) = 3$$
$$2(-15) - 4(-7.5) = 0.$$

It also has $x = 6, y = 3, z = -3$ as a solution (check this). In fact, there are infinitely many distinct numerical solutions to this system.

The major goal of this chapter is to develop a procedure for finding ALL solutions of a given linear system.

c. The linear system
$$x = 1$$
$$y = 2$$
$$z = 3$$
$$w = 4$$

clearly has exactly one solution $(x, y, z, w) = (1, 2, 3, 4)$. The procedure we develop will be so effective that, upon completion, the "general solution" of the original system will be just as obvious as in this "trivial" system. □

Recall from analytic geometry, a single linear equation $ax + by = c$ in two variables x, y has as its solution set a straight line in the Cartesian plane $\mathbb{R}^2$. When you find the intersection of two straight lines, you are actually solving a linear system with two equations and two unknowns. Its solution is the coordinates of the point of intersection of the two lines.

When the two distinct lines are parallel, they do not intersect, and you are dealing with an ***inconsistent*** linear system — that is, a system of linear equations that has no solution.

Example 1.3 (An inconsistent system.)
The solution to the separate equations in the linear system correspond to unequal parallel lines in the Cartesian plane $\mathbb{R}^2$.

$$x + y = 1$$
$$x + y = 2$$

Any solution to the linear system must be a solution to both equations and must therefore correspond to a point on both lines. Since these lines do not intersect, the linear system has no solution and is ***inconsistent***. □

SUMMARY

An equation is ***linear*** in the variables $x_1, x_2, ..., x_n$ if it can be written in the form

$$a_1 x_1 + a_2 x_2 + \cdots + a_n x_n = b,$$

where a_i and b do not depend on any x_i (usually they are numerical constants). A finite collection of ***linear equations*** is called a ***system of linear equations***. An n-tuple $(s_1, ..., s_n)$ is a ***solution*** to a system of linear equations (in n variables or unknowns) if the substitutions s_1 for x_1, s_2 for x_2, ..., s_n for x_n in each of the system's equations result in an ***identity***.

NOTE: The terms that appear in a summary in ***boldface italics*** are defined in this section. You may find their precise location by referring to the index. In addition, **boldface** UPPERCASE is used to mark warnings and for emphasis, as for example this note.

Practice Problems

1. Is it possible for an equation not to be linear in any of its variables? If not, explain why. If so, give an example.

2. Is it possible for an equation having variables r, s and t to be linear in r, linear in s and linear in t (each separately), but not linear in r, s and t (the three together)? If not, explain why. If so, give an example.

3. Is it possible for an equation having variables r, s and t to be linear in r, s and t (the three together) but not linear in r, linear in s and linear in t (each separately)? If not, explain why. If so, give an example.

4. Which of the following equations are linear in the variables x, y and z?
    (i) $2.45x - 64.03y + 28.8z = 13.22$
    (ii) $8x^2 - 12x = 64y + 7$
    (iii) $2z = 15$

5. Which of the following equations are linear in the variables p and q?
    (i) $2(3p - q) + 7 = -2(2p + 7q) - 5p$
    (ii) $p = \frac{13}{q}$
    (iii) $(2p - 4q)(2x + 7y) = 31$

6. Which of the following are linear equations (in all variables)?
    (i) $x + y + z + D = 0$
    (ii) $u + v - \frac{1}{u} - v = 3$
    (iii) $(s + t)^2 = s^2 + 2st + t^2$

7. Which of the following are linear equations?
    (i) $\frac{1}{x} + \frac{1}{xy} = \frac{1}{y}$
    (ii) $(2m + 3)(3n - 5) = 6mn + 5$
    (iii) $(3s + 4t)(2u - 5v) = 23$

8. Which of the following equations are linear in the variables x and y?
    (i) $A^3 x + A^2 y = 1$
    (ii) $(x + z)(y + z) = x(y - 7)$
    (iii) $1 = 2$

9. Identify which 4-tuples are solutions to the system of equations
    $$x + 2y + z = 6$$
    $$x + y + 2z = 4$$
    $$w + x + 5z = 5.$$
    (i) $(w, x, y, z) = (2.4, 1.1, 2.3, 0.3)$
    (ii) $(w, x, y, z) = (2, 0.8, 2.4, 0.4)$
    (iii) $(w, x, y, z) = (0.4, -3.1, 3.7, 1.7)$

10. Identify which 4-tuples are solutions to the system of equations
    $$x_1 - x_2 - x_4 = 1$$
    $$x_2 + 3x_4 = 2.$$
    (i) $(x_1, x_2, x_3, x_4) = (\frac{3}{2}, \frac{1}{2}, \frac{5}{2}, 1)$
    (ii) $(x_1, x_2, x_3, x_4) = (\frac{7}{3}, 1, \frac{7}{3}, \frac{1}{3})$
    (iii) $(x_1, x_2, x_3, x_4) = (\frac{3}{2}, -\frac{1}{4}, \frac{11}{4}, \frac{3}{4})$

11. Identify which 4-tuples are solutions to the indicated system of equations.

$$A \quad \begin{aligned} B + 2C + D &= 4 \\ -2C + 4D &= -1 \\ 3B + 5C + 4D &= 16 \end{aligned}$$

   (i) $A = -1, B = -1, C = 3, D = 1$
   (ii) $A = 3, B = 2, C = 2, D = 0$
   (iii) $A = 3, B = -7, C = 5, D = 3$

12. The Swedish chemist, Alfred Nobel (as in Nobel prize) achieved fame and fortune with his invention of dynamite. Modern dynamite is (carefully) made from TNT (trinitrotoluene), $C_7H_5N_3O_6$, by combining toluene, C_7H_8, with nitric acid, HNO_3. In this chemical reaction, the only byproduct is water, H_2O:
$$aC_7H_8 + bHNO_3 \longrightarrow cC_7H_5N_3O_6 + dH_2O$$
Assume that each atom of each of the elements N, O, C, H in the reactants is present in the products. Find four equations in the unknowns a, b, c, d from the chemical equation. Is the resulting system linear? What can you say about the number of solutions to this system?

13. Is there a three dimensional analog to the figure in Example 1.3? Suppose you had a three dimensional coordinate system with x, y and z axes. What is the shape of the set of solutions to a single linear equation? Geometrically, can you decide what the possible solution sets for a system of two linear equations in three unknowns might be?

$$* * * * * * *$$

1.2   Matrices, equivalence and row operations

Objective 1

Given a system of linear equations with numerical coefficients, write the associated augmented matrix and vice versa.

The names of the variables in a linear system are not important to the nature of the solution. The computations involved in solving a particular linear system only involve arithmetic, but they are lengthy. For this reason, notation that keeps track of only the important data (the coefficients and the constant terms) and suppresses the insignificant data (the names of the variables, the arithmetic operations and the equality signs) is used.

Any linear system of m equations relating n variables can be written:
$$\begin{aligned} a_{11}x_1 + a_{12}x_2 + \cdots + a_{1n}x_n &= b_1 \\ a_{21}x_1 + a_{22}x_2 + \cdots + a_{2n}x_n &= b_2 \\ &\cdots \\ a_{m1}x_1 + a_{m2}x_2 + \cdots + a_{mn}x_n &= b_m \end{aligned}$$

where the coefficients a_{ij} and the constant terms b_i are numerical constants (or at least do not depend on any x_i).

WARNING: The variables must appear in the same order in each of the system's equations before one can proceed.

The important information in this system comes in two parts:

$$\begin{array}{cc}
\text{coefficients} & \text{constants} \\[4pt]
\begin{array}{cccc}
a_{11} & a_{12} & \cdots & a_{1n} \\
a_{21} & a_{22} & \cdots & a_{2n} \\
& \cdots & \\
a_{m1} & a_{m2} & \cdots & a_{mn}
\end{array} &
\begin{array}{c}
b_1 \\
b_2 \\
\vdots \\
b_m.
\end{array}
\end{array}$$

Both of these parts are rectangular arrays of numbers. The first array has m rows and n columns, while the second array has m rows but only one column. When such a rectangular array of numbers is enclosed in parentheses, it is called a ***matrix***. The ***shape*** or ***size*** of a matrix is the number of rows by the number of columns. A matrix having only one column, that is, of shape m-by-1, is sometimes called a ***column vector*** or simply a ***vector***. The ***coefficient matrix*** and the ***constant vector*** of this linear system are

$$\begin{array}{cc}
\text{coefficient matrix} & \text{constant vector} \\[4pt]
\begin{pmatrix}
a_{11} & a_{12} & \cdots & a_{1n} \\
a_{21} & a_{22} & \cdots & a_{2n} \\
& \cdots & \\
a_{m1} & a_{m2} & \cdots & a_{mn}
\end{pmatrix} &
\begin{pmatrix}
b_1 \\
b_2 \\
\vdots \\
b_m
\end{pmatrix}.
\end{array}$$

You may also find matrices with brackets (that is with sharp corners), and when a small matrix is displayed on a TI-83, it will look even more peculiar:

$$\begin{bmatrix}
a_{11} & a_{12} & \cdots & a_{1n} \\
a_{21} & a_{22} & \cdots & a_{2n} \\
& \cdots & \\
a_{m1} & a_{m2} & \cdots & a_{mn}
\end{bmatrix}
\qquad
\begin{array}{l}
[[\; a_{11} \;\; a_{12} \;\cdots\; a_{1n} \;] \\
[\;\; a_{21} \;\; a_{22} \;\cdots\; a_{2n} \;] \\
[\quad\quad \cdots \quad\quad\;\;] \\
[\;\; a_{m1} \;\; a_{m2} \;\cdots\; a_{mn} \;]].
\end{array}$$

If the matrix is too big for the TI-83 display, you will have to use the direction arrows to move your window around the matrix.

The coefficient matrix of this system has shape m-by-n (also written $m \times n$). Because the coefficient matrix and the constant vector are often manipulated simultaneously, it is often convenient to combine them into a single matrix separated by nothing more than a vertical line. The resulting matrix has shape m by $n + 1$ and is called the ***augmented coefficient matrix*** or simply ***augmented matrix*** of the linear system:

$$\left(\begin{array}{cccc|c}
a_{11} & a_{12} & \cdots & a_{1n} & b_1 \\
a_{21} & a_{22} & \cdots & a_{2n} & b_2 \\
& \cdots & \\
a_{m1} & a_{m2} & \cdots & a_{mn} & b_m
\end{array}\right)$$

The entire coefficient matrix of the above system is called (upper case) A because this is the (lower case) letter used for the various coefficients. The constant vector is written with a single letter **b**, or perhaps B for the same reason. The augmented matrix is written $(A|\mathbf{b})$ or perhaps $(A|B)$.

The **entry** of A that appears in its i-th row and j-th column is written a_{ij} is said to be in **position** (ij). For example, if A is the coefficient matrix of the system in part a of Example 1.4 below, then $a_{12} = -2$ and $a_{31} = 1$.

Example 1.4 (Linear systems and augmented matrices.)

a. The indicated linear system has coefficient matrix, constant vector and augmented matrix:

$$\begin{aligned} 3x_1 - 2x_2 + 5x_3 &= 7 \\ 4x_1 + 3x_2 - x_3 &= 0 \\ x_1 - x_2 + 2x_3 &= -3 \end{aligned}$$

$$\begin{pmatrix} 3 & -2 & 5 \\ 4 & 3 & -1 \\ 1 & -1 & 2 \end{pmatrix}, \quad \begin{pmatrix} 7 \\ 0 \\ -3 \end{pmatrix} \text{ and } \left(\begin{array}{ccc|c} 3 & -2 & 5 & 7 \\ 4 & 3 & -1 & 0 \\ 1 & -1 & 2 & -3 \end{array} \right).$$

b. The indicated linear system has coefficient matrix, constant vector and augmented matrix:

$$\begin{aligned} x_1 \qquad + x_3 &= 1 \\ x_2 - x_3 &= -1 \\ 2x_1 + x_2 + x_3 &= 2 \end{aligned}$$

$$\begin{pmatrix} 1 & 0 & 1 \\ 0 & 1 & -1 \\ 2 & 1 & 1 \end{pmatrix}, \quad \begin{pmatrix} 1 \\ -1 \\ 2 \end{pmatrix} \text{ and } \left(\begin{array}{ccc|c} 1 & 0 & 1 & 1 \\ 0 & 1 & -1 & -1 \\ 2 & 1 & 1 & 2 \end{array} \right).$$

c. The indicated linear system has coefficient matrix, constant vector and augmented matrix:

$$\begin{aligned} x_1 \qquad + x_3 &= 1 \\ x_2 - x_3 &= -1 \\ x_2 - x_3 &= 0 \end{aligned}$$

$$\begin{pmatrix} 1 & 0 & 1 \\ 0 & 1 & -1 \\ 0 & 1 & -1 \end{pmatrix}, \quad \begin{pmatrix} 1 \\ -1 \\ 0 \end{pmatrix} \text{ and } \left(\begin{array}{ccc|c} 1 & 0 & 1 & 1 \\ 0 & 1 & -1 & -1 \\ 0 & 1 & -1 & 0 \end{array} \right).$$

d. The indicated linear system has coefficient matrix, constant vector and augmented matrix:

$$\begin{aligned} x_1 \qquad + x_3 &= 1 \\ x_2 - x_3 &= -1 \\ 0 &= 1 \end{aligned}$$

$$\begin{pmatrix} 1 & 0 & 1 \\ 0 & 1 & -1 \\ 0 & 0 & 0 \end{pmatrix}, \quad \begin{pmatrix} 1 \\ -1 \\ 1 \end{pmatrix} \text{ and } \left(\begin{array}{ccc|c} 1 & 0 & 1 & 1 \\ 0 & 1 & -1 & -1 \\ 0 & 0 & 0 & 1 \end{array} \right).$$

(The last equation has no solution, so the system is inconsistent. That is okay. It is still a linear equation because it can be written $0x_1 + 0x_2 + 0x_3 = 1$.) □

A linear system with all constant terms zero $b_1 = 0, \ldots b_m = 0$ is called a **homogeneous system**. Otherwise it is called **nonhomogeneous**. Homogeneous systems play an especially important role in the theory of linear systems. In working with a homogeneous system, it is standard practice to work only with the coefficient matrix, since the augmented matrix would just have a column of zeros and add no information.

Now you see the reason that the variables in all equations after the first must appear in the SAME order as they appear in the first equation. It is because matrix notation suppresses mention of the variables! If two people chose different orderings of the variables in the first equation, then they would obtain **different augmented matrices** (but one augmented matrix could be obtained from the other by rearranging the columns appropriately).

Objective 2

Given a matrix and sequence of elementary row operations, apply the row operations to obtain an equivalent matrix.

Two linear systems having the same number of equations are *equivalent* if they have exactly the same solutions. It is possible for two linear systems to be equivalent but look very different. The key to finding a solution of a complicated linear system is to find an equivalent system whose solution is obvious.

This solving of linear systems involves one of our most important *algorithms*. An algorithm is a step-by-step process that transforms *input* data to *output* in a more desirable form. Algorithms are called recipes in the culinary arts and procedures in the medical profession.

The next section develops the Gauss-Jordan elimination algorithm. It consists of a step-by-step procedure, each step of which is an elementary row operation that changes the augmented matrix so the new associated linear system is equivalent to but slightly simpler than the preceding one. The process ends when the augmented matrix has been transformed into a particularly simple form called Reduced Row Echelon Form (defined on page 17). The general solution of this final simple system (which is necessarily also the general solution of the given system) is easily written.

Parts b,c,d of Example 1.4 on the preceding page illustrate the result of this process. Each has exactly the same solutions – namely no solutions at all – and so they are equivalent. Notice also that this fact is more obvious from inspection of the last system than from the first system.

The basic steps of the Gauss-Jordan elimination algorithm are *elementary row operations.* There are three types of elementary row operations.

1. The elementary row operation that interchanges the i - th and the j - th rows of a matrix A will be denoted by: $(R_i \leftrightarrow R_j)$. The TI-83 calculator command is $rowSwap(A, i, j)$.

2. The elementary row operation that multiplies the i - th row of a matrix A by the real number $t \neq 0$ will be denoted by: $(tR_i \rightarrow R_i)$. The TI-83 calculator command is $*row(t, A, i)$.

3. The elementary row operation that replaces the i - th row of a matrix A with the sum of the i - th row and t times the j - th row $(i \neq j)$ will be denoted by: $(R_i + tR_j \rightarrow R_i)$. The TI-83 calculator command is $*row + (t, A, j, i)$.

We call these the **interchange**, **multiply** and **sum** row operations, respectively. Unfortunately, there is very little agreement concerning notation for these types. We use a rather elaborate, but visually clear, notation.

Example 1.5 (Elementary row operations.)
It is often convenient to use a more pictorial notation:

1.   a. $\begin{pmatrix} 1 & 0 & 1 & | & 0 \\ 0 & 1 & 0 & | & 0 \\ 0 & 0 & 5 & | & 0 \end{pmatrix} \overset{\times}{\longrightarrow} \begin{pmatrix} 0 & 1 & 0 & | & 0 \\ 1 & 0 & 1 & | & 0 \\ 0 & 0 & 5 & | & 0 \end{pmatrix}$ for $(R_1 \leftrightarrow R_2)$, or $rowSwap(A, 1, 2)$

     b. $\begin{pmatrix} 0 & 1 & 0 & 0 \\ 1 & 0 & 1 & 0 \\ 0 & 0 & 5 & 0 \end{pmatrix} \overset{\times}{\longrightarrow} \begin{pmatrix} 1 & 0 & 1 & | & 0 \\ 0 & 1 & 0 & | & 0 \\ 0 & 0 & 5 & | & 0 \end{pmatrix}$ for $(R_2 \leftrightarrow R_1)$ or $rowSwap(A, 2, 1)$

     (The reason parts a and b are the same is given below.)

2.   a. $\begin{pmatrix} 3 & 0 & 3 & | & 0 \\ 0 & 1 & 0 & | & 0 \\ 0 & 0 & 5 & | & 0 \end{pmatrix} \overset{(\frac{1}{3}) \rightarrow}{\longrightarrow} \begin{pmatrix} 1 & 0 & 1 & | & 0 \\ 0 & 1 & 0 & | & 0 \\ 0 & 0 & 5 & | & 0 \end{pmatrix}$ for $(\frac{1}{3}R_1 \rightarrow R_1)$, or $* \, row(1/3, A, 1)$.

     b. $\begin{pmatrix} 1 & 0 & 1 & 0 \\ 0 & 1 & 0 & 0 \\ 0 & 0 & 5 & 0 \end{pmatrix} \overset{(3) \rightarrow}{\longrightarrow} \begin{pmatrix} 3 & 0 & 3 & 0 \\ 0 & 1 & 0 & 0 \\ 0 & 0 & 5 & 0 \end{pmatrix}$ for $(3R_1 \rightarrow R_1)$, or $* \, row(3, A, 1)$.

3.   a. $\begin{pmatrix} 1 & 0 & 1 & | & 1 \\ -2 & 1 & 0 & | & 2 \\ 1 & 0 & 5 & | & 5 \end{pmatrix} \overset{(2)\neg}{\longrightarrow} \begin{pmatrix} 1 & 0 & 1 & | & 1 \\ 0 & 1 & 2 & | & 4 \\ 1 & 0 & 5 & | & 5 \end{pmatrix}$ for $(R_2 + 2R_1 \rightarrow R_2)$, or $* \, row + (2, A, 1, 2)$,

     b. $\begin{pmatrix} 1 & 0 & 1 & 1 \\ 0 & 1 & 2 & 4 \\ 1 & 0 & 5 & 5 \end{pmatrix} \overset{(-2)\neg}{\longrightarrow} \begin{pmatrix} 1 & 0 & 1 & 1 \\ -2 & 1 & 0 & 2 \\ 1 & 0 & 5 & 5 \end{pmatrix}$ for $(R_2 - 2R_1 \rightarrow R_2)$, or $* \, row + (-2, A, 1, 2)$.   □

Notice that the effect of any elementary row operation can be undone by a second *inverse* elementary row operation. Indeed, part b of Example 1.5 gives the inverse elementary row operation to part a.

The different parts of Example 1.4 on page 9 are related by elementary row operations. Check that the augmented matrix in part c is obtained from that in part b by $(R_3 - 2R_1 \rightarrow R_3)$ and that the augmented matrix in part d is obtained from that in part c by $(R_3 - R_2 \rightarrow R_3)$.

Elementary row operations and equivalence of linear systems

We pause here to show that elementary row operations are appropriate for solving systems of linear equations. We compare the solution sets of two linear systems, where the augmented matrix of one system was obtained from the augmented matrix of the other by an elementary row operation and show that the solution sets are identical. It follows that any sequence of elementary row operations will result in an equivalent linear system.

Suppose $(x_1, x_2 \ldots x_n)$ is a solution of the linear system associated with $(A|\mathbf{b})$. If $(A^*|\mathbf{b}^*)$ is obtained by an *interchange* row operation, then $(x_1, x_2 \ldots x_n)$ is certainly a solution of the new system because the equations have just been listed in a different order. In a similar way, the multiply row operation leaves the solution set unchanged because multiplying both sides of an equality by a nonzero number results in an equivalent equality.

The sum operation is a little more interesting. Consider the linear system and augmented matrix $(A|\mathbf{b})$, simultaneously:

$$
\begin{array}{cc}
\text{Linear system} & \text{Augmented matrix} \\[4pt]
\begin{array}{rcrcrcr}
x_1 & & & + & x_3 & = & 1 \\
-2x_1 & + & x_2 & & & = & 2 \\
x_1 & & & + & 5x_3 & = & 5
\end{array}
&
\left(\begin{array}{ccc|c}
1 & 0 & 1 & 1 \\
-2 & 1 & 0 & 2 \\
1 & 0 & 5 & 5
\end{array}\right)
\end{array}
$$

All you need to do is add two times the first equation to the second equation and you get $x_2 + 2x_3 = 4$. This is exactly the equation corresponding to the second row of the new augmented matrix after $(R2 + 2R1 \to R2)$ is applied to the original augmented matrix!

$$
\begin{array}{rcrcrcr}
x_1 & & & + & x_3 & = & 1 \\
& & x_2 & + & 2x_3 & = & 4 \\
x_1 & & & + & 5x_3 & = & 5
\end{array}
\qquad
\left(\begin{array}{ccc|c}
1 & 0 & 1 & 1 \\
0 & 1 & 2 & 4 \\
1 & 0 & 5 & 5
\end{array}\right)
$$

Thus, a sum elementary row operation corresponds to adding a multiple of one equation to another. This too is a system manipulation that doesn't change solution sets.

In summary, elementary row operations are a more efficient way of doing basic arithmetic manipulations with equations. The only real difference is that all the variable names (+'s, −'s and ='s) are omitted. No confusion results because the position of a number in the matrix tells us all we really need to know! For this reason, two matrices (augmented or not) are called ***equivalent*** if one can be obtained from the other by a sequence of elementary row operations.

Calculators.

The vertical line in an augmented matrix is really just a reminder that the left and right parts came from different places and have different meanings. Row and other matrix operations are oblivious to this distinction. Consequently, calculators do not distinguish between regular and augmented matrices.

As this course progresses, it will become more and more important that you work with a calculator. The reason is that the actual arithmetic required to do a problem will grow and grow. Worse still, even the slightest arithmetic error will be catastrophic. **Your best guide for your calculator is its instruction book![1]**

We will from time to time give a few comments that are specific to the TI-83 calculator series. On the original TI-83 calculator, one accesses matrix operations by the $\boxed{\text{MATRIX}}$ $\boxed{\to}$ button. But on the TI-83 plus there is no matrix button. Rather, MATRIX appears in yellow above the $\boxed{x^{-1}}$ button. **If you are working with a TI-83 plus, just substitute the key sequence $\boxed{\text{2nd}}$ $\boxed{x^{-1}}$ for $\boxed{\text{MATRIX}}$ in the following instructions.** For more detailed information, see Chapter 10 of the TI-83 or TI-83 plus manual.

---

[1]The TI-83 manual is available free at http://education.ti.com/product/tech/83/guide/83guideus.html.

Let's step through some row operations using the TI-83 series syntax. As our example we will work through part a of Example 1.7 on page 19.

Begin by entering the starting matrix into one of the matrices in the calculator's memory.

$$\boxed{\text{MATRIX}}\ \boxed{\rightarrow}\ \boxed{\rightarrow}$$

(the $\boxed{\rightarrow}$ key is in the top right-hand corner of the keyboard). At this point the display has EDIT highlighted. We will use the first matrix. It is called $[A]$, so we simply

$$\boxed{\text{ENTER}}.$$

The display will have top line:

$$\text{MATRIX}\ [A]\quad \times$$

and we must fill in the blanks so the calculator knows the shape of A. Our matrix has 3 rows and 4 columns, so we enter

$$\boxed{3}\ \boxed{\text{ENTER}}\ \boxed{4}\ \boxed{\text{ENTER.}}$$

Now we enter the matrix entries row by row: The first row is:

$$\boxed{1}\ \boxed{\text{ENTER}}\ \boxed{0}\ \boxed{\text{ENTER}}\ \boxed{1}\ \boxed{\text{ENTER}}\ \boxed{0}\ \boxed{\text{ENTER}}$$

When you have entered the final entry, the display will look like

$$
\begin{array}{cccc}
MATRIX\,[A] & 3 & \times & 4 \\
\_ & 0 & 1 & 0\ ] \\
\_ & 1 & 0 & 1\ ] \\
\_ & 3 & 2 & 2\ ]
\end{array}
$$

That is because the matrix is too large for the EDIT display, and we see only its last three columns. To finish editing and clear the screen enter $\boxed{\text{QUIT}}$. (Because the label QUIT is in yellow, you actually have to push the key sequence $\boxed{\text{2nd}}\ \boxed{\text{MODE}}$). These keys are in the top left-hand corner of the keyboard.)

We are now ready to do row operations.

$$\boxed{\text{MATRIX}}\ \boxed{\rightarrow}$$

to display the matrix operations menu labeled MATH. Because our first row operation is a sum operation, we $\boxed{\uparrow}\ \boxed{\text{ENTER}}$.

Notice the menu scrolls to the bottom when you are at the top and go up. The display now reads

$$*row + ($$

The syntax of this operation is given in part 3 of Example 1.5 on page 11.

The numerical multiplier we need is -2 and it comes first. There are two keys on the calculator that have minus signs; we want the gray (-) key that is in the bottom row (the blue $\boxed{-}$ key is for subtraction).

$$\boxed{(\text{-})}\ \boxed{2}$$

Now we must enter a comma; it is above the ⬚7⬚. The next information required is the matrix name. To display matrix names

$$\boxed{\cdot}\;\boxed{\text{MATRIX}}.$$

We want matrix $[A]$, so we just

$$\boxed{\text{ENTER}}.$$

We must still enter: comma (multiplied row) comma (target row) right parenthesis. The ⬚)⬚ key is above the ⬚9⬚ on the keyboard.

$$\boxed{\cdot}\;\boxed{1}\;\boxed{\cdot}\;\boxed{3}\;\boxed{)}.$$

The display will wrap to a second line and look like

$$*row + (-2, [A], 1, 3)$$

$$\boxed{\text{ENTER}}$$

and on the display will appear the result of your first calculator row operation!

$$\begin{bmatrix}[\,1\ 0\ 1\ 0\,] \\ [\,0\ 1\ 0\ 1\,] \\ [\,0\ 3\ 0\ 2\,]\end{bmatrix}.$$

The second row operation adds -3 times the second row to the third row.

$$\boxed{\text{MATRIX}}\;\boxed{\rightarrow}\;\boxed{\uparrow}\;\boxed{\text{ENTER}}.$$

to display

$$*row + ($$

Then

$$\boxed{(\text{-})}\;\boxed{3}\;\boxed{\cdot}.$$

Now we must enter the name of the matrix — it is NOT $[A]$ because $[A]$ remains in memory until we change it! The matrix we want is the answer of our last calculation: ⎡ANS⎤ which is also known as ⎡2nd⎤⎡(-)⎤ because ANS is yellow. Therefore, we continue with

$$\boxed{\text{ANS}}\;\boxed{\cdot}\;\boxed{2}\;\boxed{\cdot}\;\boxed{3}\;\boxed{)}\;\boxed{\text{ENTER}}.$$

The display now reads

$$\begin{bmatrix}[\,1\ 0\ 1\ \ \ 0\,] \\ [\,0\ 1\ 0\ \ \ 1\,] \\ [\,0\ 0\ 0\ -1\,]\end{bmatrix}.$$

The final row operation is done by

$$\boxed{\text{MATRIX}}\;\boxed{\rightarrow}\;\boxed{\uparrow}\;\boxed{\uparrow}\;\boxed{\text{ENTER}}.$$

to display

$$*row($$

Then

SUMMARY

Each linear system has an **augmented matrix**

$$
\begin{aligned}
a_{1,1}x_1 + a_{1,2}x_2 + \cdots + a_{1,n}x_n &= b_1 \\
a_{2,1}x_1 + a_{2,2}x_2 + \cdots + a_{2,n}x_n &= b_2 \\
&\cdots \\
a_{m,1}x + a_{m,2}x_2 + \cdots + a_{m,n}x_n &= b_m,
\end{aligned}
\qquad
\left(
\begin{array}{cccc|c}
a_{1,1} & a_{1,2} & \cdots & a_{1,n} & b_1 \\
a_{2,1} & a_{2,2} & \cdots & a_{2,n} & b_2 \\
\multicolumn{5}{c}{\cdots} \\
a_{m,1} & a_{m,2} & \cdots & a_{m,n} & b_m
\end{array}
\right).
$$

The left part of the augmented matrix is called the **coefficient matrix** and the right part is called the **constant vector**. A system is **homogeneous** if the constant vector is all zeros; otherwise it is **nonhomogeneous**.

Two linear systems are **equivalent** if they have the same solutions.
Two matrices are **equivalent** if one can be obtained from the other by a sequence of elementary row operations. There are three types of elementary row operations: **interchange**, **multiply** and **sum**.

Elementary row operations are the basic steps for the solution of linear systems.
The concepts of equivalence of linear systems and equivalence of matrices are compatible because equivalent linear systems have equivalent augmented matrices and vice versa.

Looking further
It must be emphasized that the entries of a matrix need not always be mere numbers. We sometimes (see Example 1.9 on page 21) have to work with matrices having polynomial entries. Sometimes we even take matrices as the entries of other matrices. In differential equations, one encounters matrices each of whose entries is an (unknown) function of the independent variable t.

✳ ✳ ✳ ✳ ✳ ✳ ✳

Practice Problems

For each system of linear equations in Problems 1 – 3, find the corresponding augmented matrix that has the given first row.

1. $25x_1 + 27x_2 + 76x_3 = 43$
   $79x_3 + 51x_2 + 36x_1 = 20$
   $13x_2 - 76 + 15x_1 = 9x_3$

   $(25 \quad 27 \quad 76 \mid 43)$

2. $53a - 48c + 64b = 44 \quad (-48 \quad 64 \quad 53 \mid 44)$
   $80a - 99 + 13b = 23c$
   $50 - 31b + 31c = 10a$
   $95a + 65b - 37 + 23c = 0$

3. $-14.27v + 62.781x - 4z = 0$
   $15.671w - 67.66y = 19.1v$
   $47.62 - 37.11x - 11.13z = 14.5w$

   $(62.781 \quad 0 \ -4 \ -14.27 \quad 0 \mid 0)$

Each of Problems 4 – 7 is an augmented matrix for a system of equations in variables $x_1, x_2, x_3, \ldots$. Write the corresponding system of equations, assuming the successive columns of the matrix correspond to the variables in the order listed above.

4.
$$\left(\begin{array}{ccc|c} 0 & -88 & 6 & 33 \\ 15 & -69 & 28 & 0 \\ 7 & 0 & 0 & 22 \\ 0 & 0 & 33 & 3 \end{array} \right)$$

5.
$$\left(\begin{array}{ccc|c} -23.54 & -56.02 & 0 & -54.82 \\ 0 & 42.55 & 31.02 & -0.02 \\ -15.52 & 62.1 & -90.13 & 0 \\ -82.9 & 0 & -5 & 78.51 \end{array} \right)$$

6.
$$\left(\begin{array}{cccc|c} 0 & 7 & 13 & 0 & 5 \\ 2 & 0 & 0 & 6 & 35.2 \\ 8 & 5 & 0 & 0 & 36.7 \\ 0 & -3 & 0 & -8 & 0 \end{array} \right)$$

7.
$$\left(\begin{array}{cccc|c} 67.3 & 0 & 25.9 & 6.3 & 9 \\ 0 & 0 & 56.2 & 11.9 & -44.6 \\ -23.5 & -42.6 & 0 & 28.2 & -76.2 \\ 0 & 0 & 45.2 & 61 & 18.9 \end{array} \right)$$

In Problems 8 and 9, apply the indicated row operations to the given matrix in successive order both by hand and using a calculator.

8.
$$\begin{pmatrix} 9 & 18 \\ 0 & 1 \end{pmatrix}$$
$$(1/9)R_1 \to R_1$$
$$R_1 - 2R_2 \to R_1$$

9.
$$\begin{pmatrix} 1 & 2 & 3 \\ 4 & 5 & 6 \\ 7 & 8 & 9 \end{pmatrix}$$
$$R_1 \leftrightarrow R_3$$
$$(1/7)R_1 \to R_1$$
$$R_2 - 4R_3 \to R_2$$
$$R_3 - R_1 \to R_3$$

In Problems 10 and 11, find two elementary row operations that will transform the first matrix into the second one.

10.
$$\begin{pmatrix} 4 & 8 \\ 10 & 12 \end{pmatrix} \to \begin{pmatrix} 1 & 2 \\ 0 & -8 \end{pmatrix}$$

11.
$$\begin{pmatrix} 2 & 5/2 & 3 \\ 1 & 2 & 3 \\ 4 & 5 & 6 \end{pmatrix} \to \begin{pmatrix} 1 & 2 & 3 \\ 2 & 5/2 & 3 \\ 0 & 0 & 0 \end{pmatrix}$$

12. Suppose the matrix B is obtained from the matrix A by swapping two **columns**. Now apply the same elementary row operation to both A and B obtaining matrices A_1 and B_1. Show that B_1 is obtained from A_1 by swapping the same two columns.

13. Write the augmented matrix for the given linear system and apply each of the following elementary row operations:

$$x - y = 1$$
$$x + y = 1$$

$$R_1 - R_2 \to R_1 \quad R_1 \leftrightarrow R_2 \quad 2R_1 \to R_1.$$

to obtain a total of four equivalent augmented matrices. For each of the four augmented matrices, make a sketch (as in Example 1.3 on page 5) of the separate equations in the associated linear system. Thus, obtain a geometric interpretation of these elementary row operations.

14. The section summary is (very) technically incorrect. It is possible for two linear systems to have exactly the same solutions but have inequivalent augmented matrices. Explain why and give an example.

1.3 Echelon forms and Gaussian elimination

Objective 1

Given a matrix, decide if it is in reduced row echelon form.

As already mentioned, we will use a step-by-step procedure to find a "reduced row echelon form" matrix that is equivalent to the starting matrix and for which the associated linear system is easy to solve. Each step in the process will be an elementary row operation. The key to the whole process is being really organized. Naurally, we need to know exactly what the target matrix should look like.

Two new terms are required. The *leading coefficient* of a row of a matrix is the first nonzero entry. Every row that is not all zeros has a leading coefficient. A column of a matrix that contains the leading coefficient of at least one row is called a *pivotal column*. For example, the matrix has leading coefficients in row 1 column 1, row 2 column 2 and row 3 column

$$\begin{pmatrix} 1 & 0 & 1 & 1 \\ 0 & 1 & -1 & -1 \\ 0 & 0 & 0 & 1 \end{pmatrix}$$

4, so its pivotal columns are the 1st, 2nd and 4th. The matrix A (augmented or not) is in reduced row echelon form (RREF) if all of the following conditions are met:

REDUCED ROW ECHELON FORM

1. Each leading coefficient is a 1.

2. Any rows of all zeros appear last.

3. For each pair of successive rows that are not all zeros, the leading coefficient of the first row comes in an earlier column than the leading coefficient of the following row.

4. Each pivotal column has only one nonzero entry.

NOTICE: The effect of conditions 2 and 3 is that all entries below and to the right of a leading coefficient are zero.

The matrix A is in *row echelon form* (REF) whenever the first three conditions are satisfied, whether or not the fourth holds. Thus, every matrix that is in reduced row echelon form is also in row echelon form.

Each of the echelon forms has advantages and disadvantages. An advantage of RREF is that each matrix is equivalent to ONLY one matrix in RREF. Therefore, it is easy to check your computations.

The advantage of REF is that it is easier to compute and equally useful for some purposes. The disadvantage is that the result is NOT unique. Different computations of REFs of a given matrix may result in different matrices, both of which are correct (see Example 1.7 on page 19)!

If you use a calculator (or computer) there is little reason not to work only with RREF and ignore REF. For this reason we emphasize RREF and only use REF as an intermediate step in the complete reduction process.

Example 1.6 (Echelon forms.)

a. Each of the following matrices is in reduced row echelon form (RREF):

$$\begin{pmatrix} 1 & 0 & -1 & 7 \\ 0 & 1 & 3 & 5 \\ 0 & 0 & 0 & 0 \end{pmatrix}, \quad \begin{pmatrix} 1 & 0 & 0 & 0 & 3 \\ 0 & 0 & 1 & 0 & 6 \\ 0 & 0 & 0 & 1 & 0 \end{pmatrix}, \quad \begin{pmatrix} 1 & 1 & 0 & 0 & 0 \\ 0 & 0 & 1 & 2 & 0 \\ 0 & 0 & 0 & 0 & 1 \end{pmatrix}.$$

Check that these matrices satisfy all four of the conditions.

b. The matrices

$$C = \begin{pmatrix} 1 & 2 & 3 & 1 \\ 0 & 1 & -3 & -1 \\ 0 & 0 & 1 & 3 \end{pmatrix} \text{ and } D = \begin{pmatrix} 1 & 0 & 0 & 0 & 2 \\ 0 & 1 & 0 & 1 & 1 \\ 0 & 0 & 1 & 0 & 0 \end{pmatrix}$$

are in row echelon form (REF) because they satisfy the first three conditions. Matrix C fails condition 4 and matrix D satisfies condition 4. Therefore, only matrix D is in reduced row echelon form.

c. Neither of the matrices is in row echelon form. Both fail condition 1 because the leading coefficient of the second row is not a 1. (Note, the second matrix also fails condition 4.)

$$\begin{pmatrix} 1 & 0 & 0 \\ 0 & 2 & 1 \end{pmatrix} \quad \begin{pmatrix} 1 & 2 & 3 & 1 \\ 0 & 3 & 0 & 2 \\ 0 & 0 & 1 & 3 \end{pmatrix}.$$

d. The matrices fail condition 3. The leading coefficient of the second row is in the same column as that of the first row in matrix A, while, in matrix B, the leading coefficient of the third row is in an earlier column than that of the second row.

$$A = \begin{pmatrix} 1 & 2 & 3 & 1 \\ 1 & 0 & 0 & 2 \\ 0 & 0 & 1 & 3 \end{pmatrix} \text{ and } B = \begin{pmatrix} 1 & 0 & 0 & 0 & 2 \\ 0 & 0 & 1 & 0 & 0 \\ 0 & 1 & 0 & 1 & 1 \end{pmatrix}$$

□

Objective 2

Given a matrix, apply elementary row operations to obtain an equivalent matrix in reduced row echelon form.

It is finally time to discuss the step-by-step process of reducing a matrix to REF. The good news is that each step consists of applying a single elementary row operation. The bad news is that it is by no means clear which elementary row operation will most efficiently lead to the solution. Consequently, different people may start with the same matrix and proceed in different ways, unable to compare their work until the very end.

The following algorithm (recipe or procedure) is one of the most important that we will see in this text. It gives a more or less automatic way of choosing each step and is GUARANTEED to result in a matrix in REF. It is called the ***Gaussian elimination*** in honor of one of the worlds greatest mathematicians, Karl Friedrich Gauß (1777 – 1855).[2] There is, however, evidence of a similar procedure being used by the ancient Chinese at least 2000 years ago.

GAUSSIAN ELIMINATION ALGORITHM

Step 1. Use the interchange row operation to insure that a nonzero entry in the first nonzero column appears in the first row of the working matrix. Call this entry the ***pivotal entry***.

Step 2. Use the sum row operation to produce an equivalent matrix in which zeros appear below the pivotal entry.

Step 3. Use the multiply row operation to force the pivotal entry to be a 1. (Neither the row nor the column that contain the pivotal entry will change for the rest of the process.)

Step 4. Reduce the working matrix to those entries below and to the left of the pivotal entry. If the resulting working matrix is all zeros (or empty), quit. Otherwise, return to Step 1.

Notice that input to the Gaussian elimination algorithm is a matrix A, and the output is an equivalent matrix B that is in row echelon form (NOT reduced row echelon form — that comes next!)

As the algorithm progresses, you pass through the four steps a number of times. At the end of the t-th pass, you have produced the first t rows and (at least) the first t nonzero columns of the output matrix B. The entries in the last $m - t$ rows and the remaining columns of the matrix at this point are called the ***working matrix***. (At the beginning of the algorithm, the working matrix is A.)

It may sometimes be more efficient to interchange Steps 2 and 3; the exact sequence that you use is a matter of convenience. The basic idea is to produce ever more zeros in the matrix in a systematic way by means of the sum row operation. In the process, most people (as opposed to computers) prefer to avoid fractional matrix entries for as long as possible.

Example 1.7 (Gaussian elimination.)

   a. Apply elementary row operations to the indicated matrix to obtain a matrix in row echelon form.

---

[2]Although the German letter "ß" is now archaic, it was a part of this person's name. I will try to use characters recognizable to a person when referring to that individual but use standard English transliteration when referring to something associated to that person's work.

Step 1 is unnecessary for this matrix since the $(1,1)$ entry is a 1. For Step 2, the pivotal entry of the first matrix is the 1 in position $(1,1)$, and the purpose of this row operation is the creation of the 0 in position $(3,1)$ of the second matrix.

$$\begin{pmatrix} 1 & 0 & 1 & 0 \\ 0 & 1 & 0 & 1 \\ 2 & 3 & 2 & 2 \end{pmatrix} \xrightarrow{\;(-2)\;} \begin{pmatrix} 1 & 0 & 1 & 0 \\ 0 & 1 & 0 & 1 \\ 0 & 3 & 0 & 2 \end{pmatrix}.$$

Step 3 is unnecessary for this matrix, since the $(1,1)$ entry is a 1, and Step 4 tells us to reduce the working matrix as indicated. Now circle back to Step 1 with the smaller working matrix obtained by ignoring the first row and column as indicated below.

Step 1 is unnecessary this time.

$$\begin{array}{cc} \text{Step 2} & \text{Step 3} \\ (-3)\xrightarrow{} \begin{pmatrix} 1 & 0 & 1 & 0 \\ 0 & \boxed{1} & 0 & 1 \\ 0 & 0 & 0 & -1 \end{pmatrix} & (-1)\to \begin{pmatrix} 1 & 0 & 1 & 0 \\ 0 & 1 & 0 & 1 \\ 0 & 0 & \boxed{0} & 1 \end{pmatrix}. \end{array}$$

This matrix is already in REF (but NOT RREF).

b. Any nonzero entry in the first column can be moved into position and used as a pivotal entry. The freedom to pick your pivotal entries can be used to your advantage if you prefer to avoid fractions. For example, the same matrix is reduced in two different ways in this part and in Part c below.

$$\begin{pmatrix} 4 & 3 & 1 & 6 \\ 0 & 2 & 5 & -3 \\ 1 & 0 & -3 & 4 \end{pmatrix} \xrightarrow{\;(-\frac{1}{4})\;} \begin{pmatrix} 4 & 3 & 1 & 6 \\ 0 & 2 & 5 & -3 \\ 0 & -\frac{3}{4} & -\frac{13}{4} & \frac{5}{2} \end{pmatrix} \xrightarrow{(\frac{1}{4})}$$

$$\begin{pmatrix} 1 & \frac{3}{4} & \frac{1}{4} & \frac{3}{2} \\ 0 & 2 & 5 & -3 \\ 0 & -\frac{3}{4} & -\frac{13}{4} & \frac{5}{2} \end{pmatrix} \xrightarrow{(\frac{3}{8})} \begin{pmatrix} 1 & \frac{3}{4} & \frac{1}{4} & \frac{3}{2} \\ 0 & 2 & 5 & -3 \\ 0 & 0 & -\frac{11}{8} & \frac{11}{8} \end{pmatrix} \xrightarrow[(\frac{-8}{11})]{(\frac{1}{2})\to} \begin{pmatrix} 1 & \frac{3}{4} & \frac{1}{4} & \frac{3}{2} \\ 0 & 1 & \frac{5}{2} & -\frac{3}{2} \\ 0 & 0 & 1 & -1 \end{pmatrix}.$$

c. Gaussian elimination is guaranteed to work. It is not guaranteed to be the easiest or fastest procedure. For example, starting off with an interchange operation in the preceding example leads to easier arithmetic:

$$\begin{pmatrix} 4 & 3 & 1 & 6 \\ 0 & 2 & 5 & -3 \\ 1 & 0 & -3 & 4 \end{pmatrix} \times \begin{pmatrix} 1 & 0 & -3 & 4 \\ 0 & 2 & 5 & -3 \\ 4 & 3 & 1 & 6 \end{pmatrix} \xrightarrow{\;(-4)\;} \begin{pmatrix} 1 & 0 & -3 & 4 \\ 0 & 2 & 5 & -3 \\ 0 & 3 & 13 & -10 \end{pmatrix} \xrightarrow{(-1)}$$

$$\begin{pmatrix} 1 & 0 & -3 & 4 \\ 0 & -1 & -8 & 7 \\ 0 & 3 & 13 & -10 \end{pmatrix} \xrightarrow{(3)} \begin{pmatrix} 1 & 0 & -3 & 4 \\ 0 & -1 & -8 & 7 \\ 0 & 0 & -11 & 11 \end{pmatrix} \xrightarrow[(-\frac{1}{11})\to]{(-1)\to} \begin{pmatrix} 1 & 0 & -3 & 4 \\ 0 & 1 & 8 & -7 \\ 0 & 0 & 1 & -1 \end{pmatrix}. \;\square$$

NOTE: Both the final matrices in part b and part c above are in row echelon form and are equivalent to the given matrix. The fact that two very different-looking matrices can both be equivalent to the starting matrix and still be in row echelon form can present severe problems for students and test makers alike! This ambiguity is best resolved by continuing each reduction sequence until each of the final matrices is in reduced row echelon form, at which point the final matrices will coincide (see Example 1.8).

The procedure for reducing a matrix to reduced row echelon form is to first reduce it to row echelon form and then work from RIGHT to LEFT forcing zeros above each leading coefficient. (This maximizes the amount of addition and multiplication of zeros.) The complete reduction process to reduced row echelon form is called *Gauss-Jordan elimination* and always gives a matrix in RREF.

Example 1.8 (Gauss-Jordan elimination.)
In Parts b and c of Example 1.7 on the preceding page, there were two different matrices in row echelon form that were both equivalent to the given matrix. When either of these is reduced to reduced row echelon form, the result should be the same matrix. Let's see!

$$\begin{pmatrix} 1 & \frac{3}{4} & \frac{1}{4} & \frac{3}{2} \\ 0 & 1 & \frac{5}{2} & -\frac{3}{2} \\ 0 & 0 & 1 & -1 \end{pmatrix} \xrightarrow[(-\frac{5}{2})]{} \, \Big\rfloor \xrightarrow[(-\frac{1}{4})]{} \Big\rfloor \begin{pmatrix} 1 & \frac{3}{4} & 0 & \frac{7}{4} \\ 0 & 1 & 0 & 1 \\ 0 & 0 & 1 & -1 \end{pmatrix} \xrightarrow[(-\frac{3}{4})]{} \, \Big\rfloor \begin{pmatrix} 1 & 0 & 0 & 1 \\ 0 & 1 & 0 & 1 \\ 0 & 0 & 1 & -1 \end{pmatrix}$$

On the other hand:

$$\begin{pmatrix} 1 & 0 & -3 & 4 \\ 0 & 1 & 8 & -7 \\ 0 & 0 & 1 & -1 \end{pmatrix} \xrightarrow[(3)]{} \, \Big\rfloor \begin{pmatrix} 1 & 0 & 0 & 1 \\ 0 & 1 & 8 & -7 \\ 0 & 0 & 1 & -1 \end{pmatrix} \xrightarrow[(-8)]{} \, \Big\rfloor \begin{pmatrix} 1 & 0 & 0 & 1 \\ 0 & 1 & 0 & 1 \\ 0 & 0 & 1 & -1 \end{pmatrix}. \; WOW!$$

□

It will sometimes be necessary to apply elementary row operations to matrices whose entries involve variables as well as numbers. The only difference is in the computational difficulty (even for computers!).

Example 1.9 (Reduction of matrices with polynomial entries.)
Row operations and matrix reduction can be used even when the matrix entries are not numbers. In fact, sometimes test questions have variables in a given matrix so that the problem cannot be done with the calculator alone!
Here is an example that we will return to in Chapter 4.

$$\begin{pmatrix} 8-x & 9 & 9 \\ 3 & 2-x & 3 \\ -9 & -9 & -10-x \end{pmatrix} \xrightarrow[(3)]{} \begin{pmatrix} 8-x & 9 & 9 \\ 3 & 2-x & 3 \\ 0 & -3-3x & -1-x \end{pmatrix} \xrightarrow[(-3)]{}$$

$$\begin{pmatrix} -1-x & 3+3x & 0 \\ 3 & 2-x & 3 \\ 0 & -3-3x & -1-x \end{pmatrix} \begin{matrix} \frac{1}{(x+1)} \to \\ \\ \frac{1}{(x+1)} \to \end{matrix} \begin{pmatrix} -1 & 3 & 0 \\ 3 & 2-x & 3 \\ 0 & -3 & -1 \end{pmatrix} \begin{matrix} (3)_{\raisebox{-2pt}{$\neg$}} \\ \\ (3)^{\raisebox{2pt}{$\lrcorner$}} \end{matrix}$$

$$\begin{pmatrix} -1 & 3 & 0 \\ 0 & 2-x & 0 \\ 0 & -3 & -1 \end{pmatrix} \begin{matrix} (-1) \to \\ \frac{1}{(2-x)} \to \\ (-1) \to \end{matrix} \begin{pmatrix} 1 & -3 & 0 \\ 0 & 1 & 0 \\ 0 & 3 & 1 \end{pmatrix} \xrightarrow{\;(3)^{\!\lrcorner}\;(-3)_{\raisebox{-2pt}{$\neg$}}\;} \begin{pmatrix} 1 & 0 & 0 \\ 0 & 1 & 0 \\ 0 & 0 & 1 \end{pmatrix}$$

Of course this is wrong if $x = -1$ or $x = 2$. Can you tell what the correct answer is in these cases? Do you see that they actually lead to different answers?   □

Objective 3

Solve several linear systems with the same coefficient matrix simultaneously using a generalized augmented matrix.

We will also have to solve several systems of equations with the same coefficient matrix. The solution process can be organized more efficiently by combining augmented matrices into a ***generalized augmented matrix*** in which the column vector to the right of the | is replaced by a matrix.

Suppose, for example, we have three linear systems all with the 3-by-4 coefficient matrix A. Write the constant vectors as the columns of a 3-by-3 matrix B. Then

$$(A|C) = \begin{pmatrix} a_{11} & a_{12} & a_{13} & a_{14} & b_{11} & b_{12} & b_{13} \\ a_{21} & a_{22} & a_{23} & a_{24} & b_{21} & b_{22} & b_{23} \\ a_{31} & a_{32} & a_{33} & a_{34} & b_{31} & b_{32} & b_{33} \end{pmatrix}.$$

Example 1.10 (Systems of systems.)
Suppose that both of the following systems must be solved:

$$\begin{aligned} x_1 - x_2 + x_3 &= -1 \\ 2x_1 + x_2 - x_3 &= 0 \end{aligned} \qquad\qquad \begin{aligned} x_1 - x_2 + x_3 &= 1 \\ 2x_1 + x_2 - x_3 &= -1 \end{aligned}.$$

Of course they may have different solutions — that is okay! We just need to find all solutions of the first system and find all solutions of the second system. Because the two systems have the same coefficient matrix, the two solution processes will involve exactly the same row operations in the same order. The new notation of generalized augmented matrices allows us to avoid needless repetition — that's all.
Simply combine the two augmented matrices

$$\begin{pmatrix} 1 & -1 & 1 & -1 \\ 2 & 1 & -1 & 0 \end{pmatrix} \qquad\qquad \begin{pmatrix} 1 & -1 & 1 & 1 \\ 2 & 1 & -1 & -1 \end{pmatrix}$$

into a single, generalized augmented matrix and reduce it in the usual manner.

$$\begin{pmatrix} 1 & -1 & 1 & -1 & 1 \\ 2 & 1 & -1 & 0 & -1 \end{pmatrix}$$

Each step of the process can be viewed as a step in the reduction of the first system by ignoring the last column and as a step in the reduction of the second system by ignoring the second to the last column.

We will see that the original systems have general solutions:

$$x_1 = -1/3 \qquad\qquad x_1 = 0$$
$$x_2 = 2/3 + x_3 \qquad\qquad x_2 = -1 + x_3$$
$$x_3 = x_3 \qquad\qquad\qquad x_3 = x_3. \qquad\qquad\qquad\qquad □$$

SUMMARY

A matrix A is in reduced row echelon form (RREF) if all of the following conditions are met:

1. Each leading coefficient is a 1.

2. Any rows of all zeros appear last.

3. For each pair of successive rows that are not all zeros, the leading coefficient of the first row comes in an earlier column than the leading coefficient of the following row.

4. Each pivotal column has only one nonzero entry.

A matrix A satisfying the first three conditions is in *row echelon form* (REF). *Elementary row operations* are used to reduce matrices to these forms. The systematic methods for matrix reduction called *Gaussian elimination* and *Gauss-Jordan elimination* involve working one column at a time on smaller and smaller *working matrices*.

The reduction process also applies to matrices having non-numerical entries. Simultaneous solution of several systems with the same coefficient matrix is done by using a *generalized augmented matrix*.

Looking further

There are many important variations on the algorithms of this fundamental section. These refinements have as goals either to minimize numerical errors when computing with approximations (as in Example 3.16 of Section 3.6), or to restrict the matrix entries that arise in the reduction process (as in the Smith normal form presented in Example 4.9 of Section 4.3).

Practice Problems

1. Classify each of the following matrices: REF, RREF or neither.

a.
$$\begin{pmatrix} 2 & 17 & 0 & 0 & | & -8.7 \\ 0 & 1 & 0 & -5.1 & | & 0 \\ 0 & 0 & 1 & -54 & | & -1 \\ 0 & 0 & 0 & 1 & | & 19 \end{pmatrix}$$

b.
$$\begin{pmatrix} 1 & 0 & 0 & 23 & 0 \\ 0 & 1 & -62 & 0 & 0 \\ 0 & 0 & 0 & 1 & 16 \\ 0 & 0 & 0 & 0 & 1 \end{pmatrix}$$

c.
$$\begin{pmatrix} 1 & 0 & -37 & 43 & 0 & 0 & | & 0 \\ 0 & 1 & 0 & -9 & 0 & 0 & | & 31 \\ 0 & 0 & 0 & 0 & 1 & 0 & | & 0 \\ 0 & 0 & 0 & 0 & 0 & 1 & | & 70 \end{pmatrix}$$

d.
$$\begin{pmatrix} 1 & 0 & 0 & -16 \\ 0 & 1 & 9 & 0 \\ 0 & 0 & 1 & 0 \\ 0 & 0 & 0 & 0 \end{pmatrix}$$

e.
$$\begin{pmatrix} 1 & 0 & 1 & -81 & 0 & 0 \\ 0 & 1 & -78 & 0 & 0 & 7 \\ 0 & 1 & 0 & 0 & 16 & 0 \end{pmatrix}$$

f.
$$\begin{pmatrix} 1 & 0 & 23 & -47 & 0 & 0 \\ 0 & 1 & 0 & 0 & -97 & 0 \\ 0 & 0 & 0 & 1 & 36 & 0 \\ 0 & 0 & 0 & 0 & 0 & 0 \\ 0 & 0 & 0 & 0 & 1 & 0 \end{pmatrix}$$

In Problems 2 – 7, put the matrix in RREF using elementary row operations. Identify the pivotal and nonpivotal columns of the reduced matrix.

2.
$$\begin{pmatrix} 2 & 4 & 8 & -1 & -4 \\ 1 & 2 & 4 & -1 & -3 \\ 3 & 7 & 9 & 2 & -4 \end{pmatrix}$$

3.
$$\begin{pmatrix} 2 & -2 & 2 & | & 3 \\ -3 & 1 & 5 & | & 7 \\ 1 & 0 & -2 & | & 3 \\ 4 & 0 & -8 & | & 11 \end{pmatrix}$$

4.
$$\begin{pmatrix} -3 & 2 & -8 & 4 & 8 & -13 \\ -1 & 0 & -2 & 0 & 2 & -3 \\ 4 & -2 & 10 & -4 & -9 & 16 \end{pmatrix}$$

5.
$$\begin{pmatrix} -6 & -18 & -12 & 8 & 19 & | & -2 \\ 2 & 6 & 4 & -2 & -4 & | & 2 \\ -5 & -15 & -10 & 6 & 13 & | & -5 \end{pmatrix}$$

6.
$$\begin{pmatrix} -2 & 2 & -1 & 8 & -5 & 3 \\ 4 & 9 & 0 & 10 & -5 & 13 \\ 3 & 0 & 1 & -6 & 4 & 4 \end{pmatrix}$$

7.
$$\begin{pmatrix} 2 & 4 & 5 & | & -2 \\ 3 & 7 & 9 & | & -2 \\ -1 & -2 & -3 & | & 1 \\ -4 & -11 & -14 & | & 2 \end{pmatrix}$$

In Problems 8 and 9, write the generalized augmented matrix for the two linear systems and solve the systems.

8.
$$\begin{cases} 2x_1 - 3x_2 + 5x_3 = 2 \\ x_1 + x_2 - x_3 = -1 \\ x_2 + x_3 = 3 \end{cases}$$
and
$$\begin{cases} 2x_1 - 3x_2 + 5x_3 = 0 \\ x_1 + x_2 - x_3 = -2 \\ x_2 + x_3 = 1 \end{cases}$$

9.
$$\begin{cases} 5x + 3y - 2z = 10 \\ 14x + y - z = 2 \\ x - y = 5 \end{cases}$$
and
$$\begin{cases} 5x + 3y - 2z = 2 \\ 14x + y - z = 4 \\ x - y = 6 \end{cases}$$

10. Put the matrix in reduced row echelon form using elementary row operations.
$$\begin{pmatrix} 1 & x & x^2 \\ 1 & y & y^2 \end{pmatrix}$$

    WARNING: The answer depends on whether or not $x = y$!

11. Show that no two matrices that are in RREF are equivalent.

1.4   Free variables and general solutions

Objective 1

Given an RREF augmented matrix, write the general solution of the associated linear system in terms of the variables from nonpivotal columns of the coefficient matrix, or state that there is no solution.

When a linear system has no solution (like Example 1.3 on page 5) or exactly one solution (like part c of Example 1.2 on page 4), this objective is easy. When it has more than one solution (like part b of Example 1.2 on page 4), we use $free$ variables (they are free to take any value) to obtain formulas for the other $dependent$ variables.

Sadly, exactly which variables are free and which are dependent are not forced by the mathematics. Technically there is a lot of choice. In fact, this choice is a principal topic in the theory of vector spaces. In this course, for now, we simply give an algorithm that always works and move on.

We must separate the variables into two types: independent and dependent. A variable is **dependent** if its value is (viewed as) determined by the values of other variables. A variable is called *free* or **independent** if its value is (viewed as) not determined by any other variables values. It is allowed to be any value without restriction.

You have seen free and dependent variables before. When you consider a function $y = f(x)$, you are regarding y as a dependent variable and x as a free variable. The difference now is that we have many more variables to deal with and they are related by several equations, not just one.

Whether a variable is dependent or independent depends on how the mathematical information is organized. For us, this dichotomy is determined by the order used to list the variables when the coefficient matrix is built. This ordering, together with the nature of the reduced row echelon form, determines how we separate the variables.

Here is the idea. Consider the simple linear system and its augmented matrix:

$$\begin{aligned} x_1 \quad - \quad x_3 &= 7 \\ x_2 + 3x_3 &= 5 \end{aligned}, \qquad \begin{pmatrix} 1 & 0 & -1 & 7 \\ 0 & 1 & 3 & 5 \end{pmatrix}.$$

Columns 1 and 2 are pivotal, column 3 is nonpivotal. Take x_3 as the (only) free variable and solve for the other variables in terms of it. Notice that there really is a solution for any value of x_3; so it truly is FREE.

$$\begin{aligned} x_1 &= \quad x_3 + 7 \\ x_2 &= -3x_3 + 5 \\ x_3 &= \quad x_3 \end{aligned}$$

NOTE: The equation $x_3 = x_3$ may seem unnecessary, but it must appear for formal reasons that are explained below.

The free variables are those associated with the nonpivotal columns of the coefficient matrix. Other variables are dependent.

In general, each nonzero row of an RREF coefficient matrix corresponds to an equation involving ONLY ONE dependent variable (together with some numbers and some free variables). (Because each row has only one leading coefficient!) It is a simple matter to rewrite these equations isolating the dependent variables. There results a set of formulas for each of the dependent variables in terms of numbers and free variables. When we add to this set the "dumb" equations like $x_3 = x_3$ above, we have a set of formulas for EACH of the variables. This set is called the **general solution** of the system because it satisfies the following two conditions:

- any substitution of numbers for the free variables results in a solution of the system, and

- every numerical solution of the system is obtained by this process.

Example 1.11 (From RREF to a general solution.)

a. The linear system with the indicated augmented matrix has general solution $x_1 = 3$, $x_2 = 6$, $x_3 = 0$.

$$\begin{pmatrix} 1 & 0 & 0 & | & 3 \\ 0 & 1 & 0 & | & 6 \\ 0 & 0 & 1 & | & 0 \end{pmatrix}.$$

b. The general solution of the system having the indicated matrix is also immediate: $x_1 = 3$, $x_2 = x_2$, $x_3 = 6$ and $x_4 = 0$. The variable x_2 is a free variable and can take any value. Moreover, its value is independent of the other variable values.
This solution may also be written:
BUT don't for a moment think the row form "equals" the column form. These are just two different ways to convey the same information.

$$\begin{pmatrix} 1 & 0 & 0 & 0 & | & 3 \\ 0 & 0 & 1 & 0 & | & 6 \\ 0 & 0 & 0 & 1 & | & 0 \end{pmatrix}$$

$(3, x_2, 6, 0)$ or $\begin{pmatrix} 3 \\ x_2 \\ 6 \\ 0 \end{pmatrix}.$

c. The general solution of the system having the indicated matrix is obtained by writing the associated system and putting the nonpivotal variables on the right side of the equals sign:

$$\begin{pmatrix} 1 & 0 & -1 & | & 7 \\ 0 & 1 & 3 & | & 5 \\ 0 & 0 & 0 & | & 0 \end{pmatrix}$$

$$\begin{aligned} x_1 \quad - \quad x_3 &= 7 \\ x_2 + 3x_3 &= 5 \end{aligned}$$

$$\begin{aligned} x_1 &= x_3 + 7 \\ x_2 &= -3x_3 + 5 \\ x_3 &= x_3. \end{aligned}$$

d. Here is an augmented matrix and the associated linear system:

$$\begin{pmatrix} 1 & 1 & 0 & 0 & 0 & | & 0 \\ 0 & 0 & 1 & 2 & 0 & | & 0 \\ 0 & 0 & 0 & 0 & 1 & | & 0 \end{pmatrix}$$

$$\begin{aligned} x_1 + x_2 \qquad\qquad &= 0 \\ x_3 + 2x_4 \quad &= 0 \\ x_5 &= 0 \end{aligned}$$

Because the second and fourth columns of the augmented matrix are nonpivotal, x_2 and x_4 are free variables. This linear system has general solution

$$\begin{aligned} x_1 &= -x_2 \\ x_2 &= x_2 \\ x_3 &= -2x_4 \\ x_4 &= x_4 \\ x_5 &= 0 \end{aligned} \quad \text{or} \quad \begin{pmatrix} -x_2 \\ x_2 \\ -2x_4 \\ x_4 \\ 0 \end{pmatrix}.$$

□

Objective 2

Given a system of linear equations, write an associated augmented matrix, apply elementary row operations to obtain an equivalent RREF matrix and write the general solution.

We have now presented all three parts of the process of solving a linear system and are ready to put them together.

The first part consists of replacing the system with its augmented matrix; the second part consists of reducing the augmented matrix to RREF; and the last part consists of writing the general solution of the (equivalent) system that has augmented matrix in a row echelon form. Let's put all the pieces together.

Example 1.12 (Solving a linear system.)

a. Consider the following linear system and its augmented matrix:

$$\begin{array}{rl} x_3 + 2x_4 &= 3 \\ 2x_1 + 4x_2 - 2x_3 &= 4 \\ 2x_1 + 4x_2 - x_3 + 2x_4 &= 7 \end{array} \quad \left(\begin{array}{cccc|c} 0 & 0 & 1 & 2 & 3 \\ 2 & 4 & -2 & 0 & 4 \\ 2 & 4 & -1 & 2 & 7 \end{array}\right).$$

While reducing this matrix to reduced row echelon form, we choose to keep integer entries as long as possible.

$$(1/2) \rightarrow \left(\begin{array}{cccc|c} 1 & 2 & -1 & 0 & 2 \\ 0 & 0 & 1 & 2 & 3 \\ 2 & 4 & -1 & 2 & 7 \end{array}\right) \begin{array}{c}(-2)\\ \\ (-1)\end{array} \quad \left(\begin{array}{cccc|c} 1 & 2 & -1 & 0 & 2 \\ 0 & 0 & 1 & 2 & 3 \\ 0 & 0 & 0 & 0 & 0 \end{array}\right)$$

This matrix is in row echelon form but not in reduced row echelon form. We proceed to reduce the matrix **all the way to its reduced row echelon form** using Gauss-Jordan reduction. Then we write the general solution. The variables labeling the nonpivotal columns of the RREF coefficient matrix are free.

$$(1) \quad \left(\begin{array}{cccc|c} 1 & 2 & 0 & 2 & 5 \\ 0 & 0 & 1 & 2 & 3 \\ 0 & 0 & 0 & 0 & 0 \end{array}\right) \qquad \begin{array}{rl} x_1 &= 5 - 2x_2 - 2x_4 \\ x_2 &= x_2 \\ x_3 &= 3 - 2x_4 \\ x_4 &= x_4 \end{array}$$

b. The indicated linear system has augmented matrix that transforms to the REF matrix:

$$\begin{array}{rl} x_1 + 2x_2 - x_3 + 3x_4 + x_5 &= 2 \\ 2x_1 + 4x_2 - 2x_3 + 6x_4 + 3x_5 &= 6 \\ -x_1 - 2x_2 + x_3 - x_4 + 3x_5 &= 4 \end{array},$$

$$\left(\begin{array}{ccccc|c} 1 & 2 & -1 & 3 & 1 & 2 \\ 2 & 4 & -2 & 6 & 3 & 6 \\ -1 & -2 & 1 & -1 & 3 & 4 \end{array}\right) \begin{array}{c}(1)\\(-2)\\ \end{array} \left(\begin{array}{ccccc|c} 1 & 2 & -1 & 3 & 1 & 2 \\ 0 & 0 & 0 & 0 & 1 & 2 \\ 0 & 0 & 0 & 2 & 4 & 6 \end{array}\right) \begin{array}{c}(\frac{1}{2})\end{array} \left(\begin{array}{ccccc|c} 1 & 2 & -1 & 3 & 1 & 2 \\ 0 & 0 & 0 & 1 & 2 & 3 \\ 0 & 0 & 0 & 0 & 1 & 2 \end{array}\right).$$

Continuing to RREF by the method of Gauss-Jordan gives:

$$\xrightarrow{}\;\underset{(-1)}{\overset{}{\xrightarrow{}}}\;\underset{(-2)}{\overset{}{\xrightarrow{}}}\;\left(\begin{array}{ccccc|c} 1 & 2 & -1 & 3 & 0 & 0 \\ 0 & 0 & 0 & 1 & 0 & -1 \\ 0 & 0 & 0 & 0 & 1 & 2 \end{array}\right)\;\underset{(-3)}{\xrightarrow{}}\;\left(\begin{array}{ccccc|c} 1 & 2 & -1 & 0 & 0 & 3 \\ 0 & 0 & 0 & 1 & 0 & -1 \\ 0 & 0 & 0 & 0 & 1 & 2 \end{array}\right),$$

and the general solution:
$$x_1 = 3 - 2x_2 + x_3, \quad x_2 = x_2, \quad x_3 = x_3, \quad x_4 = -1, \quad x_5 = 2. \quad \Box$$

Because all that is involved in going from RREF to the general solution is isolating the variables associated with pivotal columns in each equation of the linear system, it is clear that:

If every column of an RREF coefficient matrix is pivotal, then the system has AT MOST ONE solution. If the final column of an RREF augmented matrix is pivotal too, then it has NO SOLUTION.

The number of independent or free variables in the general solution is the number of nonpivotal columns in RREF(A) (NOT necessarily the number of nonpivotal columns in A itself!). The number of pivotal columns in RREF(A) is called the **rank** of A. **The concept of rank is central in linear algebra and matrix theory, as you will see in other courses.** We use it only to summarize the solutions of an arbitrary linear system.

THEOREM A linear system in n variables and augmented matrix $(A|\mathbf{b})$ has no solution if rank(A) $\neq$ rank($A|\mathbf{b}$). Otherwise the general solution has $n - rank(A)$ free variables.

SUMMARY

Solving a system of linear equations involves almost all of the objectives so far. Form an augmented matrix, reduce it using ***elementary row operations*** to reduced row echelon form and then write the ***general solution***. The shape and ***rank*** of the coefficient and augmented matrices of a linear system determine the nature of its solutions. In fact, the fundamental theorem of linear systems is:

THEOREM A linear system in n variables with augmented matrix $(A|\mathbf{b})$ has no solution if rank(A) $\neq$ rank($A|\mathbf{b}$). Otherwise the general solution has $n - rank(A)$ free variables.

Looking further
The question of which variables could be "freed" by reordering the variables has to do with building a **basis** for the **solution space** of the associated homogeneous system. This abstact theory of vector spaces appears the Appendix, page 217, and in sequel linear algebra courses.

Practice Problems

In Problems 1 – 4, decide which variables are free variables.

1.
$$\begin{cases} x + y + z = 1 \\ x - 2y - z = -4 \\ 3x + 7y + z = 3 \end{cases}$$

2.
$$\begin{cases} a - b + 3c + d = 4 \\ 3a - 2b + c + 4d = 10 \\ a + 10b - 7c - 3d = -1 \end{cases}$$

3.
$$\begin{cases} r + s + t = 1 \\ 10r + 4s + 2t = 20 \\ 3r \quad\ + 2t = 9 \\ 5r + 2s + t = 10 \end{cases}$$

4.
$$\begin{cases} p - q + 3r + 2s = 3 \\ -p + 2q - 6r - 2s = -2 \\ p + q - 3r + 2s = 5 \end{cases}$$

5. Write a corresponding system of equations and its general solution.

a.
$$\left(\begin{array}{cccc|c} 1 & 0 & 0 & 0 & 2 \\ 0 & 1 & 2 & 0 & 3 \\ 0 & 0 & 0 & 1 & -1 \end{array}\right)$$

b.
$$\left(\begin{array}{cccc|c} 1 & 0 & 2 & 0 & 0 & 2 & -2 \\ 0 & 1 & 4 & 0 & 0 & 3 & 2 \\ 0 & 0 & 0 & 1 & 0 & 4 & -1 \end{array}\right)$$

c.
$$\left(\begin{array}{cccc|c} 1 & 0 & 2 & 0 & 2 \\ 0 & 1 & -3 & 0 & 1 \\ 0 & 0 & 0 & 1 & 3 \\ 0 & 0 & 0 & 0 & 0 \end{array}\right)$$

d.
$$\left(\begin{array}{cccccc|c} 1 & 3 & 4 & 0 & 0 & 0 & 1 \\ 0 & 0 & 0 & 1 & 0 & 0 & 2 \\ 0 & 0 & 0 & 0 & 1 & 0 & 4 \\ 0 & 0 & 0 & 0 & 0 & 1 & 3 \end{array}\right)$$

For Problems 6 – 11, use matrix reduction to solve the linear system.

6.
$$\begin{cases} -10a - 6b + 23c + 6d = 1 \\ -6a + 6b + 9c + 10d = -9 \\ -8a \quad\ + 16c + 8d = -4 \\ 2a \quad\ - 4c - 2d = 1 \end{cases}$$

7.
$$\begin{cases} 3v + 6w \qquad\qquad = -1 \\ 6v + 12w + 2x + 3y = 1 \\ \qquad\quad 2x + 3y + 3z = 2 \\ 6v + 12w \quad\ + 2y = -2 \\ 3v + 6w - 2x = -4 \end{cases}$$

8.
$$\begin{cases} 2p + 3q + r - 2s = -3 \\ -3p - 5q - 2r = -5 \\ 3p + 3q + r - 2s = -4 \\ 2p + 3q + r - s = 0 \end{cases}$$

9.
$$\begin{cases} 2a + 3b + c - 2d = -3 \\ -3a - 5b - 2c = -5 \\ 3a + 3b + c - 2d = -4 \\ 2a + 3b + c - d = 0 \end{cases}$$

10.
$$\begin{cases} 6v - 2w - 3y = 6 \\ -3v + 4w + x + 6y = -2 \\ -3v + 2w + x + 5y = 0 \\ 3v - 2w - 2y = 4 \end{cases}$$

11.
$$\begin{cases} 3a + 2b + c + 2d + 5e = 0 \\ -4a - 2b + c - 9e = 5 \\ -2a - b + c - 5e = 2 \\ 2a + b + d + 4e = -1 \\ -a - 2b - 3c - 2d + e = -7 \end{cases}$$

12. Suppose two people started with the same system of linear equations but chose different orders for listing the variables when they wrote augmented matrices. Give an example to show that they can have different free variables in their general solutions.

13. Suppose a linear system is given and the general solution has been computed. Also suppose that a free variable f appears in the formula for the dependent variable d. Show that d can be "freed" by swapping the f and d columns. What happens to f in this case?

1.5   The vector form of the general solution

Objective 1

Given a matrix A and a vector $\mathbf{x}$, compute $A\mathbf{x}$ if it exists. Translate between a linear system and the associated matrix equation.

We have used matrices to increase efficiency in solving linear systems and have obtained general solutions as a set of formulas. We now extend arithmetic from numbers to matrices and vectors to obtain a simpler way to express general solutions. As earlier, a single letter A (the coefficient matrix) is used to denote all the coefficients of a linear system of equations, the entire system

$$\begin{aligned} a_{1,1}x_1 + a_{1,2}x_2 + \ldots + a_{1,n}x_n &= b_1 \\ a_{2,1}x_1 + a_{2,2}x_2 + \ldots + a_{2,n}x_n &= b_2 \\ &\cdots \\ a_{m,1}x + a_{m,2}x_2 + \ldots + a_{m,n}x_n &= b_m \end{aligned}$$

can be expressed as a single matrix equation. As before, the entire list of b's (respectively x's) is written as a column vector $\mathbf{b}$ (respectively $\mathbf{x}$):

$$\mathbf{b} = \begin{pmatrix} b_1 \\ \vdots \\ b_m \end{pmatrix} \text{ and } \mathbf{x} = \begin{pmatrix} x_1 \\ \vdots \\ x_n \end{pmatrix}.$$

NOTE: We use $\mathbb{R}^n$ to denote the set of all vectors with n real entries. Although the geometric interpretation of vectors in $\mathbb{R}^2$ and $\mathbb{R}^3$ is presented in the next section, officially $\mathbb{R}^n$ has no particular geometric or physical content.

Sometimes the notation $\underline{b}$, $\vec{b}$ or $\tilde{b}$ is used for $\mathbf{b}$. These notations distinguish vectors from mere numbers. Notice that none of the vector notations conveys the shape of the vector; they just remind you "this is a vector" rather like the upper case notation for a matrix doesn't include its shape.

Now the whole linear system

$$\begin{aligned} a_{1,1}x_1 + a_{1,2}x_2 + \cdots + a_{1,n}x_n &= b_1 \\ a_{2,1}x_1 + a_{2,2}x_2 + \cdots + a_{2,n}x_n &= b_2 \\ &\cdots \\ a_{m,1}x_1 + a_{m,2}x_2 + \cdots + a_{m,n}x_n &= b_m \end{aligned}$$

is written as the matrix equation:

$$\begin{pmatrix} a_{1,1} & a_{1,2} & \cdots & a_{1,n} \\ a_{2,1} & a_{2,2} & \cdots & a_{2,n} \\ & & \cdots & \\ a_{m,1} & a_{m,2} & \cdots & a_{m,n} \end{pmatrix} \begin{pmatrix} x_1 \\ x_2 \\ \vdots \\ x_n \end{pmatrix} = \begin{pmatrix} b_1 \\ b_2 \\ \vdots \\ b_m \end{pmatrix} \text{ or simply } A\mathbf{x} = \mathbf{b}.$$

That last equation has TWO new constructs. The equal sign is used (for the very first time) with matrices. As you might expect, two matrices are **equal** if they have the same shape and all corresponding entries are equal. The other new usage is $A\mathbf{x}$, or matrix times vector multiplication. The product of the m-by-n matrix A and the vector $\mathbf{x}$ of length n is DEFINED to be the vector $\mathbf{b}$ of length m whose coordinates are given by the above linear system.

That's all, there isn't any more. The simple-looking equation $A\mathbf{x} = \mathbf{b}$ has, by definition, exactly the same meaning as the entire system of equations. The only clue to its grand meaning is the uppercase A and the boldface $\mathbf{x}$ and $\mathbf{b}$.

Notice that $A\mathbf{x}$ is NOT defined unless the number of columns of A equals the number of rows of $\mathbf{x}$. Because the i-th coordinate of the product $A\mathbf{x}$ is the expression:

$$a_{i1}x_1 + a_{i2}x_2 + ... + a_{in}x_n.$$

This compatibility condition insures that you run out of a_{ij}'s and x_j's together!

Example 1.13 (Linear systems in matrix form, matrix vector multiplication.)
   a. The following linear system and matrix equation mean the same thing:

$$\begin{array}{rcl} x + 2y + z &=& 6 \\ x + y + 2z &=& 4 \\ w + x \quad + 5z &=& 5 \end{array} \qquad \begin{pmatrix} 0 & 1 & 2 & 1 \\ 0 & 1 & 1 & 2 \\ 1 & 1 & 0 & 5 \end{pmatrix} \begin{pmatrix} w \\ x \\ y \\ z \end{pmatrix} = \begin{pmatrix} 6 \\ 4 \\ 5 \end{pmatrix}$$

b.
$$\begin{pmatrix} 1 & 2 & 3 & 4 \\ 6 & 7 & 8 & 9 \end{pmatrix} \begin{pmatrix} 5 \\ 10 \\ 15 \\ 20 \end{pmatrix} = \begin{pmatrix} 1*5 + 2*10 + 3*15 + 4*20 \\ 6*5 + 7*10 + 8*15 + 9*20 \end{pmatrix} = \begin{pmatrix} 150 \\ 400 \end{pmatrix}$$

c. Also
$$\begin{pmatrix} 1 & 2 & 3 & 1 \\ 2 & 4 & 6 & 3 \\ -1 & -2 & -1 & 3 \end{pmatrix} \begin{pmatrix} 3 \\ 0 \\ -1 \\ 2 \end{pmatrix} = \begin{pmatrix} 2 \\ 6 \\ 4 \end{pmatrix}$$

d. However,
$$\begin{pmatrix} 1 & 2 & 3 \\ 6 & 7 & 8 \end{pmatrix} \begin{pmatrix} 5 \\ 10 \\ 15 \\ 20 \end{pmatrix}$$
DOES NOT EXIST because the number of columns (3) of the matrix is not equal to the number of rows (4) of the vector. □

Objective 2

Compute expressions and verify identities using matrix-vector multiplication and linear combinations.

If A and B are matrices of the same shape, then $A \pm B$ (sum or difference) is defined to be the matrix whose (i, j)-th entry is $a_{ij} \pm b_{ij}$. This is written

$$(A \pm B)_{ij} = a_{ij} \pm b_{ij}.$$

A vector with all coordinates equal to zero (and any shape) is called a **zero vector** and written $\mathbf{0}$. Note that $\mathbf{x} + \mathbf{0} = \mathbf{x}$ for any $\mathbf{x}$ (of the same shape as this particular $\mathbf{0}$).

A second arithmetic operation involving matrices of all shapes is **scalar multipli-cation**. Here, the number k is combined with the matrix A to form the matrix kA whose entries are k times each of the entries of A.

Example 1.14 (Addition of vectors and matrices, scalar multiplication.)
   a. Addition is "coordinate-wise":

$$\begin{pmatrix} 3 \\ 4 \\ 5 \end{pmatrix} + \begin{pmatrix} 4 \\ 5 \\ 3 \end{pmatrix} = \begin{pmatrix} 7 \\ 9 \\ 8 \end{pmatrix} \text{ and } \begin{pmatrix} 4 \\ 3 \end{pmatrix} - \begin{pmatrix} -2 \\ 1 \end{pmatrix} = \begin{pmatrix} 6 \\ 2 \end{pmatrix}, \text{ but } \begin{pmatrix} 6 \\ 2 \end{pmatrix} + \begin{pmatrix} 6 \\ 1 \\ 2 \end{pmatrix}$$

   DOES NOT EXIST because the vectors have different shapes.

   b. Scalar multiplication is easy: $2 \begin{pmatrix} 1 \\ 2 \\ 3 \end{pmatrix} = \begin{pmatrix} 2 \\ 4 \\ 6 \end{pmatrix}$; $0 \begin{pmatrix} 4 \\ 3 \end{pmatrix} = \begin{pmatrix} 0 \\ 0 \end{pmatrix} = \mathbf{0}.$

   c. Also $\begin{pmatrix} 1 & 2 \\ 7 & 3 \end{pmatrix} + \begin{pmatrix} 9 & 5 \\ 8 & 4 \\ 7 & 3 \end{pmatrix}$ DOES NOT EXIST because the two sum-mands have different shapes.

   d. However, $\begin{pmatrix} 9 & 5 \\ 8 & 4 \\ 7 & 3 \end{pmatrix} + \begin{pmatrix} 9 & 5 \\ 8 & 4 \\ 7 & 3 \end{pmatrix} = \begin{pmatrix} 18 & 10 \\ 16 & 8 \\ 14 & 6 \end{pmatrix} = 2 \begin{pmatrix} 9 & 5 \\ 8 & 4 \\ 7 & 3 \end{pmatrix}$ □

A particularly important formula that combines matrix multiplication and addition of vectors is the **distributive law:**

$$A(\mathbf{x} + \mathbf{y}) = A\mathbf{x} + A\mathbf{y}.$$

It holds for **all** matrices and vectors where the indicated sums and products exist. To check the correctness of this formula, just work out both sides and compare cor-responding entries. This is easy if you first notice that the (ij) entry of the matrix times vector product really only depends on one row of the matrix. Therefore, it is enough to consider the case when A is only one row. Suppose $A = (a_{11} \cdots a_{1n})$ (even the shortest matrices have two subscripts for their entries) and $\mathbf{x}, \mathbf{y} \in \mathbb{R}^n$ have i-th coordinate x_i, y_i respectively. Then by the distributive law of **arithmetic**

$$A\mathbf{x} + A\mathbf{y} = (a_{11}x_1 + a_{12}x_2 + \cdots + a_{1n}x_n) + (a_{11}y_1 + a_{12}y_2 + \cdots + a_{1n}y_n)$$
$$= a_{11}x_1 + a_{11}y_1 + a_{12}x_2 + a_{12}y_2 + \cdots a_{1n}x_n + a_{1n}y_n$$
$$= a_{11}(x_1 + y_1) + a_{12}(x_2 + y_2) + \cdots a_{1n}(x_n + y_n)$$
$$= A(\mathbf{x} + \mathbf{y}).$$

There are lots of other familiar-looking formulas relating these operations:

$$0\,\mathbf{x} = \mathbf{0} \qquad\qquad A + B = B + A$$
$$(\alpha\beta)\mathbf{x} = \alpha(\beta\mathbf{x}) \qquad\qquad \alpha(\mathbf{x} + \mathbf{y}) = \alpha\mathbf{x} + \alpha\mathbf{y},$$

for $\alpha, \beta \in \mathbb{R}$, $\mathbf{x}, \mathbf{y} \in \mathbb{R}^n$ and A, B of the same shape. But these all have new meanings because the symbols are vectors and matrices as well as numbers.

In addition, there may be some values of the variables for which the indicated operation is not defined. (Also, wasn't that a wonderful pun?) For example, the last formula doesn't make sense unless $\mathbf{x}$ and $\mathbf{y}$ have the same shape. *Any such formula's correctness is checked by comparing each side entry by entry.*

When vector addition (or subtraction) and scalar multiplication are used together to produce a new vector from (possibly several) old vectors, the new vector is called a *linear combination* of the old ones. Thus, the vector $\mathbf{v}$ is, by definition, a linear combination of the set of vectors $\mathbf{x}_1 \ldots, \mathbf{x}_s$ if $\mathbf{v}$ can be written as a sum of scalar multiples of vectors in the set $\mathbf{x}_1 \ldots \mathbf{x}_s$.

Example 1.15 (Linear combinations of vectors and matrices.)

a. $2\begin{pmatrix} 3 \\ -1 \\ -2 \end{pmatrix} + 3\begin{pmatrix} -1 \\ 3 \\ -5 \end{pmatrix} = \begin{pmatrix} 3 \\ 7 \\ -19 \end{pmatrix}$ and $2\begin{pmatrix} 2 \\ 3 \end{pmatrix} - 3\begin{pmatrix} 3 \\ 4 \end{pmatrix} = \begin{pmatrix} -5 \\ -6 \end{pmatrix}$.

b. If $A = \begin{pmatrix} 2 & -3 & 4 \\ 1 & 5 & 6 \end{pmatrix}$, $B = \begin{pmatrix} 5 & 9 & -6 \\ 8 & -2 & -5 \end{pmatrix}$ and $C = \begin{pmatrix} 1 & 0 \\ 0 & 1 \end{pmatrix}$, then

$$2A - 3B = \begin{pmatrix} -11 & -33 & 26 \\ -22 & 16 & 27 \end{pmatrix},$$

but $A + C$ doesn't exist. The matrices have incompatible shapes.   □

When a set of vectors $\mathbf{v}_1, \ldots \mathbf{v}_t$ has $\mathbf{0}$ as a nontrivial linear combination:

$$\alpha_1 \mathbf{v}_1 + \cdots + \alpha_t \mathbf{v}_t = \mathbf{0} \text{ where some } \alpha_i \neq 0$$

the vectors $\mathbf{v}_1, \ldots \mathbf{v}_t$ are called *linearly dependent*. Any set of vectors that is not linearly dependent is *linearly independent.*

These terms play a pivotal role in mathematical theory underlying matrices. It is therefore important that you understand their meaning. As an illustration, we show that whenever you have a linearly dependent set of vectors, you could throw at least one of them away, and not lose any vectors as linear combinations. More formally:

THEOREM If vectors $\mathbf{v}_1 \ldots \mathbf{v}_t$ are linearly dependent, then some one of them, say $\mathbf{v}_i$, can be written as a linear combination of the others. If $\mathbf{w}$ is expressible as a linear combination of $\mathbf{v}_1 \ldots \mathbf{v}_t$ then it is possible to do this without $\mathbf{v}_i$.

Indeed, if the given vectors are linearly dependent, then there are numbers $\alpha_1 \ldots \alpha_t$ **not all zero**, such that

$$\alpha_1 \mathbf{v}_1 + \cdots + \alpha_t \mathbf{v}_t = \mathbf{0}.$$

Just choose i so that $\alpha_i \neq 0$ and multiply both sides of this equation by $\frac{1}{\alpha_i}$, then isolate $\mathbf{v}_i$ to obtain

$$\mathbf{v}_i = \frac{-\alpha_1}{\alpha_i}\mathbf{v}_1 + \cdots + \frac{-\alpha_{i-1}}{\alpha_i}\mathbf{v}_{i-1} + \frac{-\alpha_{i+1}}{\alpha_i}\mathbf{v}_{i+1} + \cdots + \frac{-\alpha_t}{\alpha_i}\mathbf{v}_t.$$

If $\mathbf{w}$ is expressed as a linear combination of the $\mathbf{v}$s and it involves $\mathbf{v}_i$, then substitute the above expression for $\mathbf{v}_i$ and simplify.

Objective 3

Given $Ax = b$, with A in an echelon form, write the general solution as a linear combination of basic solutions and the distinguished solution.

Vector arithmetic provides an important alternative way to express the general solution of linear systems. Two special types of solutions are used.

Consider the reduced row echelon form augmented matrix $(A|b)$ and the associated vector form of the general solution x, $(A|b) =$

$$\begin{pmatrix} 1 & 2 & -1 & 0 & 0 & | & 3 \\ 0 & 0 & 0 & 1 & 0 & | & -1 \\ 0 & 0 & 0 & 0 & 1 & | & 2 \end{pmatrix}, \quad x = \begin{pmatrix} -2x_2 + x_3 + 3 \\ x_2 \\ x_3 \\ -1 \\ 2 \end{pmatrix} = \begin{pmatrix} -2x_2 + x_3 + 3 \\ x_2 \\ x_3 \\ -1 \\ 2 \end{pmatrix}.$$

(The last step just spreads things out so that the different parts of the vector x are aligned vertically.) Now use vector arithmetic to rewrite the general solution as a linear combination of three vectors, breaking x into pieces according to the above grouping. Finally, factor out scalar multiples of the free variables from the pieces:

$$x = \begin{pmatrix} -2x_2 \\ x_2 \\ 0 \\ 0 \\ 0 \end{pmatrix} + \begin{pmatrix} x_3 \\ 0 \\ x_3 \\ 0 \\ 0 \end{pmatrix} + \begin{pmatrix} 3 \\ 0 \\ 0 \\ -1 \\ 2 \end{pmatrix} = x_2 \begin{pmatrix} -2 \\ 1 \\ 0 \\ 0 \\ 0 \end{pmatrix} + x_3 \begin{pmatrix} 1 \\ 0 \\ 1 \\ 0 \\ 0 \end{pmatrix} + \begin{pmatrix} 3 \\ 0 \\ 0 \\ -1 \\ 2 \end{pmatrix}.$$

The vector with no free variables is the system's ***distinguished solution***.

Notice that the left-hand portion of an augmented matrix in RREF is the RREF of the associated homogeneous system. Therefore, the first two vectors involving free variables are the same no matter what the constant column of the RREF augmented matrix. For this reason, the vectors that are multiplied by free variables are called ***basic solutions of the associated homogeneous system***. We abbreviate this hopelessly long name as BSAHS.

DISTINGUISHED AND BASIC SOLUTIONS

Obtain RREF(A) and write the general solution.
Set all free variables equal to 0 in the system's general solution. The vector that remains is the **distinguished solution**. For each free variable, write the vector of its coefficients in the general solution. Each vector so obtained is a **basic solution** (of the associated homogeneous system).

Every consistent linear system has exactly one ***distinguished solution***. BUT only homogeneous systems have ***basic solutions***. Each basic solution of a homogeneous system corresponds to a free variable in its general solution. When we wish to refer to the basic solutions in this order, we use the term ***standard list*** of basic solutions.

Example 1.16 (Matrix form of the general solution of a linear system.)

a. The linear system in part c of Example 1.11 on page 26 has general solution as indicated: It can be written in vector notation as:

$$x_1 = x_3 + 7$$
$$x_2 = -3x_3 + 5.$$

$$\begin{pmatrix} x_3 + 7 \\ -3x_3 + 5 \\ x_3 \end{pmatrix} = x_3 \begin{pmatrix} 1 \\ -3 \\ 1 \end{pmatrix} + \begin{pmatrix} 7 \\ 5 \\ 0 \end{pmatrix}.$$

The first vector on the right-hand side is the BSAHS, and the second is the **distinguished** solution to the original nonhomogeneous system.

b. The system in part a of Example 1.12 on page 27 has general solution:

$$x_1 = 5 - 2x_2 - 2x_4$$
$$x_2 = \qquad x_2$$
$$x_3 = 3 \qquad -2x_4$$
$$x_4 = \qquad x_4$$

It can be written:

$$\begin{pmatrix} -2x_2 - 2x_4 + 5 \\ x_2 \\ -2x_4 + 3 \\ x_4 \end{pmatrix} = x_2 \begin{pmatrix} -2 \\ 1 \\ 0 \\ 0 \end{pmatrix} + x_4 \begin{pmatrix} -2 \\ 0 \\ -2 \\ 1 \end{pmatrix} + \begin{pmatrix} 5 \\ 0 \\ 3 \\ 0 \end{pmatrix}.$$

The first two vectors on the right-hand side are the BSAHS, and the third is the **distinguished** solution to the original nonhomogeneous system.

c. The indicated general solution can be written in vector notation as:

$$x_1 = -x_2$$
$$x_2 = x_2$$
$$x_3 = -2x_4$$
$$x_4 = x_4$$
$$x_5 = 0$$

$$\begin{pmatrix} -x_2 \\ x_2 \\ -2x_4 \\ x_4 \\ 0 \end{pmatrix} = x_2 \begin{pmatrix} -1 \\ 1 \\ 0 \\ 0 \\ 0 \end{pmatrix} + x_4 \begin{pmatrix} 0 \\ 0 \\ -2 \\ 1 \\ 0 \end{pmatrix}$$

The distinguished solution of this system is the zero vector 0 and is omitted. There are two basic solutions. □

Consider Parts b and c of the preceding example. Notice that the free variables are x_2, x_4. Moreover, only the corresponding BSAHSs has a 1 in the second and fourth position, respectively. Therefore, the only way to get a linear combination of the BSAHSs to equal **0** would be to have both free variables equal to zero. **In other words, the BSAHSs are linearly independent!** Problem 6 addresses the general case of this important fact.

Distinguished and basic solutions are used repeatedly in the coming chapters. You must be able to find them correctly with little effort, because they will be just a small part of much more complicated algorithms. For this reason, we pause to notice some of their most important properties.

The distinguished solution of a nonhomogeneous system has zeros in the positions of the free variables and its other entries come from the last column of the RREF augmented matrix – in order. The distinguished solution of a homogeneous system is ALWAYS the zero vector **0** and is usually not written explicitly.

Suppose the j-th column of the RREF matrix is nonpivotal, so the j-th variable is free and let **s** be the associated basic solution. Then the j-th entry of **s** equals one and the other entries of **s** that correspond to free variables all equal zero. Finally, the remaining entries of **s** (that correspond tp the dependent variables) are the negatives of the j-th column of the RREF matrix — in order.

SUMMARY

The ***product*** $A\mathbf{x}$ is the vector whose i-th entry is:

$$a_{i,1}x_1 + a_{i,2}x_2 + \cdots + a_{i,n}x_n$$

and only defined if the number of columns of A equals the number of rows in **x**. In this section, matrix ***addition*** and ***subtraction*** are defined in terms of the addition and subtraction of the entries:

$$(A \pm B)_{ij} = a_{ij} \pm b_{ij}.$$

Scalar multiplication of a matrix and a number is also introduced. Anything obtained from a set of matrices by addition and scalar multiplication is called a ***linear combination*** of the set of matrices.

If a set of vectors $\mathbf{v}_1 \ldots \mathbf{v}_t$ has **0** as a nontrivial linear combination, then it is ***linearly dependent***, otherwise the set is ***linearly independent***.

A linear system $A\mathbf{x} = \mathbf{b}$ is ***homogeneous*** if $\mathbf{b} = \mathbf{0}$. Each consistent linear system has one ***distinguished solution***. Each homogeneous system has as many ***basic solutions*** as nonpivotal columns of RREF(A). They fall in a natural order given by the ***standard list*** of basic solutions. The BSAHS are linearly independent.

Vector arithmetic is a special case of matrix arithmetic that provides an important new way of writing the general solution of a linear system as a ***linear combination*** of distinguished and basic solutions.

Looking further
Matrix times vector multiplication marks a turning point. With it we move from using matrices as an "efficiency tool" to viewing matrices as operators. This new view is further explained and developed in the next section with linear functions and transform plots. While this move is pivotal in developing more powerful use of matrices, you will be amazed how many different times and ways we use this section. The concepts of linear combination/dependent/independent are fundamental in more abstract settings, as discussed in the Appendix, page 217.

* * * * * * *

Practice Problems

1. Compute the following matrix-vector products or state that the product is not defined. When it does exist, write the linear system (in the variables $x_1, x_2, \ldots$) that has the indicated vector as a solution.

a. $\begin{pmatrix} 1 & 2 & 0 \\ 4 & -3 & 1 \\ 6 & 0 & 7 \end{pmatrix} \begin{pmatrix} 3 \\ -2 \\ -4 \end{pmatrix}$ b. $\begin{pmatrix} 2 & -3 \\ 1 & 4 \\ 3 & -2 \end{pmatrix} \begin{pmatrix} 7 \\ 1 \end{pmatrix}$ c. $\begin{pmatrix} 0 & 1 & 3 \\ -4 & 2 & 6 \end{pmatrix} \begin{pmatrix} 3 \\ 9 \end{pmatrix}$

d. $\begin{pmatrix} 5 & 4 & 7 & 0 \\ 3 & -1 & 2 & -4 \end{pmatrix} \begin{pmatrix} 1 \\ -1 \\ 0 \\ 5 \end{pmatrix}$ e. $\begin{pmatrix} 1 & -4 & -9 \\ 3 & 2 & -1 \\ 0 & 4 & 2 \\ 6 & 3 & 1 \end{pmatrix} \begin{pmatrix} 1 \\ 4 \\ 9 \\ 0 \end{pmatrix}$

2. The following equations give the general solution of a linear system. Write each one as a linear combination of the distinguished solution and BSAHS.

a. $\begin{aligned} x_1 &= x_3 - 5 \\ x_2 &= 7x_3 + 2 \end{aligned}$ b. $\begin{aligned} x_1 &= 3x_2 - 9x_4 + 8 \\ x_3 &= \quad\;\; - 7x_4 - 5 \end{aligned}$

c. $\begin{aligned} x_1 &= \quad 3x_4 + 7 \\ x_2 &= \;\; -x_4 \\ x_3 &= -5x_4 - 1 \end{aligned}$ d. $\begin{aligned} x_1 &= 2x_2 - 3x_4 \qquad\quad + 7 \\ x_3 &= \qquad\; - 4x_4 - 8x_5 + 2 \end{aligned}$

3. Find the general solution of each equation and write it as a linear combination of the distinguished solution and BSAHS. Give the standard list of basic solutions.

a. $\begin{pmatrix} 1 & 3 & -8 & 0 & 8 \\ 0 & 0 & 0 & 1 & -1 \end{pmatrix} \mathbf{x} = \begin{pmatrix} 2 \\ 6 \end{pmatrix}$ b. $\begin{pmatrix} 1 & -8 & 0 & 7 & 0 \\ 0 & 0 & 1 & -2 & 0 \\ 0 & 0 & 0 & 0 & 1 \end{pmatrix} \mathbf{x} = \begin{pmatrix} 7 \\ 6 \\ 1 \end{pmatrix}$

c. $\begin{pmatrix} 1 & 5 & 0 & 0 & 8 \\ 0 & 0 & 1 & 0 & 6 \\ 0 & 0 & 0 & 1 & -3 \end{pmatrix} \mathbf{x} = \begin{pmatrix} 2 \\ -1 \\ 7 \end{pmatrix}$ d. $\begin{pmatrix} 1 & -6 & 5 & 0 & -8 \\ 0 & 0 & 1 & -6 \end{pmatrix} \mathbf{x} = \begin{pmatrix} 1 \\ 7 \end{pmatrix}$

In Problems 4 and 5, decide which of the vectors is the distinguished solution or a basic solution to the given equation.

4. $\begin{pmatrix} 1 & -6 & 3 & 0 & -3 \\ 0 & 0 & 0 & 1 & 1 \end{pmatrix} \mathbf{x} = \begin{pmatrix} 5 \\ 0 \end{pmatrix}$ (i) $\begin{pmatrix} -6 \\ 1 \\ 0 \\ 0 \\ 0 \end{pmatrix}$ (ii) $\begin{pmatrix} -3 \\ 0 \\ 1 \\ 0 \\ 0 \end{pmatrix}$ (iii) $\begin{pmatrix} 3 \\ 1 \\ 1 \\ 0 \\ 0 \end{pmatrix}$

5. $\begin{pmatrix} 1 & 6 & 0 & 0 & 7 \\ 0 & 0 & 1 & 0 & -2 \\ 0 & 0 & 0 & 1 & -7 \end{pmatrix} \mathbf{x} = \begin{pmatrix} 3 \\ -1 \\ 4 \end{pmatrix}$ (i) $\begin{pmatrix} -6 \\ 1 \\ 0 \\ 0 \\ 0 \end{pmatrix}$ (ii) $\begin{pmatrix} -7 \\ 0 \\ 2 \\ 7 \\ 0 \end{pmatrix}$ (iii) $\begin{pmatrix} 3 \\ 0 \\ -1 \\ 4 \\ 0 \end{pmatrix}$

6. Show that the basic solutions of a homogeneous linear system are linearly independent.

7. Show that a nonempty subset of a set of linearly independent vectors is itself, linearly independent.

8. Suppose the matrix B is obtained from A by removing some rows and columns of A. Show that the rank of A is greater than or equal to the rank of B.

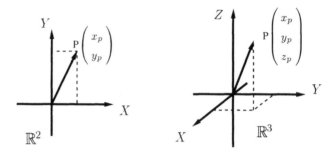

1.6 Geometric vectors and linear functions

Objective 1

Decide if two directed line segments represent the same geometric vector.
Interpret vector arithmetic geometrically.
Compute distances and angles in $\mathbb{R}^2$ and $\mathbb{R}^3$.

We now make an important visual connection between linear systems and geometry. Spatial coordinate systems make this connection. We use standard rectangular right-handed Cartesian coordinate systems, as in the figure. They are based on mutually perpendicular coordinate axes. **Notice** that we list the coordinates of a point in a column rather than a row. It is critical that we do this so that we will be able to understand matrix-times-vector multiplication in a geometric way.

The **origin**, O of a coordinate system is the point with all coordinates zero. Suppose R and P are two points. The directed line segment from R to P is written $\overrightarrow{RP}$. Each directed line segment $\overrightarrow{OP}$ **represents** the **geometric vector P** having coordinates x_P, y_P if we are working in $\mathbb{R}^2$ and x_P, y_P, z_P if we are working in $\mathbb{R}^3$.

The relationship between **geometric vectors** and directed line segments representing them is a little like rational numbers and fractions. A single rational

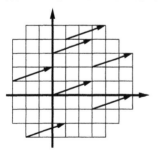

number can be represented by several different, but equivalent, fractional expressions ($\frac{1}{4} = \frac{16}{64}$). For each rational number, there is a unique fraction in "reduced" form. In a similar way, the directed line segment $\overrightarrow{OP}$, is a "reduced form" representative of the geometric vector **P**. For this reason **P** is called the **position vector** of the point P. Actually, however, the geometric vector **P** is represented by ANY directed line segment $\overrightarrow{QR}$ in the same direc-

tion and of the same length as $\overrightarrow{OP}$. Two directed line segments that represent the same geometric vector are called **equivalent.** The directed line segments in the figure are all equivalent.

The **length** or **magnitude** $|\mathbf{v}|$ of a geometric vector $\mathbf{v}$ is the distance between the end points of any representing directed line segment. Thus,

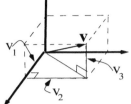

$$|\mathbf{v}| = \sqrt{\left(\sqrt{v_1^2 + v_2^2}\right)^2 + v_3^2} = \sqrt{v_1^2 + v_2^2 + +v_3^2}$$

by two applications of the Pythagorean theorem as indicated in the figure. A vector with length 1 is called a **unit vector**. The **basic unit vectors** $\mathbf{i}$, $\mathbf{j}$, $\mathbf{k}$ are the unit vectors in the directions of the positive x-axis, y-axis, z-axis, respectively.

Geometric vectors have direction and magnitude or length, but NOT position. Two directed line segments $\overrightarrow{AB}$, $\overrightarrow{CD}$ represent the same geometric vector if the quadrilateral $ABDC$ is a parallelogram. This condition is expressed algebraically by

$$x_B - x_A = x_D - x_C \quad \text{and}$$
$$y_B - y_A = y_D - y_C \quad \text{and}$$
$$z_B - z_A = z_D - z_C.$$

NOTE: Because any formula in $\mathbb{R}^2$ can be obtained from the corresponding formula in $\mathbb{R}^3$ by removing z-coordinates, we often omit the $\mathbb{R}^2$ formula.

The geometric interpretation of scalar multiplication is easy. The vector $t\mathbf{v}$ has length equal to $|t|$ times the length of $\mathbf{v}$ and the same or opposite direction as $\mathbf{v}$ depending on whether t is positive or negative.

The geometric interpretation of vector addition uses equivalent directed line segments. To find $\mathbf{v} + \mathbf{w}$ geometrically, arrange directed line segments representing the vectors "tip to tail." This can be done in two ways, each of which ends up halfway, around the parallelogram.

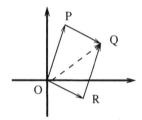

$$\overrightarrow{OP} + \overrightarrow{PQ} = \overrightarrow{OQ} = \overrightarrow{OR} + \overrightarrow{RQ}$$

Linear combinations are more interesting. Start with two position vectors $\overrightarrow{OA}$ and $\overrightarrow{OB}$ in $\mathbb{R}^3$ and let $\mathbf{A}$ and $\mathbf{B}$ be the geometric vectors that they represent. Consider the points P where

$$\overrightarrow{OP} = \mathbf{A} + t\mathbf{B} \text{ for } t \text{ in } \mathbb{R}$$

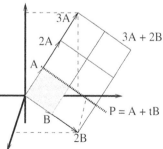

As the figure indicates, the points P form a straight line through A and parallel to $\mathbf{B}$. Replace $\mathbf{A}$ with $s\mathbf{A}$, for s in $\mathbb{R}$, to get a second line parallel to the first. Thus, the points P of the form

$$\overrightarrow{OP} = s\mathbf{A} + t\mathbf{B} \text{ for } s, t \text{ in } \mathbb{R}$$

form the plane containing O, A and B.

As you know, the law of cosines generalizes the Pythagorean theorem and relates the angle θ between two sides of a triangle to the lengths of the sides:

$$|\mathbf{u} - \mathbf{v}|^2 = |\mathbf{u}|^2 + |\mathbf{v}|^2 - 2|\mathbf{u}|\,|\mathbf{v}|\cos(\theta).$$

Solve for $|\mathbf{u}|\,|\mathbf{v}|\cos(\theta)$ in the law of cosines using the above formula for the lengths of the three vectors involved:

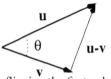

$$|\mathbf{u}|\,|\mathbf{v}|\cos(\theta) = u_1 v_1 + u_2 v_2 + u_3 v_3.$$

The expression on the right-hand side is often called the *dot product* $\mathbf{u} \cdot \mathbf{v}$ of the vectors $\mathbf{u}$ and $\mathbf{v}$. We can express the "dot product" as a "matrix-times-vector product" by flipping the first column vector on its side. This "flipping" extends to matrices and is called "transposing." In general, the *transpose* M^T of a matrix M is obtained from M by interchanging the rows and columns of M. In particular, the transpose of a column vector is a row vector and $\mathbf{u} \cdot \mathbf{v} = \mathbf{u}^T\mathbf{v} = \mathbf{v}^T\mathbf{u}$.

The law of cosines thus leads to the facts:

$$\cos(\theta) = \frac{\mathbf{u}^T\mathbf{v}}{|\mathbf{u}|\,|\mathbf{v}|} \qquad \mathbf{u}, \mathbf{v} \text{ are perpendicular if and only if } \mathbf{u}^T\mathbf{v} = 0.$$

WARNING: It is geometrically clear that the angle θ between two vectors satisfies $0 \le \theta \le \pi$, in radian measure. Fortunately, this is exactly the domain of the standard *arcos* $(\cos^{-1})$ function. Therefore, this function on a calculator will give you the correct answer. Do not be lulled into a false sense of security. Things will not be so nice if you try to use *arctan* $(\tan^{-1})$ in Section 2.1.

Example 1.17 (Distance, geometric vectors and angles in $\mathbb{R}^3$.)
Suppose P, Q and R have coordinates:

$$x_P = 2, y_P = 1, z_P = -2$$
$$x_Q = 1, y_Q = 2, z_Q = 2$$
$$x_R = 3, y_R = 3, z_R = 0.$$

Then the position vectors $\mathbf{P}$, $\mathbf{Q}$ and $\mathbf{R}$ are:

$$\mathbf{P} = \begin{pmatrix} 2 \\ 1 \\ -2 \end{pmatrix}, \mathbf{Q} = \begin{pmatrix} 1 \\ 2 \\ 2 \end{pmatrix}, \mathbf{R} = \begin{pmatrix} 3 \\ 3 \\ 0 \end{pmatrix}$$

The geometric vector $\mathbf{P} = 2\mathbf{i} + \mathbf{j} - 2\mathbf{k}$ is represented by both $\overrightarrow{OP}$ and $\overrightarrow{QR}$ because

$$x_R - x_Q = 3 - 1 = 2$$
$$y_R - y_Q = 3 - 1 = 1 \text{ and}$$
$$z_R - z_Q = -2.$$

The lengths of these vectors are:

$$|\mathbf{P}| = \sqrt{2^2 + 1^2 + (-2)^2} = 3, \ |\mathbf{Q}| = \sqrt{1^2 + 2^2 + 2^2} = 3, \ |\mathbf{R}| = \sqrt{3^2 + 3^2 + 0^2} = 3\sqrt{2}.$$

The triangle $\triangle OQR$ has sides representing the three vectors $\mathbf{Q} = \overrightarrow{OQ}$, $\mathbf{P} = \overrightarrow{QR}$ and $\mathbf{R} = \overrightarrow{OR}$. The cosines of the angles of the triangle $\triangle OQR$ are:

$$\cos(\angle OQR) = \frac{\overrightarrow{QO}^T \overrightarrow{QR}}{|\overrightarrow{QO}||\overrightarrow{QR}|} = \frac{-\mathbf{Q}^T \mathbf{P}}{|-\mathbf{Q}||\mathbf{P}|} = ((-1)(2) + (-2)(1) + (-2)(-2))/3 \cdot 3 = 0$$

$$\cos(\angle QOR) = \frac{\overrightarrow{OQ}^T \overrightarrow{OR}}{|\overrightarrow{OQ}||\overrightarrow{OR}|} = \frac{\mathbf{Q}^T \mathbf{R}}{|\mathbf{Q}||\mathbf{R}|} = ((1)(3)+(2)(3)+(2)(0))/3 \cdot 3\sqrt{2} = 1/\sqrt{2}$$

$$\cos(\angle QRO) = \frac{\overrightarrow{RQ}^T \overrightarrow{RO}}{|\overrightarrow{RQ}||\overrightarrow{RO}|} = \frac{-\mathbf{P}^T (-\mathbf{R})}{|-\mathbf{P}||-\mathbf{R}|} = \cdots = 1/\sqrt{2}$$

The angles are therefore $\pi/2, \pi/4$, and $\pi/4$, respectively. Thus, the vectors $\mathbf{P}$ and $\mathbf{Q}$ are perpendicular, and the triangle $\triangle OQR$ is an isosceles right triangle that just happens to live in a skew plane of $\mathbb{R}^3$.   □

<hr>

Objective 2

Use transform plots to present linear functions and linear systems geometrically.

<hr>

Suppose an m-by-n matrix A. The distributive property of matrix $\times$ vector multiplication implies that the *vector function*

$$f_A : \mathbb{R}^n \to \mathbb{R}^m \text{ DEFINED BY } f_A(\mathbf{x}) = A\mathbf{x}$$

satisfies the conditions:

$$f_A(\mathbf{x} + \mathbf{y}) = A(\mathbf{x} + \mathbf{y}) = A\mathbf{x} + A\mathbf{y} = f_A(\mathbf{x}) + f_A(\mathbf{y}) \text{ for all } \mathbf{x}, \mathbf{y} \in \mathbb{R}^m,$$
$$f_A(\alpha \mathbf{x}) = A(\alpha \mathbf{x}) = \alpha(A\mathbf{x}) = \alpha f_A(\mathbf{x}) \qquad\qquad \text{ for all } \alpha \in \mathbb{R}, \ \mathbf{x} \in \mathbb{R}^m.$$

(The symbol $\in$ means "in" or "is a member of.")

Any vector function $f : \mathbb{R}^m \to \mathbb{R}^n$ satisfying the properties

$$f(\mathbf{x} + \mathbf{y}) = f(\mathbf{x}) + f(\mathbf{y}) \text{ for all } \mathbf{x}, \mathbf{y} \in \mathbb{R}^m,$$
$$\text{and} \qquad f(\alpha \mathbf{x}) = \alpha f(\mathbf{x}) \qquad\qquad \text{ for all } \alpha \in \mathbb{R}, \ \mathbf{x} \in \mathbb{R}^m$$

is called a *linear* function. We will show at the end of this section that **every** linear function is f_A for some matrix A, but for now we simply wish to use linear functions to interpret linear systems in a NEW geometrical way. In contrast to the past emphasis on solutions, our goal this time is to study the set of vectors $\mathbf{b}$ for which $A\mathbf{x} = \mathbf{b}$ has a solution (see Part c of Example 1.19).

Graphs of functions of a single variable $y = f(x)$ are very familiar from other courses. It is natural to try to extend this idea to vector functions. But can you imagine what such a graph would have to look like even for $n = m = 2$? Both the new $\mathbf{x}$-axis and the new $\mathbf{y}$-axis would be planes, and they would be perpendicular, crossing only at the origin. This is impossible in $\mathbb{R}^3$.

It is still helpful to make a **table of values** to understand a given linear vector function. And it is possible to get a pretty good idea of what the given linear vector function does. A useful geometric model for vector functions is a **transform plot**. Transform plots have two coordinate systems: one for the *domain* or "before" coordinate system, and the other for the the *range* or "after" coordinate system.

Here is how to use transform plots. Simply draw a figure in the before coordinate system. Then compute a table of values for key points of the figure. Finally, sketch the image of the figure from the computed values. We will show that the images of the points on a straight line in the before coordinate system form a straight line in the after coordinate system. To "track" the **standard unit square**

$$S = \{(x,y)^T : 0 \le x \le 1,\ 0 \le y \le 1\}$$

all we need are the images of its four corners, AND one of these is trivial because $f_A((0\ 0)^T) = (0\ 0)^T$.

Example 1.18 (Transform plots.)

a. Let $A = \begin{pmatrix} 2 & 0 \\ 0 & 1 \end{pmatrix}$. Then a simple transform plot that tracks S is:

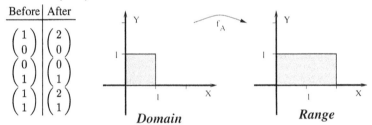

f_A fattens everything by a factor of 2 with no change in height or orientation. Linear functions associated with diagonal matrices are combinations of fattening (horizontal scaling) and stretching (vertical scaling).

b. Let $A = \frac{1}{4} \begin{pmatrix} 5 & 3 \\ 3 & 5 \end{pmatrix}$ and again track the standard unit square S.

Compute the image of each corner of the square separately to see that the figure is stretched along the line $y = x$ by a factor of 2 and shrunk along the line $y = -x$ by a factor of $\frac{1}{2}$.

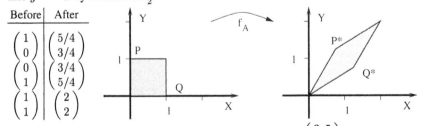

NOTE: It is important to label the points because $B = \frac{1}{4} \begin{pmatrix} 3 & 5 \\ 5 & 3 \end{pmatrix}$ has the same transform plot except that P^* and Q^* are flipped.

c. Let $A = \begin{pmatrix} 0 & 1 \\ -1 & 2 \end{pmatrix}$. Then a transform plot tracking S is:

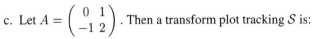

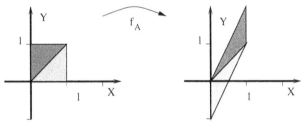

Sometimes "tracking" a more elaborate figure or figures is necessary to understand the function. This function moves points above the line $y = x$ up and to the right while moving points below the line $y = x$ down and to the left. The amount of the shift depends on the distance from the point to the line $y = x$. Such functions are called *shears*.   □

Straight lines and linear functions

The points on the straight line $y = mx + b$ can be expressed as:

$$\begin{pmatrix} x \\ y \end{pmatrix} = \begin{pmatrix} x \\ b + mx \end{pmatrix} = \begin{pmatrix} 0 \\ b \end{pmatrix} + x \begin{pmatrix} 1 \\ m \end{pmatrix}, x \in \mathbb{R}.$$

Apply the function f_A "multiply by A." Then use properties of matrix arithmetic to obtain

$$f_A \begin{pmatrix} x \\ y \end{pmatrix} = A \begin{pmatrix} x \\ y \end{pmatrix} = A \begin{pmatrix} 0 \\ b \end{pmatrix} + A \left(x \begin{pmatrix} 1 \\ m \end{pmatrix} \right)$$

$$= A \begin{pmatrix} 0 \\ b \end{pmatrix} + xA \begin{pmatrix} 1 \\ m \end{pmatrix} = f_A \begin{pmatrix} 0 \\ b \end{pmatrix} + xf_A \begin{pmatrix} 1 \\ m \end{pmatrix}, x \in \mathbb{R}.$$

Provided $f_A \begin{pmatrix} 1 \\ m \end{pmatrix} \neq \mathbf{0}$, the line through $f_A \begin{pmatrix} 0 \\ b \end{pmatrix}$ in the direction of $f_A \begin{pmatrix} 1 \\ m \end{pmatrix}$ consists of points corresponding to a values of x, as described when we discussed vector linear combinations in the preceding objective. This shows a (nonvertical) straight line in the domain is transformed to either a straight line or a point in the range by a linear function.

In making transform plots, it is also convenient to use the fact that any two parallel lines in the domain map to parallel lines in the range. Be aware that angles between lines are NOT necessarily preserved by linear functions. In fact, one of the best ways to understand a linear function transform plot is to select some line segments of domain "graph paper" that form a "grid." Then treat the "grid" as a figure and track it.

Example 1.19 (Transform plots of grids and linear systems.)

a. Consider the linear function appearing in part a of Example 1.18 on page 42 and the somewhat silly linear system:

$$A\mathbf{x} = \begin{pmatrix} 2 & 0 \\ 0 & 1 \end{pmatrix} \begin{pmatrix} x \\ y \end{pmatrix} = \begin{pmatrix} -2 \\ 3 \end{pmatrix} = \mathbf{b}.$$

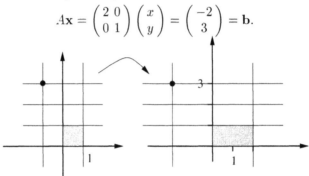

The solution to the linear system is the point in the domain that maps to the specified range point. Of course the point with coordinates $\mathbf{x} = \begin{pmatrix} -1 \\ 3 \end{pmatrix}$ in the domain works. Indeed $f_A(\mathbf{x}) = \mathbf{b}$.

b. Consider the linear system and solution associated with the linear function appearing in part b of Example 1.18 on page 42:

$$A\mathbf{x} = \frac{1}{4} \begin{pmatrix} 5 & 3 \\ 3 & 5 \end{pmatrix} \begin{pmatrix} x \\ y \end{pmatrix} = \begin{pmatrix} 1 \\ 3 \end{pmatrix} = \mathbf{b}.$$

The point with coordinates $(x, y)^T = (-1, 3)^T$ in the domain maps to the point $(1, 3)^T$ in the range. Therefore, this transform plot presents the linear system and its solution in a geometric way.

c. This is a transform plot from $\mathbb{R}^2$ to $\mathbb{R}^3$.
It shows the linear system:
The image of the figure is the grid
lying in the skew plane.

$$A\mathbf{x} = \begin{pmatrix} 1 & -1 \\ 1 & 0 \\ 1 & 1 \end{pmatrix} \begin{pmatrix} x \\ y \end{pmatrix} = \begin{pmatrix} 1 \\ 0 \\ 2 \end{pmatrix} = \mathbf{b}.$$

Notice that $\mathbf{b}$ is NOT in the skew plane that is the image of the "before grid." This implies that there is no possible $\mathbf{x}$ such that $f_A(\mathbf{x}) = \mathbf{b}$! Put another way, the original linear system is inconsistent!

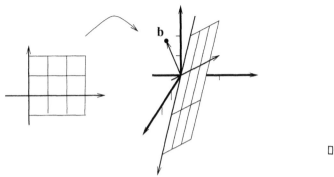

□

Notice that the image of a linear function f_A is exactly the set of all **b** for which $A\mathbf{x} = \mathbf{b}$ has a solution. This set can also be described as all possible linear combinations of the columns of A, and it is the prototypical example of what is called a *subspace* in more advanced courses.

Given a transform plot, it is possible to find the matrix of the associated linear function (or show that no such matrix exists). All that is necessary is to deduce the images of two points having linearly independent position vectors. If additional information is given, then you should check that the matrix you have computed really leads to the indicated transform plot.

Example 1.20 (From transform plots to matrices.)

  a. Find the matrix of the linear function with the indicated transform plot.

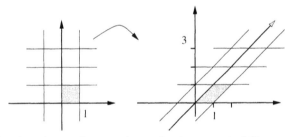

Because the domain graph paper has only three vertical lines and they are symmetric about the y axis, you can figure out the image orientation and see that:

$$f_A\begin{pmatrix}1\\0\end{pmatrix} = \begin{pmatrix}1\\0\end{pmatrix} \text{ while } f_A\begin{pmatrix}0\\1\end{pmatrix} = \begin{pmatrix}1\\1\end{pmatrix}.$$

These facts are expressed by the matrix equations

$$\begin{pmatrix}a_{11}\\a_{21}\end{pmatrix} = \begin{pmatrix}a_{11} & a_{12}\\a_{21} & a_{22}\end{pmatrix}\begin{pmatrix}1\\0\end{pmatrix} = \begin{pmatrix}1\\0\end{pmatrix} \quad \begin{pmatrix}a_{12}\\a_{22}\end{pmatrix} = \begin{pmatrix}a_{11} & a_{12}\\a_{21} & a_{22}\end{pmatrix}\begin{pmatrix}0\\1\end{pmatrix} = \begin{pmatrix}1\\1\end{pmatrix}.$$

Compare coordinates of the first and last vectors in these equations to see that
The third point of the standard unit square provides a check:

$$A = \begin{pmatrix}1 & 1\\0 & 1\end{pmatrix}.$$

$$A\begin{pmatrix}1\\1\end{pmatrix} = \begin{pmatrix}2\\1\end{pmatrix}.$$

b. This method is only a little more complicated for a more exotic figure.

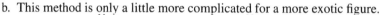

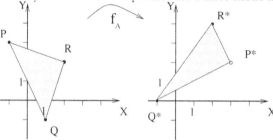

The transformed points indicate that

$$f_A\begin{pmatrix} -1 \\ 3 \end{pmatrix} = \begin{pmatrix} 3 \\ 2 \end{pmatrix}, \quad f_A\begin{pmatrix} 1 \\ -1 \end{pmatrix} = \begin{pmatrix} -1 \\ 0 \end{pmatrix} \text{ and } f_A\begin{pmatrix} 2 \\ 2 \end{pmatrix} = \begin{pmatrix} 2 \\ 4 \end{pmatrix}.$$

Each of these equations involves multiplication of a vector by A and as in Example 1.10 on page 22, we can combine them into a generalized augmented matrix:

$$\begin{pmatrix} a_{11} \; a_{12} & -1 \; 1 \; 2 \\ a_{21} \; a_{22} & 3 \; -1 \; 2 \end{pmatrix} = \begin{pmatrix} 3 \; -1 \; 2 \\ 3 \; 0 \; 4 \end{pmatrix},$$

but the variables are in the wrong part! Nonetheless, this is a system of six linear equations in the four variables $a_{11}, a_{12}, a_{21}, a_{22}$ that is really two linear systems of three equations in two sets of variables. (The first system having variables a_{11}, a_{12} and the second system having variables a_{21}, a_{22}.)
The resulting system of linear systems has a generalized augmented matrix **just** as in Example 1.10 on page 22!
This generalized augmented matrix has RREF:

$$
\begin{aligned}
-a_{11} + 3a_{12} &&&= 3 \\
&& -a_{21} + 3a_{22} &= 3 \\
a_{11} \;\; - \;\; a_{12} &&&= -1 \\
&& a_{21} \;\; - \;\; a_{22} &= 0 \\
2a_{11} + 2a_{12} &&&= 2 \\
&& 2a_{21} + 2a_{22} &= 4
\end{aligned}
$$

$$\begin{pmatrix} -1 & 3 & 3 & 3 \\ 1 & -1 & -1 & 0 \\ 2 & 2 & 2 & 4 \end{pmatrix}!$$

$$\begin{pmatrix} 1 & 0 & 0 & 1 \\ 0 & 1 & 1 & 1 \\ 0 & 0 & 0 & 0 \end{pmatrix}. \text{ Thus, } \begin{matrix} a_{11} = 0 \\ a_{12} = 1 \end{matrix}; \quad \begin{matrix} a_{21} = 1 \\ a_{22} = 1 \end{matrix} \text{ so } A = \begin{pmatrix} a_{11} \; a_{12} \\ a_{21} \; a_{22} \end{pmatrix} = \begin{pmatrix} 0 & 1 \\ 1 & 1 \end{pmatrix}.$$

The process we have just gone through will seem very much simpler when we can use matrix multiplication as introduced in Section 2.2. For example, the pattern of a_{ij}s in the middle of the last displayed equation is switched from the pattern of the matrix entries. We actually solved a matrix equation $CX = B$ where the variable part $X = A^T$ was the **transpose** of A, as pointed out in Problem 10 of Section 2.2.

Notice also that if one of the data points was wrong, then at least one of the two linear systems would have been inconsistent, and at least one of the augmented columns would have been pivotal.   □

All linear functions come from matrices

This is nothing more than a general version of the last example.

Suppose a linear function $f : \mathbb{R}^n \to \mathbb{R}^m$ is given. Let $\mathbf{e}_j$ be the column vector of length n that has all zeros except for a 1 in the j-th position. Then any $\mathbf{x}$ in $\mathbb{R}^n$ can be written

$$\mathbf{x} = x_1 \mathbf{e}_1 + x_2 \mathbf{e}_2 + \cdots + x_n \mathbf{e}_n,$$

where x_j is the j-th coordinate of $\mathbf{x}$.

Build an m-by-n matrix A whose j-th column is $f(\mathbf{e}_j)$. Then

$$A\mathbf{x} = \left(\begin{array}{cccc} f(\mathbf{e}_1) & f(\mathbf{e}_2) & \cdots & f(\mathbf{e}_n) \end{array} \right) \begin{pmatrix} x_1 \\ x_2 \\ \vdots \\ x_n \end{pmatrix} = f(\mathbf{e}_1)x_1 + f(\mathbf{e}_2)x_2 + \cdots + f(\mathbf{e}_n)x_n.$$

But this is the same $f(\mathbf{x})$ because of the defining properties of linear functions:

$$f(\mathbf{e}_1)x_1 + f(\mathbf{e}_2)x_2 + \cdots + f(\mathbf{e}_n)x_n = f(x_1\mathbf{e}_1) + f(x_2\mathbf{e}_2) + \cdots + f(x_n)\mathbf{e}_n$$
$$= f(x_1\mathbf{e}_1 + x_2\mathbf{e}_2 + \cdots + x_n\mathbf{e}_n) = f(\mathbf{x}).$$

This shows that $f(\mathbf{x}) = A\mathbf{x}$ for any $\mathbf{x}$ in $\mathbb{R}^n$, as desired. **It also shows that the $j - th$ column of A is the images of the $\mathbf{e}_j$ under the associated linear function.**

SUMMARY

Geometric vectors are represented by directed line segments. Two directed line segments represent the same geometric vector if they have the same *length* and *direction*. The angle θ between two geometric vectors is given by the formula

$$|\mathbf{u}|\,|\mathbf{v}|\cos(\theta) = \mathbf{u}^T\mathbf{v}.$$

Here we use the *transpose* operator that interchanges the rows and columns of a matrix.

For each matrix A, we define the *linear function* $f_A(\mathbf{x})$ by $f_A(\mathbf{x}) = A\mathbf{x}$. Transform plots can be useful in understanding these functions geometrically. A key property of a transform plot is that straight lines are mapped to straight lines.

Solving the linear system $A\mathbf{x} = \mathbf{b}$ amounts to computing the inverse image of $\mathbf{b}$ under the linear function f_A. Inconsistent linear systems occur when $\mathbf{b}$ is not in the image of f_A.

Looking further

The extension of familiar geometric concepts, like "distance" and "angle" to more than three dimensions is amazingly useful, even if initially counterintuitive. Einstein's theory of relativity is based a non-Euclidean "metric" in the four dimensions of "Spacetime."[3]

---

[3] *The Geometry of Spacetime*, J. Callahan, Springer-Verlag, New York, 1999.

Practice Problems

1. On a single coordinate system, plot the directed line segments $\overrightarrow{PQ}$ for each point pair. For which pairs do $\overrightarrow{PQ}$ represent the same geometric vector? Find the distance between each pair.

   a. (i) $P\begin{pmatrix}1\\1\end{pmatrix}$, $Q\begin{pmatrix}-2\\1\end{pmatrix}$ (ii) $P\begin{pmatrix}0\\1\end{pmatrix}$, $Q\begin{pmatrix}-3\\1\end{pmatrix}$ (iii) $P\begin{pmatrix}-1\\-1\end{pmatrix}$, $Q\begin{pmatrix}2\\-1\end{pmatrix}$

   b. (i) $P\begin{pmatrix}1\\1\\1\end{pmatrix}$, $Q\begin{pmatrix}0\\2\\1\end{pmatrix}$ (ii) $P\begin{pmatrix}-1\\1\\0\end{pmatrix}$, $Q\begin{pmatrix}0\\0\\0\end{pmatrix}$ (iii) $P\begin{pmatrix}0\\0\\0\end{pmatrix}$, $Q\begin{pmatrix}-1\\1\\0\end{pmatrix}$

2. Decide if the following vectors are perpendicular.
   a. $i+j+k, -2i+j+k$
   b. $3i+5j, 5i-3j$
   c. $(a,b,c)^T, (b,-a,0)^T$
   d. $(3,3,5,1,6)^T, (2,-5,1,1,-3)^T$

3. Find the angle between the given vectors and sketch.
   a. $(1,1)^T, (0,1)^T$
   b. $j, 2j+k$
   c. $(1,2,3)^T, (4,5,6)^T$

4. For each matrix, make a table of values for the indicated points and draw the transform plot of the given square.

   a. $\begin{pmatrix}0 & 2\\2 & 0\end{pmatrix}$
   b. $\begin{pmatrix}1 & 1/2\\-1/2 & 1\end{pmatrix}$

   c. $\begin{pmatrix}1 & 0\\0 & 1\end{pmatrix}$
   d. $\begin{pmatrix}1 & 1\\1 & 1\end{pmatrix}$

5. Find the matrices that transform $\triangle ABC$ into the specified $\triangle A^* B^* C^*$.
   WARNING: Check all three points!

   a.
   A: (-3,-1)
   B: (-2, 2)
   C: (0, 0)

   b.
   A: (1,-3)
   B: (-2,-2)
   C: (-1,1)

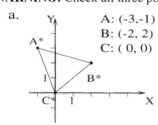

   c.
   A: (1/4,5/4)
   B: (3/2,-1/2)
   C: (-1/4,-1/4)

   d.
   A: (-1,3)
   B: (1,1)
   C: (-1,0)

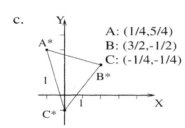

6. The unit vector x is perpendicular to $(1\ 1\ 1)^T$ and to $(-1\ 0\ 1)^T$. Solve for x?

7. Prove that parallel lines are mapped to parallel lines by linear functions.

1.7 Polynomial interpolation

Objective 1

Determine the general form of a polynomial function having degree at most n and having m specified functional values, $n, m \leq 3$.

We know that two points uniquely determine a straight line and that a function of the form $f(x) = mx + b$ has a straight line as its graph. In this section we USE linear systems to develop an important generalization of this well-known fact.

Recall that a **polynomial function** $f(x)$ of **degree** n has the form

$$f(x) = a_0 + a_1 x + a_2 x^2 + \cdots + a_n x^n$$

for constants $a_0, \ldots, a_n$ and $a_n \neq 0$. Recall also that $f(x)$ is determined by its graph or table of values and that its graph is a straight line whenever $f(x)$ has degree 1.

Warning: We generally write polynomials with the constant term on the **left** and leading coefficient on the **right**. This turns out to be important in Chapter 3.

Suppose $f(x)$ is known to have degree at most 2, and the graph $y = f(x)$ passes through the 3 points $(0,0), (1,0), (2,2)$. This means that the functional values $f(0) = 0$ $f(1) = 0$ and $f(2) = 2$ are known.

Since $f(x)$ is a polynomial of degree at most 2, we know that the formula for $f(x)$ has the form $f(x) = a_0 + a_1 x + a_2 x^2$ for some numbers a_0, a_1, a_2. Our problem is to determine the a_i's for which $f(0) = 0$, $f(1) = 0$ and $f(2) = 2$. By writing out this information

$$f(0) = a_0 + a_1(0) + a_2(0^2) = 0$$
$$f(1) = a_0 + a_1(1) + a_2(1^2) = 0$$
$$f(2) = a_0 + a_1(2) + a_2(2^2) = 2$$

there emerges a linear system of three equations in three unknowns, BUT the unknowns are the coefficients a_i. This may be unexpected because the **coefficients** are not normally thought of as **variables**. (**This shift in focus from x as the variable to the coefficients as variables is an important new twist!**)

The augmented matrix of this linear system is easy to reduce to RREF

$$\begin{pmatrix} 1 & 0 & 0 & 0 \\ 1 & 1 & 1 & 0 \\ 1 & 2 & 4 & 2 \end{pmatrix} \xrightarrow{RREF} \begin{pmatrix} 1 & 0 & 0 & 0 \\ 0 & 1 & 0 & -1 \\ 0 & 0 & 1 & 1 \end{pmatrix} \Rightarrow \begin{pmatrix} a_0 \\ a_1 \\ a_2 \end{pmatrix} = \begin{pmatrix} 0 \\ -1 \\ 1 \end{pmatrix}$$

and has the indicated vector as its only solution.

It follows that the ONLY polynomial function $y = f(x)$ of degree at most 2 whose graph goes through the three data points $(0,0), (1,0), (2,2)$ actually has degree 2 and is $y = -x + x^2$. (You can check that the three points $(0,0), (1,1), (2,2)$ lead to the unique solution $y = x$. For this reason, we cannot assume more than $y = f(x)$ had degree **at most** 2.)

In general, if $m + 1$ data points $(x_0, y_0), \ldots, (x_m, y_m)$ are given for a polynomial of degree at most n, then there arises a linear system $V\mathbf{u} = \mathbf{b}$ where

$$
V = \begin{pmatrix} 1 & x_0 & x_0^2 & \cdots & x_0^n \\ 1 & x_1 & x_1^2 & \cdots & x_1^n \\ & \cdots & & \cdots & \\ 1 & x_m & x_m^2 & \cdots & x_m^n \end{pmatrix} \text{ and } \mathbf{b} = \begin{pmatrix} y_0 \\ \vdots \\ y_m \end{pmatrix}.
$$

WARNING: V is an $m + 1$-by-$n + 1$ matrix whose row and column indices start at 0 not 1 as is normal; this is essentially forced because we want to have the subscripts of a polynomial's coefficients match the powers of x.

Matrices in which each row consists of the successive powers of a fixed number x_i are quite important and are called ***Vandermonde matrices***. We will eventually show that a Vandermonde matrix $V(x_0, x_1, \ldots, x_m)$ has rank $m+1$ whenever the numbers $x_0, x_1, \ldots, x_m$ are distinct and it has at least $m + 1$ columns. This theorem implies an important fact:

> A polynomial function $f(x)$ having degree less than or equal to n
> is uniquely determined by its values at $n + 1$ distinct data points.

Since there is but one such function, you must expect that there might be no solution if more than $n+1$ data points are specified. On the other hand, if fewer ($m+1 < n+1$) data points are specified, we can determine ALL polynomial functions of degree less than or equal to n whose graph passes through these data points. This process of computing coefficients of a function from data points is called ***interpolation***.

Polynomial Interpolation

The general form of polynomial functions $f(x)$ of degree n passing through $m + 1$ distinct data points is determined by the solution of a linear system having a Vandermonde matrix as a coefficient matrix and a constant vector made up of the corresponding functional values.

Example 1.21 (Polynomial interpolation.)

    a. Suppose the graph of the polynomial function $y = f(x)$ goes through the points: $(x,\ y) = (-1,\ 1),\ (1,\ 0),\ (2,\ 1)$. If $f(x)$ has degree at most three, what is its general form?

    Since $f(x)$ has the form $f(x) = a_0 + a_1 x + a_2 x^2 + a_3 x^3$, we need a Vandermonde matrix with four columns. Since there are three data points, our Vandermonde matrix will have three rows. We are led to the linear system:

$$
\begin{array}{l} \text{powers of } x_0 = -1 \rightarrow \\ \text{powers of } x_1 = 1 \rightarrow \\ \text{powers of } x_2 = 2 \rightarrow \end{array} \begin{pmatrix} 1 & -1 & 1 & -1 \\ 1 & 1 & 1 & 1 \\ 1 & 2 & 4 & 8 \end{pmatrix} \begin{pmatrix} a_0 \\ a_1 \\ a_2 \\ a_3 \end{pmatrix} = \begin{pmatrix} 1 \\ 0 \\ 1 \end{pmatrix} \begin{array}{l} \leftarrow f(-1) \\ \leftarrow f(1) \\ \leftarrow f(2) \end{array}
$$

The augmented matrix reduces to:

$$\begin{pmatrix} 1 & -1 & 1 & -1 & | & 1 \\ 1 & 1 & 1 & 1 & | & 0 \\ 1 & 2 & 4 & 8 & | & 1 \end{pmatrix} \xrightarrow{RREF} \begin{pmatrix} 1 & 0 & 0 & -2 & | & 0 \\ 0 & 1 & 0 & 1 & | & -1/2 \\ 0 & 0 & 1 & 2 & | & 1/2 \end{pmatrix}.$$

The general solution is
$$\begin{pmatrix} a_0 \\ a_1 \\ a_2 \\ a_3 \end{pmatrix} = a_3 \begin{pmatrix} 2 \\ -1 \\ -2 \\ 1 \end{pmatrix} + \begin{pmatrix} 0 \\ -1/2 \\ 1/2 \\ 0 \end{pmatrix}.$$

Therefore, $f(x)$ has the form:

$$f(x) = 2a_3 + (-a_3 - 1/2)x + (-2a_3 + 1/2)x^2 + a_3 x^3.$$

b. Suppose the graph of the polynomial function $y = f(x)$ goes through the points: $(x, y) = (0, 0), (1, 0), (2, 0), (3, 0)$. If $f(x)$ has degree at most 4, what is its general form?

Since $f(x)$ has the form $f(x) = a_0 + a_1 x + a_2 x^2 + a_3 x^3 + a_4 x^4$, we need a Vandermonde matrix with five columns. Since there are four data points our Vandermonde matrix will have four rows. Thus, we are led to the linear system:

$$\begin{matrix} \text{powers of } x_0 = 0 \rightarrow \\ \text{powers of } x_1 = 1 \rightarrow \\ \text{powers of } x_2 = 2 \rightarrow \\ \text{powers of } x_3 = 3 \rightarrow \end{matrix} \begin{pmatrix} 1 & 0 & 0 & 0 & 0 \\ 1 & 1 & 1 & 1 & 1 \\ 1 & 2 & 4 & 8 & 16 \\ 1 & 3 & 9 & 27 & 81 \end{pmatrix} \begin{pmatrix} a_0 \\ a_1 \\ a_2 \\ a_3 \\ a_4 \end{pmatrix} = \begin{pmatrix} 0 \\ 0 \\ 0 \\ 0 \end{pmatrix} \begin{matrix} \leftarrow f(0) \\ \leftarrow f(1) \\ \leftarrow f(2) \\ \leftarrow f(3) \end{matrix}$$

The augmented matrix reduces to:

$$\begin{pmatrix} 1 & 0 & 0 & 0 & 0 & | & 0 \\ 1 & 1 & 1 & 1 & 1 & | & 0 \\ 1 & 2 & 4 & 8 & 16 & | & 0 \\ 1 & 3 & 9 & 27 & 81 & | & 0 \end{pmatrix} \xrightarrow{RREF} \begin{pmatrix} 1 & 0 & 0 & 0 & 0 & | & 0 \\ 0 & 1 & 0 & 0 & 6 & | & 0 \\ 0 & 0 & 1 & 0 & -11 & | & 0 \\ 0 & 0 & 0 & 1 & 6 & | & 0 \end{pmatrix}$$

so the general solution to this system is
$$\begin{pmatrix} a_0 \\ a_1 \\ a_2 \\ a_3 \\ a_4 \end{pmatrix} = \begin{pmatrix} 0 \\ -6 \\ 11 \\ -6 \\ 1 \end{pmatrix}.$$

Therefore, $f(x)$ has the form $f(x) = -6a_4 x + 11a_4 x^2 - 6a_4 x^3 + a_4 x^4$.

Notice that the factorization $f(x) = a_4 x(x - 1)(x - 2)(x - 3)$ could have been recognized more easily. □

SUMMARY

Given $m + 1$ functional values points $f(x_0), \ldots, f(x_m)$ for a polynomial $f(x) = a_0 + a_1 x + a_2 x^2 + \cdots a_n x^n$ of degree at most n, the coefficients satisfy the linear system $V\mathbf{u} = \mathbf{b}$ where V is the Vandermonde matrix:

$$V = \begin{pmatrix} 1 & x_0 & x_0^2 & \cdots & x_0^n \\ 1 & x_1 & x_1^2 & \cdots & x_1^n \\ & \cdots & & \cdots & \\ 1 & x_m & x_m^2 & \cdots & x_m^n \end{pmatrix} \text{ and } \mathbf{b} = \begin{pmatrix} f(x_0) \\ \vdots \\ f(x_n) \end{pmatrix}.$$

If $m > n$ there may be no solution.

Looking further
If data doesn't quite fit a given degree polynomial, we seek a good approximation to the data. The method of least squares in Chapter 2 is used for this purpose.

An important variation on this theme is used to "fill in" a smooth curve when an interval has been erased. A *spline* is determined by as few as two x values BUT the values of the function AND its derivatives $f(x_i), f'(x_i), \ldots$ are specified as data. This leads to a system of linear equations in the spline polynomial's coefficients.[4]

Practice Problems
Sketch the data and describe the general form of the indicated function whose graph passes through the given points, if it exists.

1. $y = f_0 + f_1 x + f_2 x^2$ $(-2, 4), (0, 0), (1, 1).$

2. $y = f_0 + f_1 x + f_2 x^2$ $(0, 0), (1, 1), (-2, 4), (3, 4).$

3. $y = f_0 + f_1 x + f_2 x^2 + f_3 x^3$ $(-1, 3), (0, 0), (1, -3), (3, 15).$

4. $y = f_0 + f_1 x + f_2 x^2 + f_3 x^3$ $(-1, 3), (0, 2), (2, 10), (3, 29).$

5. $y = f_0 + f_1 x + f_2 x^2 + f_3 x^3$ $(-1, 9), (1, -2), (2, 9).$

6. This problem points out that our methods extend beyond polynomials. Suppose $y = g_0 + g_1 \sin(x) + g_2 \cos(x) + g_3 \tan(x)$ passes through the points $(-\pi/4, 0), (0, 2), (\pi/4, 2 + \sqrt{2})$ and $(\pi/3, 3(1 + \sqrt{3})/2)$. Set up a linear system for g_0, g_1, g_2, g_3 and solve it.

7. The methods of this section also extend to functions of several variables. Suppose $z = A + Bx + Cy$ is a function of x AND y and that $(x, y, z) = (1, 1, 1), (1, 2, 3)$ and $(0, 0, 1)$ are three data points. Set up and solve a system of equations in A, B, C.

---

[4]*The Theory of Splines and Their Applications*, J. H. Ahlberg, E. N. Nelson and J. L. Walsh, Academic Press, New York, 1967.

Chapter 2

Matrix number systems

Calculations that we now recognize as matrix arithmetic appear in the work of Gauß and especially Cauchy in 1826. You have seen that matrices and vectors can be useful to focus attention on the significant aspects of a linear system and its solution. But they have many more sophisticated mathematical uses. We develop the arithmetic of matrices as a realization of the algebra of linear functions. The power and significance of this view was not recognized until the 1850's by Kronecker, Weierstraß and Cayley – the originator of the notation and mathematical term "matrix."

The fact that the position of a number in a matrix is just as important as its value allows matrices to carry a great deal more information than most of the other mathematical symbols. In this way, the step to matrix notation might be compared to the step from, say, Roman numerals to the modern place value decimal number system. In the Roman system, V means five and L means fifty no matter where they appear in a number. In contrast, 5 may mean five, fifty, five hundred ... in the decimal notation, depending upon the position 5 occupies within the number being considered.

We are not often aware of the fact that our language influences our thought processes. We are even less aware of the manner in which our mathematical notation influences our mathematical work, but it is nonetheless incredibly important. A good illustration of this fact is the following exercise: translate 1683 and 1826 into Roman numeral notation and then compute their product as a Roman would have been forced to do without any decimal notation. It is safe to say that the inventors of the Roman numeration system did not multiply large numbers often.

In fact, anthropologists and archaeologists have noticed that the type of numeration scheme used by a culture correlates well with the general level of commerce and science it possesses.[1]

The point being made is that matrix notation opens up a whole new mathematical perspective. Situations that are incredibly complicated without such notation become manageable with it, and the process of extending familiar arithmetic operations to matrices leads to new uses and deeper understanding of the original arithmetic.

In this chapter we extend the arithmetic of numbers in several ways. First we recall/introduce the most important number systems culminating with the **complex**

---

[1] *Evolution of Mathematical Concepts*, Raymond L. Wilder, John Wiley & Sons, New York, 1968.

numbers. Then we introduce multiplication of matrices and explore some properties of matrices analogous to the number systems. Many new features of matrix
arithmetic arise. Of particular importance are certain matrices called symmetric projectors. We then explore two applications: least squares fitting of data and changing plane coordinates. The fact that each of these relies on matrix arithmetic and
symmetric projectors is a good example of the paradox that the great utility of mathematics derives from its apparent abstraction. The final optional section develops
some properties of polynomials needed later and uses them to explain the famous
fast Fourier transform factorization of certain Vandermonde matrices.

Throughout this chapter we use the trigonometric functions $sin(x)$ and $cos(x)$. If
the basic properties of these functions are not familiar, some review may be needed.

2.1   Complex numbers

Objective 1

Compute sums, differences, products and quotients in $\mathbb{C}$. Compute the zeros of
quadratic polynomials.

You know many number systems. The number system known to all human cultures,
and to our dog Serena, is the **natural** number system $\mathbb{N}$ used for counting (1 dog
biscuit, 2 dog biscuits, etc.). The "give and take" of commerce requires that $\mathbb{N}$ be
extended to the **integers** $\mathbb{Z}$ by introducing negative numbers.

A number system is **closed** under an operation if whenever you apply the operation to
members of the system, the result is again in the system. The natural numbers $\mathbb{N}$ are
not closed under subtraction ($1 - 3$ is not in $\mathbb{N}$). The integers $\mathbb{Z}$ are the "subtraction
closure" of $\mathbb{N}$. Similarly, the integers $\mathbb{Z}$ are not closed under division and the **rational
numbers** $\mathbb{Q}$ are the "division closure" of $\mathbb{Z}$. If a linear system involves only rational
numbers then so does its general solution. In other words, $\mathbb{Q}$ is closed under linear
system solution.

The ancient Greeks thought that $\mathbb{Q}$ was given by the creator and sufficient for all
purposes. They developed geometry and used numbers to measure concepts of length
and area. The Pythagoreans (a school of ancient Greek mathematicians) were thrown
into turmoil by the realization that the length ($\sqrt{2}$) of the diagonal of a 1 by 1 square
is not in $\mathbb{Q}$. Eventually, there evolved the "number line" of **constructible numbers**.

When the concept of limits, essential in the development of calculus, was rigorously
developed, the "constructible number line" was again found inadequate and the real
numbers $\mathbb{R}$ emerged. Today $\mathbb{R}$ is informally presented as the numbers having "non-
terminating decimal expansions." It is a technically subtle construct that few use
with complete understanding (not unlike an automatic transmission car).

We need a number system bigger than $\mathbb{Q}$, but closed in a different way. Our number system has to be "algebraically closed." This means that for each polynomial $f(x)$ of positive degree, there must be at least one number c that is a "zero of f," that is, $f(c) = 0$. That such a number system actually exists was not proven until the 18th century by Karl Friedrich Gauß. This theorem, often called the fundamental theorem of algebra, has a rich history and many published proofs, including three different proofs by Gauß. An accessible treatment can be found in *A Concrete Introduction to Higher Algebra* by L. Childs, Springer-Verlag, New York, 1984, pages 138–140.

The *complex* numbers $\mathbb{C}$ are the algebraic closure of $\mathbb{R}$. This number system is fundamental to modern mathematics and an essential tool for modern technology. In fact, Nobel laureate E. Wigner cites the appearance of $\mathbb{C}$ in the laws of quantum mechanics as a prime example of "the enormous usefulness of mathematics in the natural sciences," concluding that "it is difficult to avoid the conclusion that a miracle confronts us here."[2]

Suppose $f(x) = ax^2 + bx + c$ is a quadratic polynomial. The *quadratic formula* is a formula for the *zeros of* $f(x)$:

$$x = (-b \pm \sqrt{b^2 - 4ac})/2a.$$

This formula has no solution in $\mathbb{R}$ when the *discriminant* $b^2 - 4ac$ is negative. Therefore any "algebraic closure" of $\mathbb{R}$ must include square roots of negative numbers. The *imaginary unit* i is defined by

$$i^2 = -1.$$

Put another way, all that is postulated about i is that i is a zero of the polynomial $x^2 + 1$. By including the imaginary unit in the number system, we guarantee that the quadratic formula will work whenever $a, b, c \in \mathbb{R}$.

We build the complex number system $\mathbb{C}$ from the $\mathbb{R}$ and i using addition, subtraction, multiplication and division. All the ordinary rules of arithmetic remain valid in $\mathbb{C}$. It turns out that any complex number z can be expressed uniquely as

$$z = a + bi \text{ where } a, b \text{ are in } \mathbb{R}.$$

The number a is the *real part* of z, and b is the *imaginary part* of z.
NOTE: Every complex number can be expressed in this standard form.

The *complex conjugate* of $z = a + bi$ is defined to be $\bar{z} = a - bi$. Notice that

$$z\bar{z} = (a + bi)(a - bi) = a^2 + abi - abi - b^2 i^2 = a^2 + b^2$$

is always a non-negative real number. The *absolute value* or *modulus* of z is defined:

$$|z| = \sqrt{z\bar{z}} = \sqrt{a^2 + b^2}.$$

---

[2]"The unreasonable effectiveness of mathematics in the physical sciences," by E. Wigner *Communications in pure and applied mathematics*, vol. 13 (1960).

Any complex number $z = a + bi \neq 0$ has a (unique) multiplicative inverse

$$z^{-1} = \bar{z}/|z|^2.$$

We will see that both the complex conjugate and absolute value have very natural meanings in the geometric representation of $\mathbb{C}$.

Example 2.1 (Complex arithmetic.)

a. $f(x) = x^2 - x + 1$ has zeros given by the quadratic formula

$$x = \frac{1 \pm \sqrt{1-4}}{2} = \frac{1}{2} \pm i\frac{\sqrt{3}}{2}.$$

b. $f(x) = 1 + x + x^2 + x^3$ factors as

$$f(x) = (x + 1)(x^2 + 1) = (x + 1)(x - i)(x + i).$$

Therefore, $f(x)$ has three zeros $x = -1, i, -i$ in $\mathbb{C}$. In general, a polynomial with real coefficients having a complex zero z must also have $\bar{z}$ as a zero. It follows that very polynomial with real coefficients and degree 3, has either three real zeros, or one real zero and two complex zeros.

c. Addition and subtraction in $\mathbb{C}$ works easily by combining coefficients:

$$(1 - 2i) - (-5 + i) = (1 + 5) + (-2 - 1)i = 6 - 3i.$$

d. Multiplication in $\mathbb{C}$ is easy. Just use the fundamental property of i.

$$(1 + 2i)(1 - i) = 1 + 2i - i + 2i(-i) = 1 + i - 2i^2 = 1 + i - (-2) = 3 + i.$$

e. Division in $\mathbb{C}$ can be difficult unless you "rationalize" the denominator.

$$\frac{1 - i}{3 + 2i} = \frac{1 - i}{3 + 2i}\frac{3 - 2i}{3 - 2i} = \frac{(1 - i)(3 - 2i)}{(3 + 2i)(3 - 2i)}$$

$$= \frac{(1 - i)(3 - 2i)}{13} = \frac{3 - 3i - 2i + 2i^2}{13} = \frac{1}{13} - \frac{5i}{13}.$$

f. It bears repeating that any complex number, no matter how it is presented, can be expressed in standard form

$$(1 + i)^7 = (1 + 2i + i^2)^3(1 + i) = (2i)^3(1 + i) = -8i(1 + i) = 8 - 8i. \quad \Box$$

In the next section we see that complex arithmetic can be expressed as matrix arithmetic for certain special two by two matrices.

Objective 2

Interpret complex arithmetic in the geometric representation using both the rectangular and polar forms.

The geometric representation of $\mathbb{C}$ was first given by Aargand in 1806. In it, each complex number labels a point of a Cartesian plane.

In Aargand's **geometric representation** the real numbers appear as the "X" axis. The "Y" axis consists of real multiples of the imaginary unit i, sometimes called "purely imaginary numbers." Every complex number $z = a + bi \in \mathbb{C}$ has a real part a and imaginary part bi. (The symbol $\in$ means "in" or "is a member of.")

The point with coordinates (a, b) are labeled by $z = a + bi$. Complex addition uses geometric vectors in the Aargand diagram. Arrange them "tip to tail" as in vector addition in $\mathbb{R}^2$:

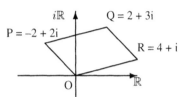

$$\overrightarrow{OP} + \overrightarrow{OR} = \overrightarrow{OP} + \overrightarrow{PQ} = \overrightarrow{OQ}.$$

The geometric interpretation of complex multiplication requires a change to polar coordinates. The complex number $z = a + bi$ may also be written in its **polar form**:

$$z = |z| \, (cos(\theta) + i \, sin(\theta)),$$

where $|z| = \sqrt{a^2 + b^2}$ is the magnitude of z and $tan(\theta) = b/a$. The angle θ — measured counterclockwise from the positive X-axis — called the **argument** of z. The absolute value of a complex number is just the magnitude of the associated Aargand diagram vector. Doesn't it seem a very natural generalization of absolute value for real numbers?

Let y be a second complex number, expressed in its polar form using its absolute value and its argument:

$$y = |y|w \quad \text{where } w = \cos(\alpha) + i\sin(\alpha).$$

Note that $|w| = \cos^2(\alpha) + \sin^2(\alpha) = 1$. The effect of multiplication by y on other complex numbers can be explained in two parts based on this **polar factorization**.

The first part, multiplication by $|y|$ is a *scaling* — moving each complex number z away from 0 by a factor $|y|$ just like scalar multiplication of geometric vectors.

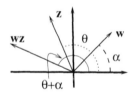

But what about the second part, multiplication by w — a complex number of absolute value one? If you just blast it out algebraically, using the formal properties of the imaginary unit, you wind up with some "familiar" trigonometric identities:

$$w \, z = 1(\cos(\alpha) + i \, \sin(\alpha)) \, z = |z| \, [\cos(\alpha) + i \, \sin(\alpha] \, [\cos(\theta) + i \, \sin(\theta)]$$
$$= |z| \, [\cos(\alpha)\cos(\theta) + i \, \sin(\alpha)\cos(\theta) + i \, \cos(\alpha)sin(\theta) - \sin(\alpha)\sin(\theta)]$$
$$= |z| \, [\cos(\alpha)\cos(\theta) - \sin(\alpha)\sin(\theta)] + i \, [\sin(\alpha)\cos(\theta) + \cos(\alpha)\sin(\theta)]$$

for the sin and cos of the sum of two angles — remember? This shows that the effect of multiplication by a complex number w of absolute value 1 is **counterclockwise rotation** through an angle $arg(w)$. In particular,

$$y \, z = |y|wz = |y||z| \, [\cos(\alpha + \theta) \, + i \, \sin(\alpha + \theta)].$$

Notice that the complex conjugate $\bar{z} = a - bi$ is symmetric with $z = a + bi$ about the "real axis" and $arg(\bar{z}) = -arg(z)$. Therefore, $arg(|z|^2) = arg(z\bar{z}) = arg(z) + -argz = 0$ and $|z|$ is a non-negative real number, this way too.

Example 2.2 (Polar form and multiplication geometrically.)

a. Consider $z = -1 + i\sqrt{3}$. $|z| = \sqrt{(-1)^2 + (\sqrt{3})^2} = 2$. The argument of θ z satisfies the following equations:

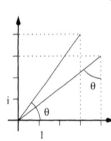

$\tan(\theta) = -\sqrt{3}$, $\cos(\theta) = \frac{-1}{2}$ and $\sin(\theta) = \frac{\sqrt{3}}{2}$.

Because the cosine is negative and the sin is positive, $\pi/2 < \theta < \pi$. The "reference angle" is $\pi/3$ or $60°$. Therefore, $\theta = \pi - \pi/3 = 2\pi/3$, or $120°$.

WARNING: If this seems unfamiliar PLEASE review your trigonometry NOW! It is tempting BUT WRONG to use the first equation and the inverse tangent function on your calculator. The reason is that the inverse tangent function only returns angles between $-\pi/2$ and $\pi/2$, while complex numbers can have any angle as an argument.

b. Compute $(4 + 3i)(3 + 4i)$ in two ways, first using the geometric polar from model and second algebraically.

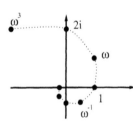

Each factor has absolute value $\sqrt{16 + 9} = 5$. Moreover each factor determines a 3-4-5 right triangle with the origin and a point on a coordinate axis, so the arguments of the two factors sum to $\pi/2 = 90°$. It follows from the multiplication formula that

$(4+3i)(3+4i) = 5\,5(\cos(\pi/2) + i\sin(\pi/2)) = 25i$.

Algebraically:
$(4 + 3i)(3 + 4i) = 12 + 9i + 16i + 12(i^2) = 25i$.

c. Suppose $w = 1 + i$. Plot the powers w^n for $-4 \le n \le 4$. What powers of z are real?

Because $w = \sqrt{2}(\cos(\pi/4) + i\,\sin(\pi/4))$, each of its powers has an argument that is a multiple of $\pi/4$. As you can see, the powers of w form a spiral in $\mathbb{C}$. Because $\sin(4n(\pi/4)) = 0$, only its 4-th powers are real. Thus, the real powers of w are $w^{-4} = 1/4, w^0 = 1, w^4 = 4, \dots$.

$\square$

There is a famous mathematical identity called ***Euler's formula*** that expresses complex numbers of absolute value 1 in **exponential** form:
$$\cos(\theta) + i\,\sin(\theta) = e^{i\theta}.$$

Sadly, justification of this fabulous formula requires calculus, but is hinted at in Problem 9. It is mentioned here because the TI-83's notation for the polar form of complex numbers is motivated by Euler's formula — "show offs."

Calculators

A TI-83 is "born" knowing about complex numbers. In order to have it display complex numbers, you must use the MODE key that is in the top row. There are two modes to consider.

$$\boxed{\text{MODE}}\,\boxed{\downarrow}\,^6$$

(that is $\boxed{\downarrow}$ repeated 6 times) takes you to a line that reads

$$\text{Real} \qquad a + bi \qquad re\hat{\ }\theta\, i.$$

Here you may select whether you want the calculator to display only real numbers, complex numbers in rectangular form or complex numbers in their polar form. The second mode is

$$\boxed{\text{MODE}}\,\boxed{\downarrow}\,^2$$

.

This time you can choose between Radian and Degree angle measurement. **It is a good idea to make sure you and your calculator are in the same mode**, especially if you have strange answers with a TEST calculator.

At the bottom of page 2-16 in the TI-83 manual is a small note that should be a BIG WARNING. It essentially says that conversions between modes are not exact, so radian angle measure will work better with complex numbers in polar form. This raises the touchy subject that **calculator arithmetic is not exact**.

Put more dramatically, calculators do not work with a number system! This is because the calculator has only finite storage and numbers have infinite variety. Therefore it must approximate numbers. Most calculators choose to do this by working only with "most significant" digits using, "floating point" notation. Some computer packages, like MAPLE and MATHEMATICA, make more serious efforts to work in a proper mathematical number system. Perhaps in the future floating point numbers will be regarded as curious ancient artifacts rather like Roman numerals — perhaps. There is another unfortunate fact about the TI-83 and complex numbers. It is not mentioned in the manuals but has been confirmed to me by "official sources." A TI-83 will not accept a complex number as a matrix entry!

In order for $\mathbb{C}$ to be algebraically closed, the quadratic formula has to work for the equation coefficients a, b, c in $\mathbb{C}$. This implies that $\mathbb{C}$ must be closed under square roots! Here is how to geometrically take n-th roots in $\mathbb{C}$. Suppose we are given $z \in \mathbb{C}$. Write z in polar form and observe that

$$w = |z|^{1/n}(\cos(k\, arg(z)/n) + i\sin(k\, arg(z)/n))$$

where $k = 1\ldots n - 1$ satisfies $w^n = z$.

SUMMARY

The imaginary unit i is, by definition, a ***zero*** of $f(x) = x^2 + 1$. Each complex number z may be written $z = a + bi$ for $a, b \in \mathbb{R}$. The numbers a and b respectively are called the ***real and imaginary*** parts of z. The ***complex conjugate*** of z is $\bar{z} = a - bi$.

The complex numbers $\mathbb{C}$ are algebraically closed. This means that every polynomial with coefficients in $\mathbb{C}$ has a zero in the number system.

The geometric representation of $\mathbb{C}$ labels the points of a Cartesian plane with coordinates (a,b) with $z = a + bi$. The polar form of

$$z(= r\,\cos(\theta) + ir\,\sin(\theta), r, \theta \in \mathbb{R}, r > 0)$$

arises from polar coordinates in the plane. Addition and multiplication in $\mathbb{C}$ each have simple interpretation in the geometric representation of $\mathbb{C}$. The absolute value $|z|$ of the complex number z is the distance of its geometric representation from the origin.

$$|z| = \sqrt{a^2 + b^2}.$$

Looking further

We can thank Charles Steinmetz[3] (1865–1923) for introducing[4] complex numbers to electrical engineering. Steinmetz, the "Wizard of Schenectady," was General Electric's chief engineer when Nikola Tesla's[5] (1856–1943) 1887 patents of an alternating current power distribution system were exploited to develop the electric power potential of Niagara Falls. The "hopelessly complex" AC theory of the day was based on "empirical" and graphical methods. Steinmetz drew on his broad mathematical background to introduce complex numbers, "phasors." This both simplified the conceptual framework and made precise computations feasible. Together these two geniuses are responsible for the inefficient "Edison" DC power distribution system being replaced with the AC system universally used today.

Steinmetz also served as a professor at Union College and was a very colorful character. To him the purpose of education was "to make a man able to make the best use of himself and for human society at large." For this reason "a man needs an extensive knowledge of history, language, literature and science, mathematics and engineering. Strictly technical training, therefore, is not education. It is a very small part of it."

---

[3]**Steinmetz: Engineer and Socialist.** by R. Kline, Johns Hopkins University Press, Baltimore 1992.

[4]"Die Anwendung complexer Größen in der Elektrotechnik" in Electrotechnisch Zeitschrift, XIV, pp 597-599, 631-635, 641-643, 653–654. Berlin 1893.

[5]http://www.pbs.org/tesla/ll/index.html.

He also had a clear understanding of mathematics as the engineer's "tool kit" saying "Mathematics is the most exact science, and its conclusions are capable of absolute proof. But this is only because mathematics does not attempt to draw absolute conclusions. All mathematical truths are relative, conditional."

* * * * * *

Practice Problems

1. Compute the indicated expressions.

   a. $(2 + 3i) + (9 - 2i)$ b. $(4 + 7i) - (10 - 9i)$
   c. $(1 + i)(2 - i)$ d. $(3 + 4i)(3 - 4i)$
   e. $(3 + i)/(3 - 4i)$ f. $|z|$ for $z = (i + 3)$

2. Find the argument and modulus and express in polar form, sketch.

   a. 4   b. $(2 + 5i)$
   c. $(4 + 3i)$ d. $(8i - 1)$
   e. $(i - 2)$ f. $(9 - 7i)$

3. In each case $f(x)$ has one real zero r. Graphically locate r and use the quadratic formula on $f(x)/(x - r)$ to obtain complex zeros. Compute the absolute value of each of the zeros of $f(x)$ and sketch their geometric representation.

   a. $f(x) = x + x^3$ b. $f(x) = x + x^2 + x^3$
   c. $f(x) = x^3 - 2x^2 + 3x - 6$

4. Solve $x^3 - i = 0$.

5. Geometrically describe the solutions to the complex equation $|z - 1| = 1$. Verify your solution algebraically.

6. Verify the trigonometric identity
   $$\sin(3\theta) = 3\sin(\theta) - 4\sin^3(\theta) \text{ by computing } (\cos(\theta) + i\sin(\theta))^3.$$

7. Use Euler's formula to show
   $$\cos(\theta) = \frac{e^{i\theta} + e^{-i\theta}}{2} \text{ and } \sin(\theta) = \frac{e^{i\theta} - e^{-i\theta}}{2i}.$$

8. Show that $\cos(n\theta)$ is a polynomial in $\cos(\theta)$ for any natural number n.
   HINT: Let n be the first value of n for which this fails. Expand Euler's formula for $\cos(\theta)^n$ and simplify.

9. (For Calculus students) Compare the MacLaurin series for the left- and right-hand side of Euler's formula.

2.2   Matrix multiplication

Objective 1

Given matrices, compute sums, scalar multiples and products as indicated, or decide the indicated expression does not exist.

We saw in Example 1.10 on page 22 that several linear systems with the same coefficients can be solved simultaneously.

Confusion between the systems is avoided by using variables with two subscripts. The first subscript tells which variable and the second subscript tells which system. The same scheme is used to keep track of the constants so $b_{i,j}$ is the i-th constant in the j-th system of equations and the vector $\mathbf{b}$ is replaced with a matrix B.

Just fatten up $\mathbf{b}$ of the augmented matrix to a generalized augmented matrix having right-hand part the matrix B. This amounts to giving different names to the variables in the different linear systems. The variables of the first system are, say, $x_{11}, x_{21}, ..., x_{n1}$; those of the second system are, say, $x_{12}, x_{22}, ..., x_{n2}$ and so on.

Example 2.3 (Systems of linear systems and matrix multiplication.)
The matrix version of the systems appearing in Example 1.10 is:

$$\begin{array}{rcr} x_1 - x_2 + x_3 & = & -1 \\ 2x_1 + x_2 - x_3 & = & 0 \end{array} \qquad \begin{array}{rcr} x_1 - x_2 + x_3 & = & 1 \\ 2x_1 + x_2 - x_3 & = & -1 \end{array}$$

$$\begin{pmatrix} 1 & -1 & 1 \\ 2 & 1 & -1 \end{pmatrix} \begin{pmatrix} x_{11} & x_{12} \\ x_{21} & x_{22} \\ x_{3,1} & x_{3,2} \end{pmatrix} = \begin{pmatrix} b_{11} & b_{12} \\ b_{21} & b_{22} \end{pmatrix} = \begin{pmatrix} -1 & 1 \\ 0 & -1 \end{pmatrix} \qquad \Box$$

In this example the second matrix is just treated as a bunch of column vectors that have been crowded together into a single matrix.
That's all there is to matrix multiplication. Just treat the second matrix as a bunch of closely packed vectors and do all the (matrix)×(vector) multiplications at once.
In general, the matrix product AX is defined whenever the number of columns of A equals the number of rows of X and is:

$$AX = \begin{pmatrix} a_{11} & \cdots & a_{1n} \\ & \cdots & \\ \boxed{a_{i1} \cdots a_{in}} \\ & \cdots & \\ a_{m1} & \cdots & a_{mn} \end{pmatrix} \begin{pmatrix} x_{11} & \cdots & \boxed{x_{1j}} & \cdots & x_{1s} \\ x_{21} & \cdots & \boxed{x_{2j}} & \cdots & x_{2s} \\ \vdots & & \vdots & & \vdots \\ x_{n1} & \cdots & \boxed{x_{nj}} & \cdots & x_{ns} \end{pmatrix} = \begin{pmatrix} b_{11} & \cdots & b_{1s} \\ & \cdots & \\ \vdots & \boxed{b_{ij}} & \vdots \\ & \cdots & \\ b_{m1} & \cdots & b_{ms} \end{pmatrix}$$

where $b_{ij} = a_{i1}x_{1j} + a_{i2}x_{2j} + \cdots + a_{in}x_{nj}$, for all i, j. (The boxes indicate the pattern of terms used to compute b_{ij} and have no other significance.)

This matrix equation is the definition of ***matrix multiplication***. Notice that the matrix AX has as many rows as A and as many columns as X.

Example 2.4 (Matrix multiplication.)

a. $\begin{pmatrix} 2 & 3 & 4 \\ 7 & 8 & 9 \end{pmatrix} \begin{pmatrix} 10 & 1 \\ 15 & 2 \\ 20 & 3 \end{pmatrix} = \begin{pmatrix} 2*10 + 3*15 + 4*20 & 2*1 + 3*2 + 4*3 \\ 7*10 + 8*15 + 9*20 & 7*1 + 8*2 + 9*3 \end{pmatrix} = \begin{pmatrix} 145 & 20 \\ 370 & 50 \end{pmatrix}.$

b. $(1 \ 2 \ 3) \begin{pmatrix} 1 \\ 2 \\ 3 \end{pmatrix} = (14)$ but $\begin{pmatrix} 1 \\ 2 \\ 3 \end{pmatrix} (1 \ 2 \ 3) = \begin{pmatrix} 1 & 2 & 3 \\ 2 & 4 & 6 \\ 3 & 6 & 9 \end{pmatrix}.$

c. $\begin{pmatrix} 3 & 4 \\ 4 & 5 \\ 0 & 3 \end{pmatrix} \begin{pmatrix} 2 & 6 & -1 & 0 \\ 3 & -2 & 7 & 1 \end{pmatrix} = \begin{pmatrix} 18 & 10 & 25 & 4 \\ 23 & 14 & 31 & 5 \\ 9 & -6 & 21 & 3 \end{pmatrix}.$

d. $\begin{pmatrix} 2 & 3 & 4 \\ 6 & 7 & 8 \end{pmatrix} \begin{pmatrix} 2 & 6 & -1 & 0 \\ 3 & -2 & 7 & 1 \end{pmatrix}$ DOES NOT EXIST because they have incompatible shapes! □

Matrix multiplication is a fundamental new operation. We will use it more and more, treating it almost like numerical multiplication. **Make sure you can do matrix multiplication flawlessly before going on.**

Objective 2

Given an adjacency matrix, sketch the associated labeled digraph and vice versa.
Use matrix multiplication to compute the number of walks of specified type.

A **digraph** $\mathcal{D}$ is a set of **vertices** and **arcs**. Each arc has a vertex as its **tail** and a vertex as its **tip**. No two arcs have the same tip and tail. Digraphs are a basic discrete mathematical model type that we will revisit in Section 3.3.

You might imagine a digraph as a street map in which two-way streets are identified as two one-way streets in opposite directions, except that "scenic loops" are allowed. (Arcs having the same vertex as tail and tip are allowed and are called **loops**.)

To draw a digraph, start with a bunch of dots (one for each vertex) and add a bunch of arrows (one for each arc). An arrow from one dot to another dot is drawn to indicate that the digraph has an arc with the first vertex v as tail and the second vertex w as tip. When this occurs, say w is an **out neighbor** of v. For simplicity we will always label the vertices by the numbers $1, 2, 3, \ldots$.

The **adjacency matrix**, $A(\mathcal{D})$, of a digraph $\mathcal{D}$ having vertices $\{1, \ldots, n\}$ is a n by n matrix of zeros and ones. There is a 1 in position $i \ j$ whenever there is an arc from vertex i to vertex j.

$\begin{pmatrix} 1 & 0 & 1 & 0 \\ 1 & 0 & 1 & 1 \\ 0 & 1 & 0 & 1 \\ 0 & 0 & 0 & 0 \end{pmatrix}$

The indicated digraph has the given matrix as adjacency matrix.

A **walk** of **length** t in a digraph $\mathcal{D}$ is a sequence of t arcs $a_1 a_2 \ldots a_t$ where the tip of each arc a_i is the same vertex as the tail of a_{i+1}, $i = 12 \ldots t - 1$. The walk $a_1 a_2 \ldots a_t$ is said to *start* at the tail of a_1 and *terminate* or end at the tip of a_t and is called ***closed*** if it starts and terminates at the same vertex. All of this terminology is also used for ***graphs***, which differ from digraphs only in that the arcs are not directed.

Example 2.5 (Digraph walks.)

Let's compute the $1, 2$ entry of $A(\mathcal{D})^2$ for the preceding digraph. $A(\mathcal{D})^2_{12}$ is obtained by summing the products of the successive entries in the first row of A and the second column of A.

These entries are always either 0 or 1. Notice that the product of two such numbers is 0 unless they are both 1, in which case the product is 1. Therefore $A(\mathcal{D})^2_{12}$ is just the number of times there is a 1 in a position of the first row of A and, at the same time there is a 1 in **the same** position of the second column of $A(\mathcal{D})$.

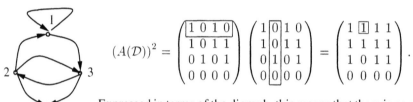

$$(A(\mathcal{D}))^2 = \begin{pmatrix} 1 & 0 & 1 & 0 \\ 1 & 0 & 1 & 1 \\ 0 & 1 & 0 & 1 \\ 0 & 0 & 0 & 0 \end{pmatrix} \begin{pmatrix} 1 & 0 & 1 & 0 \\ 1 & 0 & 1 & 1 \\ 0 & 1 & 0 & 1 \\ 0 & 0 & 0 & 0 \end{pmatrix} = \begin{pmatrix} 1 & 1 & 1 & 1 \\ 1 & 1 & 1 & 1 \\ 1 & 0 & 1 & 1 \\ 0 & 0 & 0 & 0 \end{pmatrix}.$$

Expressed in terms of the digraph, this means that there is an arc from 1 to some vertex AND an arc from that vertex to 2. Such a tip to tail arc arrangement is called a walk (of length 2) from 1 to 2. The only walk of length 2 from 1 to 2 goes through 3. Thus, $A(\mathcal{D})^2_{12} = 1$.   □

Check that the $A(\mathcal{D})^2$ in the example is itself the adjacency matrix of the indicated digraph:

In general, the number of walks from vertex i to vertex j having length k is $A(\mathcal{D})^k_{ij}$, the i, j entry of the k-th power of the adjacency matrix.

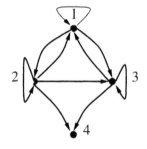

For example, the diagonal entries of $(A(\mathcal{D})^3)_{ii}$ count the number of closed walks of length 3, starting and ending at vertex i.

Objective 3
Simplify and evaluate matrix expressions using properties of matrix arithmetic.

Recall from page 40 that the ***transpose*** of an m-by-n matrix M is the n-by-m (note the subscript switch!) matrix M^T obtained from M by putting the (i, j) entry of M in the (j, i) position. An n-by-n matrix is called ***square***.

Since matrix multiplication generalizes ordinary numerical multiplication, you won't be surprised to learn that matrix analogs to the numbers 0 and 1 are of particular importance. A matrix, all of whose entries are zero, is called a **zero matrix** and written 0 or perhaps **0** (no matter what its shape!). Zero matrices are analogous to the number 0 because, for example:

$$0 + A = A \text{ and } \mathbf{0} \, A = 0.$$

(Since 0 is interpreted to have appropriate shape, the left- and right-hand sides of these equations are defined, but the different 0 matrices may have unequal shapes.) The matrix analog to the number 1 is a square matrix called an **identity matrix** and is written I, or perhaps I_n (and is located as the fifth item in the calculator's matrix math menu). I satisfies equations like:

$$IA = A \, I = A$$

for all matrices A (provided shapes are compatible so that all indicated matrix operations are actually defined). The subscript n notes the size:

$$I_2 = \begin{pmatrix} 1 & 0 \\ 0 & 1 \end{pmatrix} , I_3 = \begin{pmatrix} 1 & 0 & 0 \\ 0 & 1 & 0 \\ 0 & 0 & 1 \end{pmatrix} .$$

In fact, identity matrices are so much like 1 that we sometimes add numbers to square matrices with the understanding that the number is to be interpreted as a multiple of the appropriate identity matrix.

Matrix arithmetic is a really new system! There are important differences from ordinary arithmetic! Here are a few surprises:

1. It is possible for AB to be different from BA. Thus, matrix multiplication is **noncommutative.**. Try $A = \begin{pmatrix} 0 & 0 \\ 1 & 0 \end{pmatrix}$ and $B = \begin{pmatrix} 0 & 1 \\ 0 & 0 \end{pmatrix}$.

2. It is possible for the product AB of two matrices to equal a zero matrix while neither $A = 0$ nor $B = 0$. Try $A = B$ equal to either of the above matrices.

3. The transpose of a product is the product of the transposes with order reversed!

$$(AB)^T = B^T A^T.$$

   (To see this, just remember that the (j, i) entry of $(AB)^T$ is the (i, j) entry of AB, which is the dot product of the i-th row of A and the j-th column of B. But this is just the same as the dot product of the j-th row of B^T and the i-th column of A^T; which, in turn, is the (j, i) entry of the right-hand side.)

WARNING: The fact that matrix multiplication is noncommutative is especially hard to get used to, and constant care must be taken.

There is some good news. Matrix arithmetic does have a number of important properties in common with ordinary arithmetic. For example:

(Distributive Law) $A(B + C) = AB + AC$

(Associative Law) $A(BC) = (AB)C$

provided all indicated products exist.

The 1's of an identity matrix occupy positions in a square matrix that is called the *main diagonal*. A matrix all of whose nonzero entries are on the main diagonal is called a *diagonal* matrix. The arithmetic of diagonal matrices is especially easy; for this reason one of the major themes of linear algebra is to find ways that any square matrix is similar to a diagonal matrix.

A square matrix A is called **upper triangular** if all entries below and to the left of the main diagonal are zeros (that is $a_{ij} = 0$ for $i > j$). A matrix is **lower triangular** if its transpose is upper triangular. Notice that any square matrix in REF is upper triangular and that the sum or product of two upper triangular matrices is again upper triangular.

Example 2.6 (Matrix arithmetic and the transpose.)

a. Let $v = \begin{pmatrix} 1 \\ 2 \\ 3 \end{pmatrix} = (1\ 2\ 3)^T$. Then $vv^T - 2 = \begin{pmatrix} 1 & 2 & 3 \\ 2 & 4 & 6 \\ 3 & 6 & 9 \end{pmatrix} - 2I_3 = \begin{pmatrix} -1 & 2 & 3 \\ 2 & 2 & 6 \\ 3 & 6 & 7 \end{pmatrix}$,

but $v^T v - 2 = (1 \cdot 1 + 2 \cdot 2 + 3 \cdot 3) - 2 = 12$.

b. $\begin{pmatrix} 5 & 2 & 7 & 3 \\ 1 & 3 & 9 & 6 \end{pmatrix}^T \begin{pmatrix} 2 & 3 \\ 4 & 1 \end{pmatrix} + 3 \begin{pmatrix} 0 & 1 \\ 1 & 0 \\ 0 & 1 \\ 1 & 0 \end{pmatrix} = \begin{pmatrix} 14 & 16 \\ 16 & 9 \\ 50 & 30 \\ 30 & 15 \end{pmatrix} + \begin{pmatrix} 0 & 3 \\ 3 & 0 \\ 0 & 3 \\ 3 & 0 \end{pmatrix} = \begin{pmatrix} 14 & 19 \\ 19 & 9 \\ 50 & 33 \\ 33 & 15 \end{pmatrix}$. □

Objective 4

Verify complex identities in the matrix representation of $\mathbb{C}$.

We have seen that the 2-by-2 identity matrix is a matrix analog to the number 1. We can make a real matrix model of $\mathbb{C}$ by finding a real matrix analogous to the imaginary unit i. This means we must find a MATRIX zero of the polynomial $f(x) = x^2 + 1$. If we further insist that complex conjugation corresponds to "transpose," then there is essentially only one choice. We have the matrix model of $\mathbb{C}$:

$$z = a + bi \quad \Leftrightarrow \quad \begin{pmatrix} a & -b \\ b & a \end{pmatrix} \text{ and } \bar{z} = a - bi \quad \Leftrightarrow \quad \begin{pmatrix} a & b \\ -b & a \end{pmatrix}.$$

You can check complex addition and multiplication equally well with the complex numbers or with the matrix representation; the results correspond! For example:

$$(a+bi)(a-bi) = a^2 + b^2 = z\bar{z} \quad \Leftrightarrow \quad \begin{pmatrix} a & -b \\ b & a \end{pmatrix} \begin{pmatrix} a & b \\ -b & a \end{pmatrix}^T = \begin{pmatrix} a^2 + b^2 & 0 \\ 0 & a^2 + b^2 \end{pmatrix}.$$

Matrices corresponding to numbers of absolute value 1 and argument θ

$$r(\theta) = \begin{pmatrix} \cos(\theta) & -\sin(\theta) \\ \sin(\theta) & \cos(\theta) \end{pmatrix}$$

are of special significance. They correspond to linear functions having transform plots that fix the origin and rotate everything else counterclockwise by an angle of θ and are used in our discussion of changing plane coordinates.

SUMMARY

The **product** AB of two matrices is not defined unless the number of columns of A equals the number of rows of B. The ij- entry of the product AB is $a_{i1}b_{1j} + a_{i2}b_{2j} + \cdots + a_{in}b_{nj}$.

If $\mathcal{D}$ is a digraph on v vertices, its **adjacency matrix** is a v-by-v matrix with a 1 in position ij if and only if there is an arc in $\mathcal{D}$ from vertex i to j. The ij entry of $A(\mathcal{D})^n$ is exactly the number of walks in $\mathcal{D}$ of length n from vertex i to j.

Matrix arithmetic satisfies many BUT NOT ALL rules of ordinary arithmetic. In particular, matrix multiplication is **noncommutative** $AB \neq BA$! There are matrix analogs of 0 and 1.

The matrix model of $\mathbb{C}$ provides an alternative way of doing complex arithmetic:

$$z = a + bi \iff \begin{pmatrix} a & -b \\ b & a \end{pmatrix}.$$

Looking further

It is hard to overemphasize the importance of the notion of multiplying matrices. Historically, Cayley was led to it by his study of linear functions in a way that appears in Section 4.1. (Gauß and Cauchy were studying something else.) By expanding the use of addition and multiplication beyond ordinary numbers, matrices open the door to "abstract algebra." Matrix arithmetic vastly expands the "mathematical tool kit," and specifically "representation theory" uses matrix arithmetic as a tool to understand more subtle mathematical structures.

$$* * * * * *$$

Practice Problems

1. Compute the matrix product, or state that the product is not defined.

a. $\begin{pmatrix} 7 & -4 & 6 \\ 3 & 1 & -9 \end{pmatrix} \begin{pmatrix} 1 & -6 \\ 2 & 0 \\ 5 & 1 \end{pmatrix}$

b. $\begin{pmatrix} 2 & 0 & 3 \\ 1 & 6 & -9 \\ 5 & -2 & 4 \end{pmatrix} \begin{pmatrix} -6 & 1 & -4 \\ 5 & -3 & 2 \\ 0 & 7 & 9 \end{pmatrix}$

c. $\begin{pmatrix} -6 & 1 & -4 \\ 5 & -3 & 2 \\ 0 & 7 & 9 \end{pmatrix} \begin{pmatrix} 2 & 0 & 3 \\ 1 & 6 & -9 \\ 5 & -2 & 4 \end{pmatrix}$ [Compare the results of parts b and c.]

d. $\begin{pmatrix} 7 & -3 & 2 & 1 & 0 \\ 5 & -6 & -4 & 8 & -1 \end{pmatrix} \begin{pmatrix} -1 & 8 \\ -7 & 2 \\ 3 & -6 \\ 4 & 5 \end{pmatrix}$

e. $\begin{pmatrix} 1 & -7 \\ -6 & 3 \\ -8 & 2 \\ 0 & 4 \end{pmatrix} \begin{pmatrix} 2 & 0 & 3 & -6 \\ -9 & 1 & 0 & 5 \end{pmatrix}$

2. Let $A = \begin{pmatrix} -1 & 3 & 2 \\ 2 & -1 & 1 \end{pmatrix}$, $B = \begin{pmatrix} 1 & -1 \\ -1 & 2 \\ 3 & -3 \end{pmatrix}$, $C = (1 \ -1 \ 1)$, $D = \begin{pmatrix} 2 \\ -1 \\ 2 \end{pmatrix}$.

Compute: a. AB b. BA c. CB d. CBA e. BAD.

3. Compute (if it exists).

a. $\frac{1}{3} \begin{pmatrix} 3 & -1 \\ -9 & 6 \end{pmatrix} - 2 \begin{pmatrix} 1 & -\frac{1}{6} \\ -9 & -1 \end{pmatrix}$

b. $\begin{pmatrix} 3 \\ -1 \\ 7 \end{pmatrix} (-1 \ 4) + 3 \begin{pmatrix} 1 & -2 \\ -2 & 3 \\ 1 & -6 \end{pmatrix}$

c. $(3 \ 2) \left[\begin{pmatrix} 2 & -1 \\ 4 & 3 \end{pmatrix} - \frac{1}{2} \begin{pmatrix} 8 & 10 \\ 2 & 18 \end{pmatrix} \right] \begin{pmatrix} 3 \\ -3 \end{pmatrix}$

d. $-3 \begin{pmatrix} 4 & -9 & 0 \\ -1 & 0 & 2 \\ 6 & -3 & 1 \end{pmatrix} + (3 \ 5 \ -7) \begin{pmatrix} 2 \\ -2 \\ 1 \end{pmatrix}$

e. $\left[\begin{pmatrix} 5 & 6 \\ 1 & 3 \end{pmatrix} \begin{pmatrix} 4 \\ 0 \end{pmatrix} - 2 \begin{pmatrix} 6 \\ 1 \end{pmatrix} \right] (1 \ 1)$

4. a.   Write the adjacency matrix of the digraph:
   b.   Count the total number of closed walks of length 5 using matrix multiplication.
   c.   What pattern arises in the successive powers of the adjacency matrix?

5. a.   Write the adjacency matrix of the digraph:
   b.   Count the total number of closed walks of length 3 using matrix multiplication. Verify your answer by finding them.
   c.   Use matrix arithmetic to count the total number of walks of length 1 or 2 or 3 in the digraph.

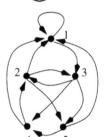

6. Given $f(x) = 3x^2 - 5x$, compute $f(A)$ for each of the following:

a. $A = \begin{pmatrix} -1 & 1 & 3 \\ 5 & -2 & 4 \\ 1 & 6 & 3 \end{pmatrix}$

b. $A = \begin{pmatrix} -1 & 1 & 3 \\ 0 & -2 & 4 \\ 0 & 0 & 3 \end{pmatrix}$

7. Make a transform plot for the linear function f_A where A is the matrix model for the complex number $1 + i$. How does this confirm the claims of the last section?

8. Use the matrix model of $\mathbb{C}$ to verify:
   a. $(1 + i)^2 = 2i$ b. $(2 + i)(2 - i) = 5$.

9. Verify the associative law $(AB)C = A(BC)$ for:

$$A = (a_1 \ a_2), \quad B = \begin{pmatrix} b_{11} & b_{12} \\ b_{21} & b_{22} \end{pmatrix} \text{ and } C = \begin{pmatrix} c_1 \\ c_2 \end{pmatrix}.$$

Use this result to establish the associative law for all 2-by-2 matrices. (Hint: Regard A as a stack of rows and C as a list of columns rather like the beginning of the section.)

10. Return to Example 1.20 and construct matrices M having the coordinates of P, Q and R as rows and C having as rows the coordinates of P*, Q* and R*. Show that we simply solved $MA^T = C$ for A^T.

2.3   Auxiliary matrices and matrix inverses

Objective 1

Given a polynomial $f(x)$, a square matrix A and a vector $\mathbf{v}$, compute $f(A)\mathbf{v}$ with an auxiliary matrix.

This objective is not just an exercise in properties of matrix arithmetic. It is crucial in Section 3.2.

We have, for some time, viewed the matrix equation $M\mathbf{x} = \mathbf{b}$ as the matrix version of a linear system having coefficient matrix M. When viewed this way, the ROWS of the matrix M are emphasized because they correspond to the separate equations of the system. This matrix equation can be expressed in a second way by concentrating on the COLUMNS of M. From this point of view, the coordinates of $\mathbf{x}$ are just coefficients used to build a linear combination of the columns of M.

Example 2.7 (Linear combinations of the columns of a matrix.)

a. Let $M = \begin{pmatrix} 4 & 1 & -3 \\ -6 & 0 & 4 \\ 2 & 0 & -3 \end{pmatrix}$, $\mathbf{b} = \begin{pmatrix} -3 \\ 6 \\ -7 \end{pmatrix}$ and $\mathbf{x} = \begin{pmatrix} 1 \\ 2 \\ 3 \end{pmatrix}$. Then the product

$$M\mathbf{x} = \begin{pmatrix} 4 & 1 & -3 \\ -6 & 0 & 4 \\ 2 & 0 & -3 \end{pmatrix}\begin{pmatrix} 1 \\ 2 \\ 3 \end{pmatrix} = \begin{pmatrix} 4 \\ -6 \\ 2 \end{pmatrix} 1 + \begin{pmatrix} 1 \\ 0 \\ 0 \end{pmatrix} 2 + \begin{pmatrix} -3 \\ 4 \\ -3 \end{pmatrix} 3 = \begin{pmatrix} -3 \\ 6 \\ -7 \end{pmatrix}.$$

b. In fact, the column linear combination point of view already appeared in transform plots. The transform plot in part b of Example 1.19 on page 44 shows that $(-1,3)^T$ in the domain is mapped to $(1,3)^T$ in the range. When expressed in terms of the image grid, this fact is:

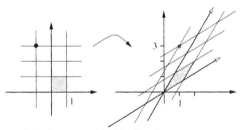

$$\frac{1}{4}\begin{pmatrix} 5 & 3 \\ 3 & 5 \end{pmatrix}\begin{pmatrix} -1 \\ 3 \end{pmatrix} = \frac{1}{4}\begin{pmatrix} 5 \\ 3 \end{pmatrix}(-1) + \frac{1}{4}\begin{pmatrix} 3 \\ 5 \end{pmatrix}(3) = \begin{pmatrix} 1 \\ 3 \end{pmatrix},$$

since the image grid consists of all integer linear combinations of the columns of the matrix. □

This objective is about matrices that are built to use the fact just illustrated. Namely, that the matrix equation $M\mathbf{x} = \mathbf{b}$ expresses $\mathbf{b}$ as a linear combination of the columns of M and that the coefficients are the entries of $\mathbf{x}$.

Let $f(x) = f_0 + f_1 x + \cdots + f_n x^n$ be a polynomial of degree n (so $f_n \neq 0$). If A is a matrix, then basic functional notation dictates that $f(A)$ be obtained by replacing every occurrence of x in the formula for $f(x)$ by the matrix A. The arithmetic operations must, of course, be interpreted as **matrix** arithmetic. This requires that A is square (as assumed in this objective). The expression $f_k A^k$ is understood as the scalar product of the number f_k by the product of k copies of the matrix A, and the $+$ is understood as matrix addition. There is one more issue. The constant coefficient must be interpreted as a multiple of the **appropriate identity matrix**, as mentioned in the preceding section. Thus,

$$f(A) = f_0 I + f_1 A + \cdots + f_n A^n,$$

and is a matrix of the same shape as A!

Example 2.8 (Evaluating a polynomial at a square matrix.)

  a. Evaluate the polynomial $f(x) = x^2 - 5x + 4$ at $A = \begin{pmatrix} 1 & 0 \\ 0 & 4 \end{pmatrix}$.

$$f(A) = \begin{pmatrix} 1 & 0 \\ 0 & 4 \end{pmatrix}^2 - 5 \begin{pmatrix} 1 & 0 \\ 0 & 4 \end{pmatrix} + 4 I_2 = \begin{pmatrix} 1 - 5 + 4 & 0 \\ 0 & 16 - 20 + 4 \end{pmatrix} = 0.$$

Of course, you realize that $x^2 - 5x + 4 = (x - 4)(x - 1)$, so it is reassuring to notice that $(A - 4I)(A - I)$

$$= \left(\begin{pmatrix} 1 & 0 \\ 0 & 4 \end{pmatrix} - 4 \begin{pmatrix} 1 & 0 \\ 0 & 1 \end{pmatrix} \right) \left(\begin{pmatrix} 1 & 0 \\ 0 & 4 \end{pmatrix} - \begin{pmatrix} 1 & 0 \\ 0 & 1 \end{pmatrix} \right) = \begin{pmatrix} -3 & 0 \\ 0 & 0 \end{pmatrix} \begin{pmatrix} 0 & 0 \\ 0 & 3 \end{pmatrix} = 0.$$

  b. Evaluate the polynomial $f(x) = x^2 - 2x - 3$ at $B = \begin{pmatrix} 1 & 1 & 1 \\ 0 & 1 & 1 \\ 0 & 0 & 1 \end{pmatrix}$.

$$B^2 - 2B + 3 = \begin{pmatrix} 1 & 2 & 3 \\ 0 & 1 & 2 \\ 0 & 0 & 1 \end{pmatrix} - 2 \begin{pmatrix} 1 & 1 & 1 \\ 0 & 1 & 1 \\ 0 & 0 & 1 \end{pmatrix} - \begin{pmatrix} 3 & 0 & 0 \\ 0 & 3 & 0 \\ 0 & 0 & 3 \end{pmatrix} = \begin{pmatrix} -4 & 0 & 1 \\ 0 & -4 & 0 \\ 0 & 0 & -4 \end{pmatrix}.$$

Since $f(x) = (x - 3)(x + 1)$, we could have computed $(B - 3I)(B + I)$. □

Now suppose $\mathbf{v} \in \mathbb{R}^n$ is a vector and A is an n-by-n matrix. If a polynomial $f(x)$ is given, then there are TWO ways to compute $f(A)\mathbf{v}$. One way is to first compute the matrix $f(A)$ (as above) and then to multiply the result by $\mathbf{v}$. The second method uses auxiliary matrices. Here is how it works:

Begin by writing out exactly what $f(A)\mathbf{v}$ is. Then use the ***distributive law*** of matrix arithmetic and regroup as follows:

$$f(A)\mathbf{v} = (f_0 + f_1 A + f_2 A^2 + \cdots + f_n A^n)\mathbf{v}$$
$$= f_0\mathbf{v} + f_1 A\mathbf{v} + f_2 A^2\mathbf{v} + \cdots + f_n A^n\mathbf{v}$$
$$= \mathbf{v} f_0 + (A\mathbf{v})f_1 + (A^2\mathbf{v})f_2 + \cdots + (A^n\mathbf{v})f_n$$

$$= \left(\mathbf{v} \middle| A\mathbf{v} \middle| A^2\mathbf{v} \middle| \cdots \middle| A^n\mathbf{v} \right) \begin{pmatrix} f_0 \\ f_1 \\ \vdots \\ f_n \end{pmatrix}.$$

The last step is not at all obvious, and to be frank, truly brilliant. It may seem more natural if you read it from bottom up? The last matrix, whose k-th column is $A^{k-1}\mathbf{v}$, is called an **auxiliary matrix** for A and $\mathbf{v}$. (The vertical lines in the auxiliary matrix are only there to emphasize that the matrix is constructed one column at a time.)

If $\mathbf{v} \in \mathbb{R}^n$ and A are given but the polynomial coefficients $f_0, \ldots, f_n$ are **unknowns** for which we must solve, then the auxiliary matrix method is perfect because it **postpones** dealing with the coefficients until the very end! All you really need is an auxiliary matrix with as many columns as there are coefficients of $f(x)$.

Example 2.9 (Computing $f(A)\mathbf{v}$ in two ways.)

Let B be as in part b of the preceding example and $\mathbf{v} = (1\ 1\ 1)^T$ and $f(x) = x^2 - 2x - 3$. $f(B)\mathbf{v} = \begin{pmatrix} -4 & 0 & 1 \\ 0 & -4 & 0 \\ 0 & 0 & -4 \end{pmatrix} \begin{pmatrix} 1 \\ 1 \\ 1 \end{pmatrix} = \begin{pmatrix} -3 \\ -4 \\ 1 \end{pmatrix}.$

The old-fashioned way to compute $f(B)\mathbf{v}$ is indicated above; simply use the computation of $f(B)$ that appeared in that example.

The new, improved auxiliary matrix way begins with computing the auxiliary matrix: this is actually pretty easy to do. You start with $\mathbf{v}$ in the first column and work your

$$\left(\mathbf{v} \middle| B\mathbf{v} \middle| B^2\mathbf{v} \right) = \begin{pmatrix} 1 & 3 & 6 \\ 1 & 2 & 3 \\ 1 & 1 & 1 \end{pmatrix}$$

way to the right. **Each new column is obtained from the last by multiplying it by B, thereby avoiding matrix powers.** Once the auxiliary matrix is computed, you need only multiply it by the coefficient vector of the the polynomial:

$$f(B)\mathbf{v} = \left(\mathbf{v} \middle| B\mathbf{v} \middle| B^2\mathbf{v} \right) \begin{pmatrix} f_0 \\ f_1 \\ f_2 \end{pmatrix} = \begin{pmatrix} 1 & 3 & 6 \\ 1 & 2 & 3 \\ 1 & 1 & 1 \end{pmatrix} \begin{pmatrix} -3 \\ -2 \\ 1 \end{pmatrix} = \begin{pmatrix} -3 \\ -4 \\ 1 \end{pmatrix}.$$

WARNING: Did you notice the order of those sneaky coefficients? It is reversed from the usual way to write polynomials. As explained in Section 1.7, the **constant** coefficient c_0 goes first for us!

Suppose that we had to FIND a polynomial $g(x)$ such that $g(B)\mathbf{v} = \mathbf{0}$. All we would do is build an auxiliary matrix M for B and $\mathbf{v}$ that for sure has a nonpivotal column when RREFed. Then find a basic solution to $M\mathbf{g} = \mathbf{0}$ and translate it into a polynomial.

$$\begin{pmatrix} 1 & 3 & 6 & 10 \\ 1 & 2 & 3 & 4 \\ 1 & 1 & 1 & 1 \end{pmatrix} \xrightarrow{RREF} \begin{pmatrix} 1 & 0 & 0 & 1 \\ 0 & 1 & 0 & -3 \\ 0 & 0 & 1 & 1 \end{pmatrix}. \text{ Basic solution: } \begin{pmatrix} g_0 \\ g_1 \\ g_2 \\ g_3 \end{pmatrix} = \begin{pmatrix} -1 \\ 3 \\ -3 \\ 1 \end{pmatrix}.$$

Therefore, $g(x) = g_0 + g_1 x + g_2 x^2 + g_3 x^3 = -1 + 3x - 3x^2 + x^3$.
Check that the polynomial $g(x)$ satisfies $g(A)\mathbf{v} = \mathbf{0}$.   □

As already plainly put, auxiliary matrices play a pivotal role later. Make sure you can work with them! They represent just one small step in the minimal polynomial algorithm.

Objective 2

Given an n-by-n matrix A, compute A^{-1} or decide that A isn't invertible.

The square matrix A is **invertible** (or **nonsingular**) if there is a matrix X such that:

$$AX = I \text{ and } XA = I.$$

When such a matrix X exists, it is unique (see Problem 7). The matrix X is called the inverse of A and is written A^{-1}. The key to computing the inverse of A is to think of the matrix equation $AX = I$ as a system of linear systems (one linear system for each column of X and I) AND all with the same coefficient matrix A, just as in Example 1.10 on page 22. The j-th of these linear systems is:

$$A \begin{pmatrix} \text{The } j\text{-th column} \\ \text{of } A^{-1} \end{pmatrix} = \begin{pmatrix} \text{The } j\text{-th column} \\ \text{of the identity} \\ \text{matrix} \end{pmatrix} = \begin{pmatrix} 0 \\ \vdots \\ 1 \\ \vdots \\ 0 \end{pmatrix} \begin{matrix} \uparrow \\ \\ j \\ \\ \downarrow \end{matrix}.$$

Now, j ranges from 1 to n and each produces a system of linear equations having the coefficient matrix A. To solve this "system of systems" all we do is form a generalized augmented matrix that has the matrix A on the left and the columns of the identity matrix on the right. Then reduce it to RREF.

$$\begin{pmatrix} a_{1,1} & a_{1,2} & \cdots & a_{1,n} & 1 & 0 & & 0 \\ a_{2,1} & a_{2,2} & \cdots & a_{2,n} & 0 & 1 & \cdots & 0 \\ & \cdots & \cdots & & & & \cdots & \\ a_{m,1} & a_{n,2} & \cdots & a_{m,n} & 0 & 0 & \cdots & 1 \end{pmatrix}.$$

If the reduced row echelon form of A has a row of zeros, then the system coming from the column of the leading coefficient in that row is inconsistent and we know that A has no inverse. If the reduced row echelon form of A is the identity matrix, then the reduced version of the above matrix has the form $(I|A^{-1})$. Actually, the reduced version of $(A|I)$ is either $(I|A^{-1})$, or A^{-1} does not exist.

Example 2.10 (Computing the inverse of a square matrix.)

a. The inverse of $A = \begin{pmatrix} 2 & 3 & 1 \\ -1 & 1 & 0 \\ 1 & 0 & 1 \end{pmatrix}$ is $A^{-1} = \begin{pmatrix} \frac{1}{4} & -\frac{3}{4} & -\frac{1}{4} \\ \frac{1}{4} & \frac{1}{4} & -\frac{1}{4} \\ -\frac{1}{4} & \frac{3}{4} & \frac{5}{4} \end{pmatrix}$.

$$\left(\begin{array}{ccc|ccc} 2 & 3 & 1 & 1 & 0 & 0 \\ -1 & 1 & 0 & 0 & 1 & 0 \\ 1 & 0 & 1 & 0 & 0 & 1 \end{array}\right) \rightarrow \left(\begin{array}{ccc|ccc} 1 & 0 & 1 & 0 & 0 & 1 \\ -1 & 1 & 0 & 0 & 1 & 0 \\ 2 & 3 & 1 & 1 & 0 & 0 \end{array}\right) \rightarrow \left(\begin{array}{ccc|ccc} 1 & 0 & 1 & 0 & 0 & 1 \\ 0 & 1 & 1 & 0 & 1 & 1 \\ 0 & 3 & -1 & 1 & 0 & -2 \end{array}\right)$$

$$\rightarrow \left(\begin{array}{ccc|ccc} 1 & 0 & 1 & 0 & 0 & 1 \\ 0 & 1 & 1 & 0 & 1 & 1 \\ 0 & 0 & -4 & 1 & -3 & -5 \end{array}\right) \rightarrow \left(\begin{array}{ccc|ccc} 1 & 0 & 0 & \frac{1}{4} & -\frac{3}{4} & -\frac{1}{4} \\ 0 & 1 & 0 & \frac{1}{4} & \frac{1}{4} & -\frac{1}{4} \\ 0 & 0 & 1 & -\frac{1}{4} & \frac{3}{4} & \frac{5}{4} \end{array}\right).$$

b. This method applies when the matrix has variable entries, too. Suppose A is as given. We show below that it has the indicated inverse whenever $a, b \in \mathbb{R}$ are not both zero.

$$A = \begin{pmatrix} a & -b \\ b & a \end{pmatrix} \text{ has inverse } A^{-1} = \frac{1}{a^2 + b^2}\begin{pmatrix} a & b \\ -b & a \end{pmatrix}.$$

Indeed, if $b = 0$, then $A = aI_2$, and the formula holds. Otherwise:

$$\left(\begin{array}{cc|cc} a & -b & 1 & 0 \\ b & a & 0 & 1 \end{array}\right) \rightarrow \left(\begin{array}{cc|cc} a & -b & 1 & 0 \\ 1 & \frac{a}{b} & 0 & \frac{1}{b} \end{array}\right) \rightarrow \left(\begin{array}{cc|cc} 0 & \frac{-a^2}{b} - b & 1 & -a/b \\ 1 & \frac{a}{b} & 0 & \frac{1}{b} \end{array}\right) \rightarrow$$

$$\left(\begin{array}{cc|cc} 0 & -b^2 - a^2 & b & -a \\ 1 & \frac{a}{b} & 0 & \frac{1}{b} \end{array}\right) \rightarrow \left(\begin{array}{cc|cc} 1 & \frac{a}{b} & 0 & \frac{1}{b} \\ 0 & 1 & \frac{-b}{b^2+a^2} & \frac{a}{b^2+a^2} \end{array}\right) \rightarrow \left(\begin{array}{cc|cc} 1 & 0 & \frac{a}{b^2+a^2} & \frac{b}{b^2+a^2} \\ 0 & 1 & \frac{-b}{b^2+a^2} & \frac{a}{b^2+a^2} \end{array}\right).$$

But you already knew that! This matrix is just the matrix representation of the complex number $z = a + bi$. We know that it is invertible whenever $z \neq 0$. In fact, we have verified the matrix representation version of $z^{-1} = \bar{z}/|z|^2$, the formula that appears on page 56.   □

There is an important formula for the inverse of the product of two matrices:
$$(AB)^{-1} = B^{-1}A^{-1}.$$

A proof is sketched in Problem 7. **We use it often in Chapters 3 and 4.**

Solving a linear system with invertible coefficient matrix A is extremely easy IF you happen to know A^{-1}. Start with $A\mathbf{x} = \mathbf{b}$. Multiply by A^{-1} and regroup $A^{-1}(A\mathbf{x}) = A^{-1}\mathbf{b}$ to obtain
$$(A^{-1}A)\mathbf{x} = A^{-1}\mathbf{b}. \text{ Then simplify to } \mathbf{x} = I\mathbf{x} = A^{-1}\mathbf{b}.$$

It is, of course, easier to solve one explicit linear system using Gaussian elimination than it is to compute A^{-1}, so you would only do this if you had to solve the same linear system repeatedly.

Inverting Vandermonde matrices: Lagrange interpolants

One of the most important situations where the linear system coefficient matrix is guaranteed invertible already appeared in Section 1.7. There we used Vandermonde matrices $V(x_0 \ldots x_n)$ to find polynomials $f(x) = f_0 + f_1 x + \cdots + f_m x^m$ that interpolate the data (x_i, y_i), $i = 0 \ldots n$ by solving the linear system $V(x_0 \ldots x_n)\mathbf{f} = \mathbf{y}$. The reason this worked is

$$V(x_0 \ldots x_n)\mathbf{f} = \begin{pmatrix} 1 & x_0 & x_0^2 & \cdots & x_0^m \\ 1 & x_1 & x_1^2 & \cdots & x_1^m \\ & & \cdots & & \\ 1 & x_n & x_n^2 & \cdots & x_n^m \end{pmatrix} \begin{pmatrix} f_0 \\ f_1 \\ \vdots \\ f_m \end{pmatrix} = \begin{pmatrix} f(x_0) \\ f(x_1) \\ \vdots \\ f(x_n) \end{pmatrix}.$$

In other words, multiplication by $V(x_0 \ldots x_n)$ transforms the coefficient vector $\mathbf{f}$ of $f(x)$ into the vector of f **values** $(f(x_0) \ldots f(x_n))^T$.

The fact is that $V = V(x_0, \ldots x_n)$ is invertible whenever it is square and the x_i are distinct. We will need this and related facts in Sections 2.5 and 3.4, so we now develop an explicit formula for V^{-1} in this case.

This may seem like a daunting task, but it actually has an elegant solution first given by Waring in 1779 (although generally attributed to Lagrange, who published it independently in 1795). The key idea is to use polynomial multiplication.

Recall that in Part b of Example 1.21 on page 51, we observed that that the solutions to the homogeneous system $V(x_0 \ldots x_n)\mathbf{f} = \mathbf{0}$ could be written in **factored** form as multiples of $Z(x) = (x - x_0) \ldots (x - x_n)$ when the Vandermonde matrix has more columns than rows. We can still use this factoring idea.

For each value of $j = 0 \ldots n$, break $Z(x)$ into two factors so that

$$Z(x) = (x - x_j)Z_j(x) \text{ for } j = 0 \ldots n.$$

Thus, $Z_j(x)$ is defined to be the product of all factors of $Z(x)$ **except** $(x - x_j)$.

Now suppose the Vandermonde matrix is square, so $m = n$, and the x_i are distinct. Imagine building $V(x_0, \ldots x_n)^{-1}$ one column at time — but don't forget that we are studying polynomials. The j-th column corresponds to the linear system:

$$V(x_0, \ldots x_n) \begin{pmatrix} f_0 \\ \vdots \\ \vdots \\ f_n \end{pmatrix} = \begin{pmatrix} 0 \\ \vdots \\ 1 \\ \vdots \\ 0 \end{pmatrix} \begin{matrix} \\ \} j \\ \\ \end{matrix}.$$

This amounts to finding the **coefficients** $f_0 \ldots f_n$ of a polynomial $f(x)$ of degree at most n that satisfies

$$f(x_j) = 1 \text{ and } f(x_i) = 0 \text{ for } i \neq j.$$

The j-th **fundamental Lagrange interpolant** $\Lambda_j(x)$ for the data $x_0 \ldots x_n$ is the polynomial corresponding to the solution to this nonhomogeneous linear system. Our job is to give a simple formula for $\Lambda_j(x)$. Then its coefficients are exactly the entries of the j-th column of $V(x_0 \ldots x_n)^{-1}$. To do this we do NOT use row operations. Rather, we think about polynomials. The fact that $\Lambda_j(x_i) = 0$ whenever

$i \neq j$ implies that it is divisible by $Z_j(x)$. Moreover, since the x_is are distinct, $Z_j(x_j) \neq 0$, and so

$$\Lambda_j(x) = \frac{Z_j(x)}{Z_j(x_j)}.$$

Example 2.11 (Vandermonde inverses and Lagrange interpolants.)
Let V be the Vandermonde matrix associated with "data" $1, i, -1, -i$. The first two fundamental Lagrange interpolants determine the first two columns of V^{-1} and are

$$\Lambda_1 = \frac{(x-i)(x+1)(x+i)}{(1-i)(1+1)(1+i)} = \frac{1+x+x^2+x^3}{4}$$

$$\Lambda_2 = \frac{(x-1)(x+1)(x+i)}{(i-1)(i+1)(i+i)} = \frac{-i-x+ix^2+x^3}{-4i} = \frac{1-ix-x^2+ix^3}{4}.$$

Miraculously, V^{-1} is almost the conjugate transpose of V:

$$V = \begin{pmatrix} 1 & 1 & 1 & 1 \\ 1 & i & -1 & -i \\ 1 & -1 & 1 & -1 \\ 1 & -i & -1 & i \end{pmatrix} \quad \text{while } V^{-1} = \frac{1}{4}\overline{V}^T = \frac{1}{4}\begin{pmatrix} 1 & 1 & 1 & 1 \\ 1 & -i & -1 & i \\ 1 & -1 & 1 & -1 \\ 1 & i & -1 & -i \end{pmatrix}.$$

□

SUMMARY

To compute $f(A)$ for a polynomial $f(x)$ and square matrix A, interpret the constant term as a scalar multiple of the **identity matrix** and interpret arithmetic operations as matrix arithmetic. You may also compute $f(A)\mathbf{v}$ by using an **auxiliary matrix** M and then multiplying M by the coefficient vector for $f(x)$. This method is especially useful if the polynomial has complicated or unknown coefficients.

To compute the inverse of a square matrix A, reduce the generalized augmented matrix $(A|I)$. The result is either $(I|A^{-1})$, or A^{-1} doesn't exist.

When $x_0 \ldots x_n$ are distinct, the square Vandermonde matrix $V(x_0 \ldots x_n)$ has as its columns the coefficients of *fundamental Lagrange interpolants*. There is a simple formula for the j-th fundamental Lagrange interpolant $\Lambda_j(x)$ that involves polynomial mutiplication.

Looking further

Our method of computing the inverse of an n-by-n matrix is equivalent to solving n linear systems. There is also a formula for the inverse of a square matrix involving determinants in Chapter 4. However, our algorithm is numerically superior.

When comparing algorithms, one must compare more than their complexity and numerical stability. The best algorithms reveal partial information as you proceed, like a picture coming into focus. Such an algorithm allows you to stop early if you run out of time or money and still have something of value. Sometimes deliberately stopping early is called "data compression."

Practice Problems

1. Compute $f(A)$, where $f(x) = (2x - 5)(x + 1)$ and $A = \begin{pmatrix} 1 & 4 & -2 \\ 3 & 3 & -5 \\ -4 & 1 & 1 \end{pmatrix}$.

2. Given $f(x) = 2x^2 - 3x - 5$, compute $f(A)$ for each of the following:

   a. $A = \begin{pmatrix} 1 & 4 & -2 \\ 0 & 3 & -5 \\ 0 & 0 & 1 \end{pmatrix}$

   b. $A = \begin{pmatrix} 1 & 0 & 0 \\ 0 & 3 & 0 \\ 0 & 0 & 1 \end{pmatrix}$

3. For the following problems, compute $f(A)\mathbf{v}$ using auxiliary matrices.

   a. $f(x) = x^2 + 5x + 4$, $A = \begin{pmatrix} 1 & 1 \\ 0 & 5 \end{pmatrix}$, $\mathbf{v} = \begin{pmatrix} 1 \\ 2 \end{pmatrix}$

   b. $f(x) = x^3 - 3x^2 + 4x - 9$, $A = \begin{pmatrix} 1 & 2 & 3 \\ 0 & 4 & 5 \\ 0 & 0 & 6 \end{pmatrix}$, $\mathbf{v} = \begin{pmatrix} -1 \\ 2 \\ -1 \end{pmatrix}$

4. For each part of Problem 3, let M be the auxiliary matrix that you used. Solve the linear system $M\mathbf{g} = \mathbf{0}$. Use this to write a polynomial $g(x)$, of degree equal to the size of A, for which $g(A)\mathbf{v} = \mathbf{0}$.

5. Compute the inverse of each of the following matrices or state that the matrix has no inverse.

   a. $\begin{pmatrix} 1 & 3 \\ 2 & 4 \end{pmatrix}$

   b. $\begin{pmatrix} 1 & 1 \\ 2 & 1 \end{pmatrix}$

   c. $\begin{pmatrix} 2 & 3 & -1 \\ 1 & 2 & 3 \\ -1 & -1 & 4 \end{pmatrix}$

   d. $\begin{pmatrix} 1 & 2 & 7 \\ 1 & -8 & 5 \\ 0 & 5 & 1 \end{pmatrix}$

   e. $\begin{pmatrix} 1 & 0 & 1 & 2 \\ 0 & 1 & 3 & 4 \\ 0 & 0 & 1 & 0 \\ 0 & 0 & 0 & 1 \end{pmatrix}$

   f. $\begin{pmatrix} 7 & 14 & 2 \\ 11 & 25 & 3 \\ 3 & 6 & 1 \end{pmatrix}$

6. Compute all the Lagrange interpolants for the indicated data set and check that they are the columns of the inverse of the appropriate Vandermonde matrix.
   a. $x_0 = 0, x_1 = 1, x_2 = 2$ b. $x_0 = 0, x_1 = 1, x_2 = 2, x_3 = 3$.

7. Suppose that two matrices X and Y satisfy $MX = I = XM$ and $MY = I = YM$. Compute XMY in two ways (using associativity of matrix multiplication) to show that $X = Y$. It follows that M has one inverse at most. Use this fact to prove $(AB)^{-1} = B^{-1}A^{-1}$.

8. Show that $(A^T)^{-1} = (A^{-1})^T$ when they both exist.

9. Suppose $\mathbf{v} \neq \mathbf{0}$. Show there is an invertible matrix B whose first column is $\mathbf{v}$.

10. Suppose $x_0 \ldots x_n$ are distinct numbers and V is an associated rectangular Vandermonde matrix with $m \leq n$ columns. Show that V has rank m.

11. Suppose a data set $x_0 \ldots x_n$ with distinct values is given. Show that the sum of the Lagrange interpolants is a function of degree $n - 1$ that takes the value 1 at each data point. Conclude that
$$\Lambda_0(x) + \Lambda_1(x) + \cdots + \Lambda_n(x) = 1.$$

2.4 Symmetric projectors, resolving vectors

Objective 1

Given an n-by-r matrix A of rank r, compute the symmetric projector P_A and the complementary projector $P_{A\perp}$. Explain their geometric meaning.

In this section we require all matrices have real entries. We repeatedly use the fact that the sum of the squares of real numbers is zero only if each term is zero.

Recall that a **symmetric** matrix M equals its transpose $M = M^T$. A matrix M is **idempotent** if $M^2 = M$. Of course, the only idempotent numbers are 0 and 1.

This section concerns symmetric idempotent matrices. The following two sections present two important applications of matrix arithmetic that build on this section.

We build a symmetric idempotent matrix P_A from any rectangular matrix A of appropriate rank. (Recall that the rank of a matrix A is the number of pivotal columns in $RREF(A)$.) The matrices A must have rank equal their number of columns. Another way of saying this is that all columns of $RREF(A)$ must be pivotal. This condition amounts to requiring that the only $\mathbf{x}$ for which $A\mathbf{x} = \mathbf{0}$ is $\mathbf{x} = \mathbf{0}$.

A bonus is that P_A actually has very useful geometric interpretation.

Given a rectangular matrix A, the **symmetric projector** P_A is defined to be:

$$P_A = A(A^T A)^{-1} A^T \qquad \text{Provided rank}(A) = \text{number of columns.}$$

(There is a general formula for P_A when A has arbitrary rank, but it more complicated and we will not use it.) The **complementary projector** relative to A is defined to be the (appropriate) identity matrix minus P_A:

$$P_{A\perp} = I - P_A.$$

Example 2.12 (Symmetric projectors.)

a. Suppose $\mathbf{v} = \begin{pmatrix} 1 & 2 & -2 \end{pmatrix}^T$. Then $\mathbf{v}^T \mathbf{v} = 1^2 + s^2 + (-2)^2 = 9$, so

$$P_{\mathbf{v}} = \begin{pmatrix} 1 \\ 2 \\ -2 \end{pmatrix} \tfrac{1}{9} \begin{pmatrix} 1 & 2 & -2 \end{pmatrix} = \tfrac{1}{9} \begin{pmatrix} 1 & 2 & -2 \\ 2 & 4 & -4 \\ -2 & -4 & 4 \end{pmatrix}$$

$$\text{and } P_{\mathbf{v}\perp} = I_3 - P_{\mathbf{v}} = \tfrac{1}{9} \begin{pmatrix} 9-1 & -2 & 2 \\ -2 & 9-4 & 4 \\ 2 & 4 & 9-4 \end{pmatrix} = \tfrac{1}{9} \begin{pmatrix} 8 & -2 & 2 \\ -2 & 5 & 4 \\ 2 & 4 & 5 \end{pmatrix}.$$

b. Suppose $A = \begin{pmatrix} 1 & -1 \\ 1 & 0 \\ 1 & 1 \end{pmatrix}$. Then $A^T A = \begin{pmatrix} 3 & 0 \\ 0 & 2 \end{pmatrix}$, so

$$P_A = \begin{pmatrix} 1 & -1 \\ 1 & 0 \\ 1 & 1 \end{pmatrix} \begin{pmatrix} 1/3 & 0 \\ 0 & 1/2 \end{pmatrix} \begin{pmatrix} 1 & 1 & 1 \\ -1 & 0 & 1 \end{pmatrix} = \frac{1}{6} \begin{pmatrix} 5 & 2 & -1 \\ 2 & 2 & 2 \\ -1 & 2 & 5 \end{pmatrix}$$

$$\text{and } P_{A\perp} = I_3 - P_A = \frac{1}{6} \begin{pmatrix} 6-6 & -2 & 1 \\ -2 & 6-2 & -2 \\ 1 & -2 & 6-5 \end{pmatrix} = \frac{1}{6} \begin{pmatrix} 1 & -2 & 1 \\ -2 & 4 & -2 \\ 1 & -2 & 1 \end{pmatrix}. \quad \square$$

Let's check the claimed algebraic properties of symmetric projectors. All it takes is the courage to write down the formulas and a little bit of "manipulation":

$$P_A^T = (A(A^T A)^{-1} A^T)^T = (A^T)^T [(A^T A)^{-1}] A]^T$$
$$= (A)[((A^T A)^{-1})^T A^T] = A(A^T A)^{-1} A^T = P_A.$$

In the middle we used the fact that $(A^T A)$ is symmetric, and the inverse of a symmetric matrix is also symmetric.

Checking that P_A is idempotent is easier

$$P_A^2 = A(A^T A)^{-1} A^T \; A(A^T A)^{-1} A^T$$
$$= A[((A^T A)^{-1}(A^T A)] \, (A^T A)^{-1} A^T = AI(A^T A)^{-1} A^T = P_A.$$

Thus, a symmetric projector is indeed a symmetric matrix and it is idempotent. PLEASE CHECK that the complementary projector to a symmetric projector is also both symmetric and idempotent. Also check that $P_A P_{A\perp} = 0 = P_{A\perp} P_A$. Like the above verification, these "proofs" are more intimidating than difficult.

Because the symmetric projector formula involves an inverse, something must be said about why the $r \times r$ matrix $S = A^T A$ is invertible. Recall from Section 1.3: If an n-by-r matrix A has rank r then $n \geq r$ and RREF(A) looks like an identity matrix stacked on top of some rows of zeros.

Proof that S=A^{T}A is invertible

We know, from the preceding section, that it is enough to show that the linear system $Sx = 0$ has only the trivial solution $x = 0$. Now, suppose $Sx = 0$. Multiply on the left by x^T and eliminate S:

$$0 = x^T 0 = x^T Sx = x^T A^T Ax = (x^T A^T)Ax = (Ax)^T Ax = b^T b,$$

where $Ax = b$ (what else?). But $b^T b$ is a sum of squares of real numbers, so it is non-negative. Thus, $b^T b = 0$ implies $b = 0$. This means that $Ax = b$ is actually a homogeneous linear system. Finally, since A has rank r, RREF(A) has only pivotal columns, and we know that such a linear system has only one solution: $x = 0$.

It can be shown that any symmetric idempotent matrix is a symmetric projector P_A for some matrix A. In particular, complementary projectors are actually symmetric projectors (but for a **different matrix** — NOT A with which we are not directly concerned). The use of the word "complementary" is therefore only meaningful relative to A. This will now be explained geometrically.

Objective 2

Given vectors $\mathbf{a}, \mathbf{w} \in \mathbb{R}^3$, resolve $\mathbf{w}$ with respect to $\mathbf{a}$.

Even more important than the algebraic properties of symmetric projectors is their geometric meaning. We only seriously attempt a geometric interpretation of symmetric projectors in $\mathbb{R}^n$, $n \leq 3$, for now. This interpretation is most important when $A = \mathbf{a}$ is a column vector. In this case $\mathbf{a}^T \mathbf{a} = |\mathbf{a}|^2$, as shown in Section 1.6. So

$$P_{\mathbf{a}}\mathbf{w} = \mathbf{a}(\mathbf{a}^T \mathbf{a})^{-1} \mathbf{a}^T \mathbf{w} = \mathbf{a}\frac{\mathbf{a}^T \mathbf{w}}{|\mathbf{a}|^2} = \frac{\mathbf{a} \cdot \mathbf{w}}{|\mathbf{a}||\mathbf{w}|}|\mathbf{w}|\frac{\mathbf{a}}{|\mathbf{a}|} = (cos\theta)|\mathbf{w}|\frac{\mathbf{a}}{|\mathbf{a}|},$$

where θ is the angle between $\mathbf{a}$ and $\mathbf{w}$.

Notice that this equation shows that $P_{\mathbf{a}}\mathbf{w}$ is a scalar multiple of $\mathbf{a}$. In addition, $P_{\mathbf{a}\perp}\mathbf{w}$ is perpendicular to $\mathbf{a}$ (because $P_A P_{A\perp} = 0$) and $\mathbf{w} = P_{\mathbf{a}}\mathbf{w} + P_{\mathbf{a}\perp}\mathbf{w}$ (because $I = P_A + P_{A\perp}$). These basic algebraic facts are summarized geometrically in the following figure. The process of writing a vector $\mathbf{w}$ as the sum of two vectors, one of which is a scalar multiple of the vector $\mathbf{a}$ and the other perpendicular to $\mathbf{a}$, is called **_resolving the vector_** with respect to $\mathbf{a}$.

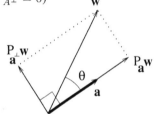

Resolving vectors is a fundamental tool in the analysis of forces experienced-by-a physical body in a complicated situation, as done in *statics* (see Problem 4 for an example).

It is important to realize that the angle between two vectors in $\mathbb{R}^n$ is defined in the PLANE geometry of the PLANE containing both vectors. Consequently, for any m-by-n matrix A (of rank n) and vector $\mathbf{v} \in \mathbf{R}^n$, we can still use the **geometric language** saying $P_A \mathbf{v}$ is perpendicular to all columns of A and $P_{A\perp}\mathbf{v}$ is perpendicular to every vector that is perpendicular to all columns of A. Therefore, $P_A \mathbf{v}$ and $P_{A\perp}\mathbf{v}$ are perpendicular to each other in $\mathbb{R}^n$!

Example 2.13 (Resolving vectors.)

   a. Resolve $\mathbf{x} = \begin{pmatrix} 1 & 1 & 1 \end{pmatrix}^T$ with respect to $\mathbf{a} = \begin{pmatrix} -1/3 & 2/3 & 2/3 \end{pmatrix}^T$.

$$P_{\mathbf{a}}\mathbf{x} = 1/9 \begin{pmatrix} 1 & -2 & -2 \\ -2 & 4 & 4 \\ -2 & 4 & 4 \end{pmatrix} \begin{pmatrix} 1 \\ 1 \\ 1 \end{pmatrix} = \begin{pmatrix} -1/3 \\ 2/3 \\ 2/3 \end{pmatrix}$$

and

$$P_{\mathbf{a}\perp}\mathbf{x} = 1/9 \begin{pmatrix} 8 & 2 & 2 \\ 2 & 5 & -4 \\ 2 & -4 & 5 \end{pmatrix} \begin{pmatrix} 1 \\ 1 \\ 1 \end{pmatrix} = \begin{pmatrix} 4/3 \\ 1/3 \\ 1/3 \end{pmatrix}.$$

Check:
$$\mathbf{x} = \begin{pmatrix} 1 \\ 1 \\ 1 \end{pmatrix} \stackrel{!}{=} \begin{pmatrix} -1/3 \\ 2/3 \\ 2/3 \end{pmatrix} + \begin{pmatrix} 4/3 \\ 1/3 \\ 1/3 \end{pmatrix} = P_\mathbf{a}\mathbf{x} + P_{\mathbf{a}\perp}\mathbf{x}$$

and
$$(P_\mathbf{a}\mathbf{x})^T P_{\mathbf{a}\perp}\mathbf{x} = (-1/3 \ 2/3 \ 2/3) \begin{pmatrix} 4/3 \\ 1/3 \\ 1/3 \end{pmatrix} \stackrel{!}{=} 0.$$

b.

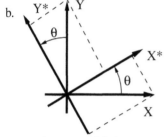

The figure shows two plane coordinate systems with the same origin. Each of the indicated vectors has length 1, and the unprimed vectors are perpendicular as are the $*$ vectors. The $*$ coordinate system is obtained from the un-$*$ by counterclockwise rotation by an angle θ. Our task is to resolve the un-$*$ vectors in terms of the $*$ vectors. Because each vector is a unit vector, this is most easily done using basic trigonometry. Virtually by definition of the sin and cos functions,

$$\mathbf{x}^* = \cos(\theta)\mathbf{x} + \sin(\theta)\mathbf{y} \qquad \mathbf{y}^* = \cos(\theta + 90°)\mathbf{x} + \sin(\theta + 90°)\mathbf{y}$$
$$\mathbf{x} = \cos(-\theta)\mathbf{x}^* + \sin(-\theta)\mathbf{y}^* \quad \mathbf{y} = \cos(-\theta + 90°)\mathbf{x}^* + \sin(-\theta + 90°)\mathbf{y}^*.$$

Numerous trigonometric identities (that follow from formulas we used on page 57) allow simplification to:

$$\mathbf{x}^* = \cos(\theta)\mathbf{x} + \sin(\theta)\mathbf{y} \qquad \mathbf{y}^* = -\sin(\theta)\mathbf{x} + \cos(\theta)\mathbf{y}$$
$$\mathbf{x} = \cos(\theta)\mathbf{x}^* - \sin(\theta)\mathbf{y}^* \qquad \mathbf{y} = \sin(\theta)\mathbf{x}^* + \cos(\theta)\mathbf{y}^*.$$

Each of these equations can be obtained by resolving the appropriate vector using a symmetric projector and its complementary projector. For example, the position vector represented by $\mathbf{y}^*$ is $(-\sin(\theta), \cos(\theta))^T$ and $|\mathbf{y}^*| = 1$, so

$$P_{\mathbf{y}^*} = \mathbf{y}^{*T}\mathbf{y}^* = \begin{pmatrix} \sin^2(\theta) & -\sin(\theta)\cos(\theta) \\ -\sin(\theta)\cos(\theta) & \cos^2(\theta) \end{pmatrix},$$

$$P_{\mathbf{y}^*\perp} = I_2 - P_{\mathbf{y}^*} = \begin{pmatrix} \cos^2(\theta) & \sin(\theta)\cos(\theta) \\ \sin(\theta)\cos(\theta) & \sin^2(\theta) \end{pmatrix}.$$

Therefore,

$$P_{\mathbf{y}^*}\mathbf{x} = P_{\mathbf{y}^*}\begin{pmatrix} 1 \\ 0 \end{pmatrix} = \begin{pmatrix} sin^2(\theta) \\ -\sin(\theta)\cos(\theta) \end{pmatrix} = -\sin(\theta)\begin{pmatrix} \sin(\theta) \\ \cos(\theta) \end{pmatrix} = -\sin(\theta)\mathbf{y}^*$$

and

$$P_{\mathbf{y}^*\perp}\mathbf{x} = P_{\mathbf{y}^*\perp}\begin{pmatrix} 1 \\ 0 \end{pmatrix} = \begin{pmatrix} \cos^2(\theta) \\ \sin(\theta)\cos(\theta) \end{pmatrix} = \cos(\theta)\begin{pmatrix} \cos(\theta) \\ \sin(\theta) \end{pmatrix} = \cos(\theta)\mathbf{x}^*$$

so at last
$$\mathbf{x} = \cos(\theta)\mathbf{x}^* - \sin(\theta)\mathbf{y}^*.$$

If we were to attempt the analogous three-dimensional problem, the tables would be turned. Symmetric projectors would be much faster than "elementary trigonometry."

SUMMARY

Given a matrix A whose RREF has only pivotal columns, the associated **symmetric projector** and **complementary projector** are

$$P_A = A(A^T A)^{-1} A^T \text{ and } P_{A^\perp} = I - P_A, \text{ respectively.}$$

These matrices are symmetric and **idempotent**.
In case $\mathbf{a} \in \mathbb{R}^3$, $P_{\mathbf{a}} \mathbf{w}$ is the geometric projection of $\mathbf{w}$ in the direction of $\mathbf{a}$, the vector $\mathbf{x}$ is **resolved** with respect to $\mathbf{a}$ with the equation

$$\mathbf{x} = P_{\mathbf{a}} \mathbf{x} + P_{\mathbf{a}^\perp} \mathbf{x}.$$

Looking further
Symmetric projectors are of extreme importance in both theory and application. Commuting idempotents are the building blocks for many "structure theories" in abstract algebra. In applications they are used to "take apart" a complicated situation so its pieces can be studied separately. Principal component analysis, briefly described in Section 3.7, is a clear example of this process.

$$* * * * * * *$$

Practice Problems

1. Compute the symmetric projectors P_A and $P_{A^\perp}$.

   a. $A = \begin{pmatrix} 1 \\ 1 \end{pmatrix}$ b. $A = \begin{pmatrix} 1 & 1 \\ 1 & 1 \end{pmatrix}$ c. $A = \begin{pmatrix} 1 & -1 \\ 1 & 1 \end{pmatrix}$

   d. $A = \begin{pmatrix} 1 \\ 1 \\ 1 \end{pmatrix}$ e. $A = \begin{pmatrix} 1 & -2 \\ 1 & -1 \\ 1 & 0 \\ 1 & 1 \\ 1 & 2 \end{pmatrix}$ f. $A = \begin{pmatrix} 1 & -2 & 4 \\ 1 & -1 & 1 \\ 1 & 0 & 0 \\ 1 & 1 & 1 \\ 1 & 2 & 4 \end{pmatrix}$

2. Resolve $\mathbf{v}$ with respect to $\mathbf{a}$ using $P_{\mathbf{a}}$.
   a. $\mathbf{v} = (10, 10)^T, \mathbf{a} = (3, 4)^T$ b. $\mathbf{v} = (1, 3)^T, \mathbf{a} = (0, 2)^T$
   c. $\mathbf{v} = \mathbf{i} + \mathbf{j} + \mathbf{k}, \mathbf{a} = \mathbf{k}$ d. $\mathbf{v} = \mathbf{i} + \mathbf{j} + \mathbf{k}, \mathbf{a} = 5\mathbf{i} - 2\mathbf{j} + \mathbf{k}$
   e. $\mathbf{v} = (1, 1, 0)^T, \mathbf{a} = (-2, 1, -1)^T$ f. $\mathbf{v} = (1, 1, 1, 1)^T, \mathbf{a} = (1, -1, -1, 1)^T$

3. Verify that $P_{A^\perp}$ is idempotent for any n-by-m matrix A of rank m. Show also that $P_{A^\perp} P_A = P_A P_{A^\perp} = 0$.

4. This problem does not use matrices. By resolving vectors it shows why sailboats can travel into the wind. Imagine the boat heading $\mathbf{b}$ at an angle θ to the wind $\mathbf{w}$ and the sail $\mathbf{s}$ at an angle ϕ to the boat's heading as in the figure.

The wind force vector is resolved by the sail and the boat only experiences $\mathbf{f} = P_{\mathbf{s}\perp}(\mathbf{w})$. Similarly, the boat's keel is designed to prevent "side slipping" and so the only force that actually contributes to the boat's motion is $P_{\mathbf{b}}(\mathbf{f})$. Compute $|P_{\mathbf{b}}(\mathbf{f})|$ as a function $|\mathbf{w}|$, θ and ϕ. For $\theta = \pi/3$, sketch this function and estimate the value of ϕ where it take its maximum value.

5. Let $\mathbf{a} \in \mathbb{R}^2$ and consider the matrix R defined to be $R = I - 2P_{\mathbf{a}}$. For each of the following vectors, use a transform plot to describe the geometric effect of the associated linear function f_R.

    a. $\mathbf{a} = (1,0)^T$ b. $\mathbf{a} = (1,1)^T$ c. $\mathbf{a} = (0,1)^T$

    d. $\mathbf{a} = (0,0,1)^T$ e. $\mathbf{a} = (1,1,0)^T$ f. $\mathbf{a} = (1,1,1)^T$

* * * * * * *

2.5   Least squares approximation

Objective 1

Given a set of up to five data points and $m \leq 4$, find the $(m-1)$-th degree least squares approximation $f(x)$ to the data.

Recall Section 1.7, where we used polynomial interpolation to "fit" an n-th degree polynomial function $f(x)$ to $n+1$ data points. In this section we discuss the more realistic situation in which there is more data and the polynomial function is only required to "best approximate" the m data points; $m > n+1$. The snag is that some choice must be made about how to resolve conflicting data, and this amounts to deciding how one might mathematically find some middle ground.

This fundamental question was first addressed by Legendre in 1806, while plotting the course of astronomical bodies, but at almost the same time Gauß developed the same methods for a more down-to-earth application — reconciling surveyor's data to build accurate maps. Today scientists use "least squares" or "regression" of experimental data when a theoretical mathematical model for a particular phenomenon is known, but the formula contains terms that must be determined by experimentation.

In Section 1.7 we saw that more than $n+1$ data points may lead to an inconsistent linear system $A\mathbf{x} = \mathbf{b}$ having no solution. The crux of the search for a middle ground is how to compare approximations quantitatively. Using the geometric machinery from Section 1.6, we present Legendre's elegant answer.

If we attempt to interpolate a linear function $f(x) = f_0 + f_1 x$ through the data points $(x_1, y_1) = (-1, 1), (x_2, y_2) = (0, 0), (x_3, y_3) = (1, 2)$, the methods of polynomial interpolation lead to the inconsistent linear system:

$$A\mathbf{f} = \begin{pmatrix} 1 & -1 \\ 1 & 0 \\ 1 & 1 \end{pmatrix} \begin{pmatrix} f_0 \\ f_1 \end{pmatrix} = \begin{pmatrix} 1 \\ 0 \\ 2 \end{pmatrix}.$$

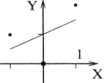

This is confirmed by the plot of the data points because it shows clearly that they are not on a straight line. Our task is to find the linear function (straight line) that comes **closest** to going through all three data points. The answer that will emerge has been plotted as well. To see that it really is the **closest** linear function to the data, we use both vector geometry and symmetric projectors BUT in a remarkably different geometric context!

Imagine the transform plot of the linear function f_A. By coincidence, it actually appears in Part c of Example 1.19 on pages 44 and 45 where we used it to give a geometric interpretation of an inconsistent linear system. The set of all linear combinations S of the columns of the coefficient matrix A form the plane (see the figure below). The inconsistency of the linear system is (still) expressed geometrically by the fact that the constant vector $\overrightarrow{OP} = (1\ 0\ 2)^T$ is not in S.

Our current situation is confusing because each point of S represents a linear function $y = F(x)$ that might be used to approximate the data. The question is "Which point, if any, in S corresponds to a function $y = F(x)$ that **best** approximates the data?"

Imagine a balloon centered at P slowly inflating. At some point it will first touch S. The point Q is the (unique) point of S that is closest to P. Doesn't it seem natural to say the solution to the linear system $A\mathbf{f} = \overrightarrow{OQ}$ is "the best approximate solution" to the inconsistent system? This is Legendre's idea of least squares approximation.

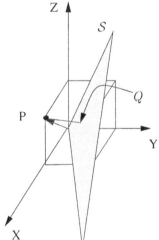

Our bonus is that we actually know how to compute Q. The vector $\overrightarrow{OQ}$ is just the symmetric projector $P_A(\overrightarrow{OP})$. By the second part of Example 2.12 on page 77

$$P_A \mathbf{b} = A(A^T A)^{-1} A^T \mathbf{b} = \cdots = \begin{pmatrix} 1/2 \\ 1 \\ 3/2 \end{pmatrix}.$$

The solution to the consistent system $A\mathbf{f} = (1/2\ 1\ 3/2)^T$ is $\mathbf{f} = (1\ 1/2)^T$, and from it we obtain $f(x) = 1 + x/2$ as the "best approximate" solution to the given data.

Recapitulating, each point of S in the figure represents a linear function $y = F(x)$. The fact that Q (corresponding to $y = f(x)$) is the point in S that is **closest** to P is equivalent to the fact that among all linear functions $y = F(x)$,

$$d^2 = (y_1 - F(x_1))^2 + (y_2 - F(x_2))^2 + (y_3 - F(x_3))^2$$

is **minimal** for $F(x) = f(x)$.

This condition may be expressed in terms of the data plot (at the top of the preceding page) by saying the sum of the squares of the **vertical** distances[6] from a data point to a graph is minimized by $y = f(x)$. The geometric interpretation using symmetric projectors occurs in $\mathbb{R}^s$ where s is the number of data points. This geometric interpretation in the **sample** space, not the data plot, makes the uniqueness of the **least squares** solution clear. Needless to say, the geometric interpretation becomes more "conceptual" when more data points are involved.

Please turn back to the preceding page and compare the two figures. Note that they represent exactly the SAME mathematical situation in two different geometric ways. It is easy to imagine more complicated interpolation problems than in the first figure, but for such problems the second figure is no longer three-dimensional. The geometric insight that the second figure gives us is a powerful reason to try to extend our geometric understanding beyond the three dimensions we physically experience. Any time you can interpret a mathematical situation in several different ways, you reach a deeper insight.

In actual practice, least squares approximation requires only the computation of the matrix $(A^T A)^{-1}$, because we can exploit properties of matrix arithmetic to "back into" the solution of $Af = P_A b$:

> The linear system $Af = P_A b$ has solution $f = (A^T A)^{-1} A^T b$. (Just multiply the indicated f by A and look!) If this linear system has solution $f = (f_0 \ f_1 \ldots f_n)^T$ then the least squares polynomial approximation is:
>
> $$y = f_0 + f_1 x + \cdots + f_n x^n.$$

The alert reader may wonder why these rectangular Vandermonde matrices A satisfy the rank condition of Section 2.4. There is a square Vandermonde matrix A^* whose first few columns are A and is invertible, by Section 2.3. Simply imagine the sequence of row operations that put A^* in REF and apply them to A stopping after the first r rows are fixed. Every column of $REF(A)$ is pivotal (compare to Problem 10 of Section 2.3 on page 76).

Example 2.14 (Least squares approximation.)
    a. Suppose we are given the points $\{(-2, 1), (-1, -1), (0, 0), (1, 2), (2, 1)\}$. Our task is to find the least squares linear (degree 1) approximation to the points.

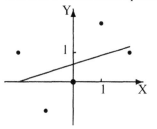

The augmented matrix of the associated (inconsistent) linear system is:

$$(A|b) = \begin{pmatrix} 1 & -2 & 1 \\ 1 & -1 & -1 \\ 1 & 0 & 0 \\ 1 & 1 & 2 \\ 1 & 2 & 1 \end{pmatrix}.$$

---

[6]The "vertical" cannot be omitted without changing the condition to a much more complicated one when the interpolating polynomial has higher degree.

The solution to the **consistent** linear system $A\mathbf{f} = P_A\mathbf{b}$ is

$$\mathbf{f} = (A^T A)^{-1} A^T \mathbf{b} = \begin{pmatrix} 5 & 0 \\ 0 & 10 \end{pmatrix}^{-1} \begin{pmatrix} 3 \\ 3 \end{pmatrix} = \begin{pmatrix} 0.6 \\ 0.3 \end{pmatrix}$$

Therefore, $f(x) = 0.6 + (0.3)x$ is the least squares linear approximation to the data points.

b. Suppose we are given the points $\{(-2, 1), (-1, -1), (0, 0), (1, 2), (2, 1)\}$. Our task is to find the least squares cubic approximation to the points.

The augmented matrix of the associated (inconsistent) linear system is:

$$(A|\mathbf{b}) = \begin{pmatrix} 1 & -2 & 4 & -8 & \vline & 1 \\ 1 & -1 & 1 & -1 & \vline & -1 \\ 1 & 0 & 0 & 0 & \vline & 0 \\ 1 & 1 & 1 & 1 & \vline & 2 \\ 1 & 2 & 4 & 8 & \vline & 1 \end{pmatrix}.$$

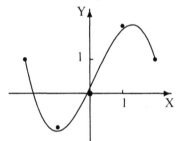

We can solve the system $A\mathbf{f} = P_A\mathbf{b}$ directly:

$$\mathbf{f} = (A^T A)^{-1} A^T \mathbf{b} = \begin{pmatrix} \frac{17}{35} & 0 & \frac{-1}{7} & 0 \\ 0 & \frac{65}{72} & 0 & -\frac{17}{72} \\ \frac{-1}{7} & 0 & \frac{1}{14} & 0 \\ 0 & -\frac{17}{72} & 0 & \frac{5}{72} \end{pmatrix} \begin{pmatrix} 3 \\ 3 \\ 9 \\ 3 \end{pmatrix} = \begin{pmatrix} \frac{6}{35} \\ 2 \\ \frac{3}{14} \\ -\frac{1}{2} \end{pmatrix}.$$

Therefore, $f(x) = \frac{6}{35} + 2x + \frac{3}{14}x^2 - \frac{1}{2}x^3$ is the least squares approximation to the data. □

SUMMARY

The *least squares* approximation of a data set of n points is based on geometry in $\mathbb{R}^n$. The least squares approximation to the data is

$$f_0 + f_1 x + + \cdots + f_x x^m \text{ where } \mathbf{f} = (A^T A)^{-1} A^T \mathbf{b}$$

and A is the Vandermonde matrix determined by the x-coordinates of the data points. This formula arises from a symmetric projector in the *sample space* $\mathbb{R}^s$ where s is the number of data points.

Looking further

Least squares data analysis, also called regression, is a fundamental statistical tool that is essential for experimental scientists of all disciplines.[7] Mathematically sophisticated readers will insist that Euclidean distance in the sample space is not the

---

[7] *Statistical Methods for Social Scientists* by Eric A. Hanushek, John E. Jackson, Academic Press, 1977.

only reasonable metric. I can only respond that it is a cornerstone of classical statistics and acknowledge that there are many variations on this basic material. Problems 3 and 4 hint at alternatives.

There is a very big question implicit in this section that is not mathematical. Suppose you have some data and would like to study the process it represents with a mathematical model. Just how do you choose the type of curve to "fit" to the data?

This is of course the essential contribution of the experimental scientist. This person usually has some sense of how the observed process works and/or knowledge of similar situations.

The choice is determined by the context or purpose of the model. If the intent to interpolate (fill in values of x between your data points) rather than to **extrapolate** (consider values of x far from the observed data), then formula choice is less critical. There is also an important distinction between prediction and projection. A prediction would require more support than merely saying "if present trends continue, then...".

The choice also depends on the mathematical tools available. There are statistical tools beyond least squares that are designed to "look for" simple patterns in mountains of data. But you must know a little about how they work before you can use them confidently. We touch briefly on one of the most important "principal components" at the end of the next chapter.

In the end, model building is an interactive process. You need to go back and "test the model" with new data, then refine the model (not the data!)

$$* \ * \ * \ * \ * \ * \ *$$

Practice Problems

1. Find and sketch the least squares linear approximation for the given points.

   a. $\{(-2,1),(-1,1),(0,-2),(1,-3),(2,-8)\}$
   b. $\{(0,1),(1,2),(2,5),(3,5),(4,7)\}$
   c. $\{(-4,2),(-2,1),(0,1),(2,0),(4,-3)\}$

2. Find and sketch the least squares cubic approximation to the given points.

   a. $\{(-2,0),(-1,0),(0,5),(1,4),(2,7)\}$
   b. $\{(-3,0),(1,4),(2,2),(4,3),(5,8)\}$
   c. $\{(-1,0),(0,0),(1,-2),(2,-1),(4,2)\}$

3. If the polynomial type being approximated is linear, then the length $|P_{A^\perp}(\mathbf{b})|$ is called the **variance** of the data. Explain why $|P_{A^\perp}(\mathbf{b})|$ is a natural measure of the quality of a least squares approximation in general.

4. Suppose that for some reason you had more confidence in some data observations than in others. How might these methods be adapted to take this into account? (This is called "generalized least squares.")

2.6 Changing plane coordinates

Objective 1

Given two plane coordinate systems specified by their origins and basic unit vectors, compute the coordinates of a point specified in one coordinate system in terms of the other.

How can two observers at different locations compare what they see? The answer requires the solution of linear systems whose coefficient matrices are derived from the relative positions of the observers. We show how to introduce intermediate coordinate systems and use matrix multiplication to build up a complicated coefficient matrix from simple pieces. Let's begin by fixing ideas with a little fiction.

Rescue at sea

Imagine it is 3 AM and a titanic disaster has just occurred at sea. By chance we are in a stationary balloon 1 km above the ocean surface. We can see both the disaster site and a search vessel. It is a cloudy night with the rising moon only faintly visible, and we don't know our own location exactly. Nonetheless, we desperately want to use our radio to tell the search vessel how to find the disaster site.

To do this we must convey some information from which the search vessel can find the way. For this reason we introduce coordinates. Imagine a three-dimensional coordinate system having our $\mathcal{B}$alloon at its origin with vertical Z_B-axis and with positive X_B-axis in the direction of the horizon nearest the moon. Since the moon is rising, this is in an easterly direction but not necessarily 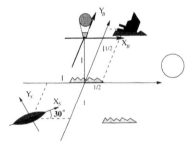 due east. Finally, we choose the Y_B-axis of our coordinate system to be perpendicular to both the X_B and Z_B axes, so positive Y_B-coordinates are to the north.

Using primitive means of angle measurement and the trig functions in our trusty calculator, we compute the coordinates (all units in km) of the search vessel and of the disaster site as indicated in the figure.

Because we are not sure that the search vessel can see the moon (and because we work so slowly, the moon could have moved before we can communicate with the search vessel), we must identify a coordinate system that will be meaningful to the $\mathcal{S}$earch vessel. To this end we observe that the search vessel is pointed in the direction at an angle of $\theta = 30°$ to the north of our x-axis. We take this as the direction of the positive X_S-axis for the search vessel and the positive Y_S axis to be perpendicular off the port (left-hand) side. We must find a formula for the search vessel's coordinates of the disaster site D.

Please note, we are NOT allowing the use of a common reference system (like latitude and longitude) because we intend to use this mathematical machinery in settings where there are none.

There are several **keys** to understanding this situation.

One is noticing exactly what is involved in choosing a coordinate system. Each coordinate system is specified by a starting point or origin O and by a set of perpendicular directions corresponding to the directions of the positive coordinate axes. As in Section 1.6 we keep track of the directions of the coordinate axes with ***basic unit vectors***. We will only work with perpendicular basic vectors oriented in the standard way. (This is a simplifying assumption more than a theoretical limitation.)

A more subtle **key** is that the observers are NOT in the water; else they could not distinguish objects on the same "line of sight." This is on the one hand obvious, but on the other hand it is a major paradigm shift. Its implementation (as perspective painting) by artists is one of the defining changes of the historic "Renaissance" period.

Mathematically, it leads to "projective geometry" and "homogeneous coordinates." In projective geometry a "point" is not a dot on graph paper, rather a "line of sight" determined by such a point and an "all seeing eyeball origin" that is OUTSIDE the situation being studied. More formally, each "point" $\langle \mathbf{v} \rangle$ of a projective plane is "represented by" a nonzero vector $\mathbf{v} \in \mathbb{R}^3$ AND two vectors "represent" the same "point" if one is a scalar multiple of the other.

A final observation is that we implicitly named each coordinate system ($\mathcal{B}$ for the balloon's and $\mathcal{S}$ for the search vessel's). We then used a compound notation for corresponding elements of the systems as a way of keeping track of all the information.

Recall that we established names for the elements of a plane coordinate system in Section 1.6. The origin is O, and the basic unit vector in the direction of the positive X-axis (respectively Y-axis) is $\mathbf{i}$ (respectively $\mathbf{j}$). When we consider several coordinate systems at once we will distinguish corresponding elements with subscripts. Thus the plane coordinate system $\mathcal{C}$ comes equipped with an origin $O_{\mathcal{C}}$ and basic unit vectors $\mathbf{i}_{\mathcal{C}}$ and $\mathbf{j}_{\mathcal{C}}$. The coordinates of the point P are $(x_{\mathcal{C}}, y_{\mathcal{C}})$ where

(key) $$\overrightarrow{O_{\mathcal{C}}P} = x_{\mathcal{C}}\mathbf{i}_{\mathcal{C}} + y_{\mathcal{C}}\mathbf{j}_{\mathcal{C}}.$$

(The label in the left margin marks this equation as the **key** equation.)

The idea of this section is to move from one coordinate system to another in two steps, changing only one element at a time.

If two coordinate systems have different origins but the same basic unit vectors, they are related by ***translation*** of coordinates. If they have different basic unit vectors but the same basic origin, they are related by ***rotation*** of coordinates.

Translation and homogeneous coordinates

Suppose two coordinate systems $\mathcal{A}, \mathcal{B}$ are related by translation of coordinates and that a point P in the plane is specified. By notational convention, P has $\mathcal{A}$-coordinates (x_A, y_A) and $\mathcal{B}$-coordinates (x_B, y_B). Our job is to find a matrix equation that moves from the $\mathcal{A}$-coordinates of P to the $\mathcal{B}$-coordinates of P.

These coordinate systems have the same basic unit vectors, so simplify notation and drop the vector subscripts. Also suppose the $\mathcal{A}$-coordinates of O_B are known, say $(O_B)_A = (s, t)$. Write the $\mathcal{A}$ **key** equation and the $\mathcal{B}$ **key**. Then take the difference:

$$\overrightarrow{O_A P} - \overrightarrow{O_B P} = (x_A \mathbf{i}_A + y_A \mathbf{j}_A) - (x_B \mathbf{i}_B + y_B \mathbf{j}_A) = (x_A - x_B)\mathbf{i} + (y_A - y_B)\mathbf{j}.$$

Recall the "tip to tail" form of vector addition to simplify the left-hand side. Then use the $\mathcal{A}$ **key** equation with $P = O_B$:

$$\overrightarrow{O_A P} - \overrightarrow{O_B P} = \overrightarrow{O_A P} + \overrightarrow{P O_B} = \overrightarrow{O_A O_B} = s\mathbf{i} + t\mathbf{j}.$$

Finally, compare coordinates in these two formulas for the same expression to obtain:

$$\begin{aligned} x_B &= x_A - s \\ y_B &= y_A - t \end{aligned} \qquad \text{where} (O_B)_A = (s, t).$$

That's the good news. The bad news is that this relationship cannot be expressed with 2-by-2 matrices. Rather than give up matrices, we introduce **homogeneous coordinates** and work with 3-by-3 matrices. This amounts to acknowledging that the observer in the balloon was not in the water. (This is revisited on page 220.)

The homogeneous coordinates of a point in $\mathbb{R}^n$ is an $n+1$-tuple whose final coordinate is not zero (and by convention is taken to be 1). Informally, this added coordinate might be regarded as playing a role similar to a decimal point when working with numbers. Usually it's just along for the ride, but sometimes you need it to make things work out correctly.

Homogeneous coordinates can indeed be affected by matrix multiplication:

$$\begin{pmatrix} x_B \\ y_B \\ 1 \end{pmatrix} = \begin{pmatrix} x_A - s \\ y_A - t \\ 1 \end{pmatrix} = \begin{pmatrix} 1 & 0 & -s \\ 0 & 1 & -t \\ 0 & 0 & 1 \end{pmatrix} \begin{pmatrix} x_A \\ y_A \\ 1 \end{pmatrix} = T_{B \leftarrow A} \begin{pmatrix} x_A \\ y_A \\ 1 \end{pmatrix}.$$

$T_{B \leftarrow A}$, (named for transition NOT translation) is called the **transition matrix** from $\mathcal{A}$-coordinates to $\mathcal{B}$-coordinates. (The weird $\leftarrow$ will be explained below.)

Example 2.15 (Translation of coordinates)

Write the transition matrix, $T_{B \leftarrow A}$, from $\mathcal{A}$ to $\mathcal{B}$ and the transition matrix, $T_{A \leftarrow B}$, from $\mathcal{B}$ to $\mathcal{A}$. Because we are asked for the transition matrix going in both directions, we need the $\mathcal{A}$-coordinates of O_B, namely $(-2, 1)^T$, and the $\mathcal{B}$-coordinates of O_A, namely $(2, -1)^T$.

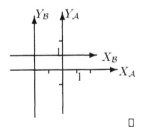

$$T_{B \leftarrow A} = \begin{pmatrix} 1 & 0 & 2 \\ 0 & 1 & -1 \\ 0 & 0 & 1 \end{pmatrix} \qquad T_{A \leftarrow B} = \begin{pmatrix} 1 & 0 & -2 \\ 0 & 1 & 1 \\ 0 & 0 & 1 \end{pmatrix}$$

Matrix number systems.

Rotation of coordinates.

Suppose two coordinate systems $\mathcal{C}, \mathcal{D}$ are related by rotation of coordinates and that a point P in the plane is specified. Again, by notational convention P has $\mathcal{C}$-coordinates (x_C, y_C) and $\mathcal{D}$-coordinates (x_D, y_D). Our job is to find a matrix equation that moves from the coordinates of P in one system to the other.

This time the two coordinate systems share the same origin $O = O_C = O_D$, but have different basic unit vectors. In view of the **key** equation, the $\mathcal{C}$-basic unit vectors must be expressed in terms of the $\mathcal{D}$-basic unit vectors. Exactly this appears in part b of Example 2.13 on page 80, but in slightly different notation (i_C was **x**, i_D was **x***).
Translating:

$$i_C = \cos(\theta)i_D - \sin(\theta)j_D$$
$$j_C = \sin(\theta)i_D + \cos(\theta)j_D$$

where θ is the angle from i_C to i_D (measured counterclockwise as always). Thus,

$$\overrightarrow{OP} = x_C i_C + y_C j_C$$
$$= x_C[\cos(\theta)i_D - \sin(\theta)j_D] + y_C[\sin(\theta)i_D + \cos(\theta)j_D]$$
$$= [\cos(\theta)x_C + \sin(\theta)y_C]i_D + [-\sin(\theta)x_C + \cos(\theta)y_C]j_D.$$

This is neatly expressed by the matrix equation:

$$\begin{pmatrix} x_D \\ y_D \\ 1 \end{pmatrix} = \begin{pmatrix} \cos(\theta) & \sin(\theta) & 0 \\ -\sin(\theta) & \cos(\theta) & 0 \\ 0 & 0 & 1 \end{pmatrix} \begin{pmatrix} x_C \\ y_C \\ 1 \end{pmatrix} = T_{D \leftarrow C} \begin{pmatrix} x_C \\ y_C \\ 1 \end{pmatrix}.$$

The matrix $T_{D \leftarrow C}$, is called the **transition matrix** associated with the rotation of coordinates from $\mathcal{C}$ to $\mathcal{D}$.

Example 2.16 (Rotation of coordinates.)

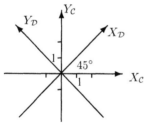

Write BOTH the transition matrix from the $\mathcal{C}$- to the $\mathcal{D}$-coordinate system AND the transition matrix from the $\mathcal{D}$- to the $\mathcal{C}$-coordinate system.

The basic unit vectors of the $\mathcal{D}$-coordinate system are obtained from those of the $\mathcal{C}$-coordinate system by a counterclockwise rotation of $45°$, AND the two coordinate systems have the same origin $O = O_C = O_D$.

The transition matrix from the $\mathcal{C}$- to the $\mathcal{D}$- uses $\theta = 45°$ while the transition matrix from the $\mathcal{D}$- to the $\mathcal{C}$- uses $\theta = -45°$:

$$T_{D \leftarrow C} \doteq \begin{pmatrix} 0.707 & 0.707 & 0 \\ -0.707 & 0.707 & 0 \\ 0 & 0 & 1 \end{pmatrix} \quad T_{C \leftarrow D} \doteq \begin{pmatrix} 0.707 & -0.707 & 0 \\ 0.707 & 0.707 & 0 \\ 0 & 0 & 1 \end{pmatrix},$$

since $\cos(45°) \doteq 0.707 \doteq \cos(-45°)$ and $\sin(45°) \doteq 0.707 \doteq -\sin(-45°)$. $\square$

General case

It is time to chain the two simple types of coordinate changes together. Suppose we have three coordinate systems $\mathcal{A}$, $\mathcal{B}$ and $\mathcal{C}$, and a point P with homogeneous $\mathcal{X}$-coordinates $(x_{\mathcal{X}}, y_{\mathcal{X}}, 1)^T$ for $\mathcal{X} = \mathcal{A}, \mathcal{B}, \mathcal{C}$. By definition of the transition matrices,

$$
\begin{pmatrix} x_C \\ y_C \\ 1 \end{pmatrix} = T_{C \leftarrow B} \begin{pmatrix} x_B \\ y_B \\ 1 \end{pmatrix} \quad \text{and} \quad \begin{pmatrix} x_B \\ y_B \\ 1 \end{pmatrix} = T_{B \leftarrow A} \begin{pmatrix} x_A \\ y_A \\ 1 \end{pmatrix}.
$$

By associativity of matrix multiplication,

$$
\begin{pmatrix} x_C \\ y_C \\ 1 \end{pmatrix} = T_{C \leftarrow B} T_{B \leftarrow A} \begin{pmatrix} x_A \\ y_A \\ 1 \end{pmatrix} \quad \text{and so } T_{C \leftarrow A} = T_{C \leftarrow B} T_{B \leftarrow A},
$$

because that is the equation defining $T_{C \leftarrow A}$. To change coordinates several times, simply **multiply** the step-by-step transition matrices to obtain the transition matrix from the first to the last coordinate system.

Since matrices multiply by (column) vectors on the left, chaining of several transitions together requires the first transition matrix to be on the right and the last to be on the left (unlike English sentences). Thus, the "backward" arrow convention; it's as easy as $\mathcal{C}\mathcal{B}\mathcal{A}$.

HINT: When combining rotations and translations, always do translations first. Because translations do not affect the basic unit vectors, the original data gives the right rotation matrix. (If you were to do rotation first, you would have to figure out the new coordinates of the origin of the final coordinate system before translating.)

Example 2.17 (Changing plane coordinates.)

a. Compute the $\mathcal{B}$-coordinates of the indicated point P. Introduce an intermediate coordinate system $\mathcal{T}$ obtained from $\mathcal{A}$ by $\mathcal{T}$ranslation and from which $\mathcal{B}$ is obtained by rotation. By inspection, the homogeneous $\mathcal{A}$-coordinates of $O_{\mathcal{B}}$ are $(1, 1, 1)^T$ and therefore,

$$
T_{\mathcal{T} \leftarrow \mathcal{A}} = \begin{pmatrix} 1 & 0 & -1 \\ 0 & 1 & -1 \\ 0 & 0 & 1 \end{pmatrix}.
$$

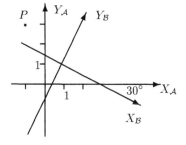

The $\mathcal{B}$-basic unit vectors are obtained from the $\mathcal{A}$-basic unit vectors by a **counterclockwise** rotation of $\theta = -30°$, so

$$
T_{\mathcal{B} \leftarrow \mathcal{T}} = \begin{pmatrix} \cos(-30°) & \sin(-30°) & 0 \\ -\sin(-30°) & \cos(-30°) & 0 \\ 0 & 0 & 1 \end{pmatrix}.
$$

Since $\cos -30° \doteq .866$ and $\sin -30° = -0.5$, the transition matrix is:

$$
T_{\mathcal{B} \leftarrow \mathcal{A}} = T_{\mathcal{B} \leftarrow \mathcal{T}} T_{\mathcal{T} \leftarrow \mathcal{A}} = \begin{pmatrix} .866 & -0.5 & 0 \\ 0.5 & .866 & 0 \\ 0 & 0 & 1 \end{pmatrix} \begin{pmatrix} 1 & 0 & -1 \\ 0 & 1 & -1 \\ 0 & 0 & 1 \end{pmatrix} \doteq \begin{pmatrix} .866 & -0.5 & -.366 \\ 0.5 & .866 & -1.37 \\ 0 & 0 & 1 \end{pmatrix}
$$

The point P has homogeneous $\mathcal{A}$-coordinates $(-1, 3, 1)^T$. Therefore its $\mathcal{B}$-coordinates are

$$T_{\mathcal{B}\leftarrow\mathcal{A}} \begin{pmatrix} -1 \\ 3 \\ 1 \end{pmatrix} = \begin{pmatrix} -2.732 \\ 0.728 \\ 1 \end{pmatrix}.$$

This is plausible geometrically because P is in the second $\mathcal{B}$ quadrant.

b. Recall the titanic disaster at sea. The $\mathcal{B}$-homogeneous coordinates of the $\mathcal{S}$earch vessel and $\mathcal{D}$isaster site that was collected in the $\mathcal{B}$alloon's coordinate system is:

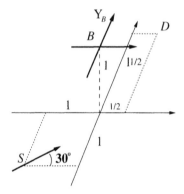

$$(O_\mathcal{S})_\mathcal{B} = \begin{pmatrix} -1 \\ -1 \\ 1 \end{pmatrix}, \quad (D)_\mathcal{B} = \begin{pmatrix} 0.5 \\ 1.5 \\ 1 \end{pmatrix}$$

Again introduce an intermediate coordinate system $\mathcal{T}$ obtained from $\mathcal{B}$ by $\mathcal{T}$ranslation and from which $\mathcal{S}$ is obtained by rotation. Then the $(O_\mathcal{S})_\mathcal{B}$ coordinates imply that

$$T_{\mathcal{T}\leftarrow\mathcal{B}} = \begin{pmatrix} 1 & 0 & 1 \\ 0 & 1 & 1 \\ 0 & 0 & 1 \end{pmatrix}.$$

Because $\mathcal{S}$-basic vectors are obtained from the $\mathcal{T}$-basic vectors (which are also the $\mathcal{B}$-basic vectors) by a counterclockwise rotation of $\theta \doteq 30°$,

$$T_{\mathcal{S}\leftarrow\mathcal{T}} = \begin{pmatrix} \cos 30° & \sin 30° & 0 \\ -\sin 30° & \cos 30° & 0 \\ 0 & 0 & 1 \end{pmatrix}.$$

Putting it all together we have $(D)_\mathcal{S} = T_{\mathcal{S}\leftarrow\mathcal{B}}(D)_\mathcal{B} = T_{\mathcal{S}\leftarrow\mathcal{T}}T_{\mathcal{T}\leftarrow\mathcal{B}}(D)_\mathcal{B}$

$$= \begin{pmatrix} \cos 30° & \sin 30° & 0 \\ -\sin 30° & \cos 30° & 0 \\ 0 & 0 & 1 \end{pmatrix} \begin{pmatrix} 1 & 0 & 1 \\ 0 & 1 & 1 \\ 0 & 0 & 1 \end{pmatrix} \begin{pmatrix} 0.5 \\ 1.5 \\ 1 \end{pmatrix}$$

$$\doteq \begin{pmatrix} 0.866 & 0.5 & 1.37 \\ -0.5 & 0.866 & .366 \\ 0 & 0 & 1 \end{pmatrix} \begin{pmatrix} 0.5 \\ 1.5 \\ 1 \end{pmatrix} \doteq \begin{pmatrix} 2.55 \\ 1.41 \\ 1 \end{pmatrix}$$

Tell the $\mathcal{S}$earch vessel to look 1.41 km ahead and $\doteq 2.55$ km to the port side from your 3 AM position and heading.

c. A coordinate system $\mathcal{D}$ is obtained from $\mathcal{C}$ by translation and rotation. The homogeneous coordinates of $(O_\mathcal{D})_\mathcal{C}$ are $(-1, -1, 1)^T$, and the positive $X_\mathcal{D}$ axis goes through $(3, 2)_\mathcal{C}$. Sketch the two coordinate systems and write the transition matrix $T_{\mathcal{D}\leftarrow\mathcal{C}}$.

This time no angle is given explicitly. But the $\mathcal{C}$-coordinates of two points on the $X_\mathcal{D}$-axis is given, and that is enough. The angle θ is not what is needed. Instead it is sufficient to calculate $\cos(\theta)$ and $\sin(\theta)$. Begin by sketching the data. It's the best way to really know what you have to work with.

The dashed box and its diagonal is the key. It has θ as an acute angle and is a 3-4-5 right triangle. Elementary trigonometry implies $\cos(\theta) = 4/5$ and $\sin(\theta) = 3/5$. From this point on, the problem is similar to the others. Introduce an intermediate coordinate system $\mathcal{T}$ obtained from $\mathcal{C}$ by $\mathcal{T}$ranslation and from which $\mathcal{D}$ is obtained by rotation.

$$T_{\mathcal{T} \leftarrow \mathcal{C}} = \begin{pmatrix} 1 & 0 & 1 \\ 0 & 1 & 1 \\ 0 & 0 & 1 \end{pmatrix}.$$

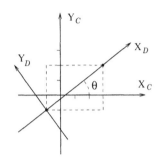

The $\mathcal{D}$-basic unit vectors are obtained from the $\mathcal{C}$-basic unit vectors by a **counterclockwise** rotation of θ, so

$$T_{\mathcal{D} \leftarrow \mathcal{T}} = \begin{pmatrix} 4/5 & 3/5 & 0 \\ -3/5 & 4/5 & 0 \\ 0 & 0 & 1 \end{pmatrix}.$$

Putting it all together we have: $T_{\mathcal{D} \leftarrow \mathcal{C}} = T_{\mathcal{D} \leftarrow \mathcal{T}} T_{\mathcal{T} \leftarrow \mathcal{C}}$

$$= \begin{pmatrix} 4/5 & 3/5 & 0 \\ -3/5 & 4/5 & 0 \\ 0 & 0 & 1 \end{pmatrix} \begin{pmatrix} 1 & 0 & 1 \\ 0 & 1 & 1 \\ 0 & 0 & 1 \end{pmatrix} = \begin{pmatrix} 0.8 & 0.6 & 1.4 \\ -0.6 & 0.8 & 0.2 \\ 0 & 0 & 1 \end{pmatrix}.$$ □

SUMMARY

Each plane coordinate system $\mathcal{C}$ is determined by an origin $O_{\mathcal{C}}$ and basic unit vectors $\mathbf{i}_{\mathcal{C}}, \mathbf{j}_{\mathcal{C}}$. We assume that $\mathbf{j}_{\mathcal{C}}$ is $90°$ counterclockwise from $\mathbf{i}_{\mathcal{C}}$.

Two such coordinate systems are related by a ***translation*** (respectively a ***rotation***) if they have different origins but the same basic unit vectors (respectively different basic unit vectors but the same origin).

For each point P, left multiplication by the transition matrix $T_{\mathcal{D} \leftarrow \mathcal{C}}$ transforms the homogeneous $\mathcal{C}$-coordinates, $(P)_{\mathcal{C}}$ of P into the homogeneous $\mathcal{D}$-coordinates, $(P)_{\mathcal{D}}$ of P

$$T_{\mathcal{D} \leftarrow \mathcal{C}} = \begin{pmatrix} \cos\theta & \sin\theta & 0 \\ -\sin\theta & \cos\theta & 0 \\ 0 & 0 & 1 \end{pmatrix} \begin{pmatrix} 1 & 0 & -s \\ 0 & 1 & -t \\ 0 & 0 & 1 \end{pmatrix}$$

where a $\mathcal{D}$-basic unit vector is obtained from the corresponding $\mathcal{C}$-basic unit vector by counterclockwise rotation through an angle of θ; and the homogeneous coordinates $(O_{\mathcal{D}})_{\mathcal{C}} = (s, t, 1)^T$.

Looking further

Computer graphics involves portraying a three-dimensional object on a flat video screen. This is done by building a mathematical model of the object and then "viewing" it from different positions. In practice, the different positions have homogeneous coordinates that are 4-tuples. In addition to translation and rotation, there are two more fundamental operations involved in the processes, **scaling** and **projection**. Scaling is used to indicate depth of field by making objects further away smaller. Projection is used to put the three-dimensional model onto the plane of the flat video screen. Each of these fundamental operations is achieved by a (generalized) transition matrix of the type we have presented.[8]

* * * * * * *

Practice Problems

1. A coordinate system $\mathcal{D}$ is obtained from the $\mathcal{C}$ by a translation. The $O_\mathcal{D}$ has $\mathcal{C}$-coordinates $(-2, -3)$. Find the $\mathcal{D}$ coordinates of P if $(P)_\mathcal{C} = (2, -1)$. Sketch.

In each of Problems 2 and 3, two plane coordinate systems are related as in the figure. Write the (homogeneous) transition matrix $T_{\mathcal{D} \leftarrow \mathcal{C}}$ in factored form. Compute the primed coordinates of the indicated point.

2.

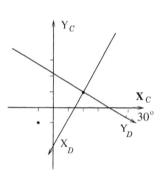

3.

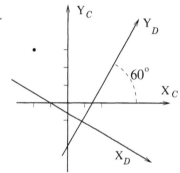

4. A coordinate system $\mathcal{D}$ is obtained from the $\mathcal{C}$ coordinate system by a translation and a rotation. The $O_\mathcal{D}$ has $\mathcal{C}$-coordinates (p,q) and the positive $X_\mathcal{D}$ axis goes through the point with $\mathcal{C}$-coordinates (a,b). Sketch the two coordinate systems and write the transition matrix $T_{(\mathcal{D} \leftarrow \mathcal{C})}$.

a. $(p, q) = (2, 3), (a, b) = (6, 0)$ b. $(p, q) = (-4, -2), (a, b) = (8, 3)$
c. $(p, q) = (-1, 1), (a, b) = (-5, -2)$ d. $(p, q) = (1, -2), (a, b) = (1, 1)$

5. Show that the coordinate changes in this section "preserve distance." To do this suppose P and Q are points in $\mathbb{R}^2$ and $\mathcal{A}$, $\mathcal{B}$ are two coordinate systems that are related by translation and/or rotation. Show that the formula for $|\overrightarrow{PQ}|$ (see Section 1.6) gives the same answer whether you use the $\mathcal{A}$-coordinates or the $\mathcal{B}$-coordinates in the computation. Why is it sufficient to just check translation and rotation separately?

---

[8]*Computer Graphics: Principles and Practice* by Foley, VanDam, Feiner and Hughes, Addison-Wesley, Reading MA, 1990.

2.7   The fast Fourier transform and the Euclidean algorithm

Objective 1

Given a polynomial $g(x)$ and an integer $n \geq deg(g)$, write the matrix R_g of the "remainder when divided by $g(x)$."

A central theme of this book is an interplay between polynomials and matrices. Sections 1.7, 2.3 and 2.5 all involve Vandermonde matrices to transfer polynomial problems to matrix problems and back. In the next chapter we use carefully chosen polynomials to decompose an arbitrary square matrix A into certain special matrices called the components of A. But to do this in the most general case requires a deeper understanding of polynomials.

We begin by recalling that division of a polynomial $f(x)$ by a polynomial $g(x)$ results in a quotient polynomial $q(x)$ and a remainder polynomial $r(x)$ and then observe that there is a matrix R_g such that the coeffcient vector of $r(x)$ is $R_g\mathbf{f}$ where $\mathbf{f}$ is the coeffcient vector of $f(x)$.

Next we introduce block matrices and use them with remainder matrices to obtain a very famous matrix factorization of certain Vandermonde matrices called the "fast Fourier transform" (see "Looking further" for some of its many applications).

We also develop the Euclidean algorithm for polynomials because it is necessary to establish key properties of matrix components in the general setting.

This section is somewhat technical and may be skipped if time is limited. Its principal use in the coming material is in some proofs of Section 3.4.

In this section $\mathbb{F}$ will be used to denote one of the number systems $\mathbb{Q}, \mathbb{R}$ or $\mathbb{C}$. Let $\mathbb{F}_n[x]$ denote the set of all polynomials with coefficients in $\mathbb{F}$, and of degree less than or equal to n. In other words, $f(x)$ is in $\mathbb{F}_n[x]$ if and only if $deg(f) \leq n$. If there is no degree restriction, this notation is simplified to $\mathbb{F}[x]$.

In fact, $\mathbb{F}[x]$ is a "number system" that shares several important properties with the integers $\mathbb{Z}$. Beyond not being closed under division, the basic shared fact is that they share long division and have a division theorem.

Example 2.18 (Long division.)

a. Most people first see long division with integers, so let's start there. Suppose $f = 169$ and $g = 12$. Then long division of f by g yields a quotient of 14 and a remainder of 1. It is presented as:

$$
\begin{array}{r}
1\ 4 \\
1\ 2\,\overline{)1\ 6\ 9} \\
\underline{1\ 2} \\
4\ 9 \\
\underline{4\ 8} \\
1
\end{array}
$$

Of course, things are more interesting if the 169 were 209. Then "borrowing" would be necessary.

b. The same structure appears with polynomials. Suppose $f(x) = x^2 + 6x + 9$ and $g(x) = x + 2$. Then division of $f(x)$ by $g(x)$ yields a quotient of $x + 4$ and a remainder of 1. It is presented as:

$$
\begin{array}{r}
x + 4 \\
x + 2 \overline{\smash{\big)}\, x^2 + 6x + 9} \\
\underline{x^2 + 2x } \\
4x + 9 \\
\underline{4x + 8} \\
1
\end{array}
$$

In "synthetic division," the $+$ signs and powers of x are suppressed. Coefficients are arranged in carefully aligned columns.

In a way, synthetic division is to polynomial division as matrices are to linear systems. Each is the bare essentials of the other. Part a is essentially a synthetic division rendering of this part.

c. If $f(x) = x^5 - x^3 + x^2 - 1$ and $g(x) = x^3 + 3x^2 + x - 5$, then division of $f(x)$ by $g(x)$ yields a quotient of $x^2 - 3x + 7$ and a remainder of $-12x^2 - 22x - 34$.

$$
\begin{array}{r}
x^2 - 3x + 7 \\
x^3 + 3x^2 + x - 5 \overline{\smash{\big)}\, x^5 - x^3 + x^2 - 1} \\
\underline{x^5 + 3x^4 + x^3 - 5x^2 } \\
- 3x^4 - 2x^3 + 6x^2 - 1 \\
\underline{- 3x^4 - 9x^3 - 3x^2 + 15x } \\
7x^3 + 9x^2 - 15x - 1 \\
\underline{7x^3 + 21x^2 + 7x - 35} \\
- 12x^2 - 22x + 34
\end{array}
$$

Polynomial long division is actually EASIER than integer long division because there is no "borrowing" (necessary to keep coefficients between 0 and 9). □

NOTICE: We have violated the convention introduced in Section 1.7 that polynomial terms be written in order of increasing degree. Long division would not look very familiar if we hadn't — to get an idea, view this page in a mirror! Be assured that we will not dwell on the polynomial long division process.

THEOREM: (Division theorem) Given polynomials $f(x)$ and $g(x)$ with the degree $deg(g)$ of $g(x)$ greater than zero, there exist **unique** $q(x)$ and $r(x)$ such that:

1. $f(x) = q(x)g(x) + r(x)$ (Think of $\frac{f(x)}{g(x)}$ equals $q(x)$ with remainder $r(x)$.)

2. $deg(r) < deg(g)$.

In this objective we fix the polynomial $g(x)$, of degree m, once and for all and study how the remainder polynomial $r(x)$ depends on the polynomial $f(x)$ in the division theorem. By the second part of the division theorem, $r(x)$ has degree less than m, so $r(x) \in \mathbb{F}_{m-1}[x]$. DEFINE the function rem_g from polynomials to polynomials by:

$$
rem_g : \mathbb{F}_n[x] \to \mathbb{F}_{m-1}[x] \begin{cases} \text{where } rem_g(f(x)) \text{ is the remainder} \\ \text{when } f(x) \text{ is divided by } g(x). \end{cases}
$$

Even though the rem_g involves polynomials, it is essentially a linear function, as defined in Section 1.6. (This is shown after the next example.) But, of course, addition and scalar multiplication must be interpreted for polynomials, not vectors.

The transition from polynomials to vectors is achieved by means of old friends, "co-efficient vectors." The matrix version R_g of rem_g simply substitutes the coefficient vector for each polynomial.

Example 2.19 (The remainder matrix R_g.)

a. Recall that the **Remainder Theorem** asserts that the value of $f(x)$ at $x = \alpha$ is the remainder when $f(x)$ is divided by $g(x) = x - \alpha$. This theorem is expressed in our current notation by the equation:

$$rem_{x-\alpha}(f) = f(\alpha).$$

It follows that $R_{x-\alpha}$ has just one row and $R_{x-\alpha} = (1, \alpha, \alpha^2 \ldots)$ — but you saw that in Section 1.7.

This simple case is actually one of the most important "remainder matrices." After all, a Vandermonde matrix is a stack of of them.

b. Consider the remainder polynomial $r(x)$ when a polynomial $f(x), deg(f) \leq 3$, is divided by $g(x) = x^2 + 1$. Our job is to find the matrix R_g

The first thing to notice is that $deg(r) \leq 1$, so $rem_g : \mathbb{F}_3[x] \to \mathbb{F}_1[x]$. This implies that R_g is a 2-by-4 matrix.

The most direct way of computing $rem_g(f)$ is by long division. It is painful, but it works:

$$
\begin{array}{r}
f_3 x + \qquad\qquad f_2 \\
\hline
x^2 + 1 \;\big|\; f_3 x^3 + f_2 x^2 + \qquad f_1 x + \qquad\qquad f_0 \\
f_3 x^3 \qquad\qquad\qquad\quad + \qquad\qquad f_3 x \\
\hline
f_2 x^2 + (f_1 - f_3)x + \qquad\qquad f_0 \\
f_2 x^2 + \qquad\qquad\qquad + \qquad\qquad f_2 \\
\hline
(f_1 - f_3)x + (f_0 - f_2)
\end{array}
$$

Comparison of coefficients reveals that the matrix of rem_g, $R_g = (r_{ij})$, satisfies:

$$
\begin{pmatrix} f_0 - f_2 \\ f_1 - f_3 \end{pmatrix} = R_g \mathbf{f} = \begin{pmatrix} r_{11} f_0 + r_{12} f_1 + r_{13} f_2 + r_{14} f_3 \\ r_{21} f_0 + r_{22} f_1 + r_{23} f_2 + r_{24} f_3 \end{pmatrix}.
$$

Since this formula holds for all polynomials $f(x)$, comparison of coefficients of $f_0 \ldots f_4$ in the two coordinates yields the values of r_{ij}.

BUT there is a much easier way. Observe that rem_g is a linear function and so its matrix can be computed one COLUMN at a time (see page 47). Each column of R_g is the remainder when a power of x is divided-by-$g(x)$. Either way,

$$
R_g = \begin{pmatrix} 1 & 0 & -1 & 0 \\ 0 & 1 & 0 & -1 \end{pmatrix}.
$$

The reason the first two columns look like an identity matrix is that each of the monomials $f(x) = 1$ and $f(x) = x$ is its own remainder when it is divided by $g(x)$. The third and fourth columns of R_g are just the coefficient vectors of the remainder when x^2 and x^3 are divided by $g(x)$. That's right, we stop with x^3 because x^4 is not in $\mathbb{F}_3[x]$!

c. To see the general pattern, suppose

$$g(x) = 4 + 3x + 2x^2 + x^3$$

$$\begin{pmatrix} 1 & 0 & 0 & -4 & 8 & -4 \\ 0 & 1 & 0 & -3 & 2 & 5 \\ 0 & 0 & 1 & -2 & 1 & 0 \end{pmatrix}$$

and consider polynomials of degree at most
5. Then R_g is a 3-by-6 matrix. Its first three
columns are an identity matrix because $x^a = rem_g(x^a)$ when $a \le 2$. The
fourth column of R_g is just the coefficient vector of the remainder when x^3 is
divided by $g(x)$, namely $(-4, -3, -2)^T$. It is a simple matter to compute the
remaining columns by dividing x^4 and x^5 by $g(x)$.

It turns out that R_g is really just an auxiliary matrix (as introduced in Section
2.3) for a very special matrix C_g, called the **companion matrix** for $g(x)$. This
shortcut is developed in Problems 3 and 4.

☐

The function rem_g really does satisfy the two "linearity conditions."   This
amounts to verifying that

$$rem_g(e + f) = rem_g(e) + rem_g(f) \quad \text{for all polynomials } e, f \in \mathbb{F}_n[x],$$
$$\text{and} \quad rem_g(\alpha f) = \alpha rem_g(f) \quad\quad\quad \text{for all } \alpha \in \mathbb{F}, \ f \in \mathbb{F}_n[x].$$

These arguments rest on the fact that there is **only one** polynomial $q(x)$ and **only one**
polynomial $r(x)$ that satisfy both parts of the division theorem.

For notational simplicity we also suppress those "x"s, writing f rather than the technically correct $f(x)$. They are only clutter, making this look harder than it is.

Let $quo_g(e), quo_g(f)$ denote the associated quotients, and apply the division theorem
to e and f separately to obtain:

$$e = quo_g(e)g + rem_g(e), \quad\quad f = quo_g(f)g + rem_g(f).$$

Now add these equations and use the fact that the sum of two polynomials has degree
no greater than either of the summands:

$$e + f = (quo_g(e) + quo_g(e)) \, g + (rem_g(e) + rem_g(f)) \,.$$

The first equation to be verified follows from the fact that $rem_g(e) + rem_g(f)$ satisfies all the conditions required of $rem_g(e + f)$ specified by the division theorem.

To obtain the second, just apply the division theorem to g and multiply the result by
α. Then observe that multiplication by α doesn't increase the degree of $rem_g(f)$.

Objective 2

Given a row partitioned Vandermonde matrix, factor it as a block diagonal matrix of
smaller Vandermonde matrices and of stacked remainder matrices.

When working with large matrices it is sometimes convenient to draw horizontal separator lines between some rows and vertical separator lines between some columns

as a way of keeping organized. Such matrices are called **partitioned** or **blocked**. A generalized augmented matrix, as in Example 1.10 on page 22, is a simple instance. It could be described as a 1-by-2 matrix, each of whose entries was itself a matrix.

The most important fact about blocked matrices is that you can add or multiply two blocked matrices together, treating the blocks like "super-numbers." All that is required is that all products of submatrices that arise meet the shape compatibility requirement discussed in Section 2.2.

Example 2.20 (Partitioned matrices.)
The matrix A is partitioned into two row blocks and two column blocks. Each "super number" is labeled by the position as indicated.

$$A = \begin{pmatrix} 4 & 2 & 2 & 3 \\ 0 & 1 & 0 & 5 \\ 0 & 1 & 1 & 5 \\ 0 & 0 & 0 & 6 \end{pmatrix} = \left(\begin{array}{c|c} A_{11} & A_{12} \\ \hline A_{21} & A_{22} \end{array} \right)$$

The (14) entry of A^2 can be computed "block wise" by first calculating:

$$A_{11}A_{12} + A_{12}A_{22} = \begin{pmatrix} 4 & 2 \\ 0 & 1 \end{pmatrix} \begin{pmatrix} 2 & 3 \\ 0 & 5 \end{pmatrix} + \begin{pmatrix} 2 & 3 \\ 0 & 5 \end{pmatrix} \begin{pmatrix} 1 & 5 \\ 0 & 6 \end{pmatrix} = \begin{pmatrix} 10 & 50 \\ 0 & 35 \end{pmatrix}$$

and then picking out the top right-hand entry to see that $(A^2)_{14} = 50$.

We could as well have blocked the same matrix into three row blocks and three column blocks as indicated (changing the name to avoid confusion between the new and old "super numbers.")

$$B = \left(\begin{array}{c|cc|c} 4 & 2 & 2 & 3 \\ \hline 0 & 1 & 0 & 5 \\ 0 & 1 & 1 & 5 \\ \hline 0 & 0 & 0 & 6 \end{array} \right) = \left(\begin{array}{c|c|c} B_{11} & B_{12} & B_{13} \\ \hline B_{21} & B_{22} & B_{23} \\ \hline B_{31} & B_{32} & B_{33} \end{array} \right)$$

Then the (14) entry of B^2 can be computed as:

$$(B^2)_{14} = B_{11}B_{13} + B_{12}B_{23} + B_{13}B_{33} = (4)(3) + \begin{pmatrix} 2 & 2 \end{pmatrix} \begin{pmatrix} 5 \\ 5 \end{pmatrix} + (3)(6) = 50.$$

In the end the matrix product does not depend on the blocking pattern. ☐

The biggest benefit from working with partitioned matrices occurs when some of the submatrices are zero. For example, this matrix is "block triangular." The product of two identically blocked matrices has the same form when there are square blocks on the "block diagonal." Check that the product of two such block triangular matrices is again block triangular.

It is common to leave blank parts of a partitioned matrix that consist of all zeros.

A matrix that has lots of well-placed zeros is called **sparse**. This rather vague term focuses attention on exactly how replacing one matrix by vector multiplication with several can speed up the process. If you multiply an n-by-n matrix by a vector, neither of which has any zero entries, then there are a total of n^2 numerical multiplications. But if the matrix is expressed as a product of several factor matrices AND if each of the factor matrices is sparse enough, then there can actually be fewer numerical multiplications required to multiply the vector by all the factor matrices! This paradox is dramatically illustrated in the next example!

Recall the key fact about **square** Vandermonde matrices $V(x_0, \ldots x_n)$ that we used in Section 1.7 and repeated in Section 2.3 on page 74. If $\mathbf{f}$ is the coefficient vector of the polynomial $f(x) = f_0 + f_1 x + \ldots f_n x^n$ then

$$V(x_0 \ldots x_n)\mathbf{f} = \begin{pmatrix} 1 & x_0 & x_0^2 & \cdots & x_0^n \\ & & & \cdots & \\ 1 & x_n & x_n^2 & \cdots & x_n^n \end{pmatrix} \begin{pmatrix} f_0 \\ \vdots \\ f_n \end{pmatrix} = \begin{pmatrix} f(x_0) \\ \vdots \\ f(x_n) \end{pmatrix}.$$

In other words, multiplication by $V(x_0, \ldots x_n)$ transforms the coefficient vector $\mathbf{f}$ of $f(x)$ into the vector of f **values** $(f(x_0), \ldots f(x_n))^T$.

This kind of transformation is actually of fundamental importance in both theory and practice. When the x_i's are the (complex) zeros of the polynomial $x^{n+1} - 1$. this transformation is called the **discrete Fourier transform.**

The discrete Fourier transform is a critical step in transmitting or recording sound digitally and has many other applications. Needless to say, even the simplest sound requires rapid sampling and an enormous amount of "multiplication by V" to achieve acceptable quality. Although the **FAST** Fourier transform has revolutionized signal processing, it is just a clever way to "multiply by V" very quickly. It is achieved by rewriting V as a product of matrices that are both sparse and blocked.

A key observation that underlies the fast Fourier transform is that computing $f(\alpha)$ doesn't require the whole coefficient vector $\mathbf{f}$. You can get by with the coefficient vector $\mathbf{r}$ of the remainder polynomial $r(x) = rem_g(f)$ for ANY polynomial $g(x)$ for which $g(\alpha) = 0$. This follows from the remainder theorem because

$$f(\alpha) = q(\alpha)g(\alpha) + r(\alpha) = q(\alpha)0 + r(\alpha) = r(\alpha).$$

Example 2.21 (Factoring Vandermonde matrices.)

    a. Let $V(-2, -1, 0, 1, 2)$ be the 5-by-5 Vandermonde matrix, so the data points are $-2, -1, 0, 1, 2$. Suppose $f(x)$ is any polynomial in $\mathbb{F}_4[x]$.

As noted above, in order to compute $f(-2)$ and $f(-1)$, it is good enough to know the remainder polynomial when $f(x)$ is divided by the **factor polynomial** $g(x) = (x+1)(x+2) = x^2 + 3x + 2$. Moreover, this remainder polynomial has coefficient vector $R_g\mathbf{f}$ as explained in the preceding objective. Consequently,

$$\begin{pmatrix} f(-2) \\ f(-1) \end{pmatrix} = V(-2, -1)R_g\mathbf{f} = \begin{pmatrix} 1 & -2 \\ 1 & -1 \end{pmatrix} \begin{pmatrix} 1 & 0 & -2 & 6 & -14 \\ 0 & 1 & -3 & 7 & -15 \end{pmatrix} \mathbf{f}.$$

Similarly, $f(0)$, $f(1)$ and $f(2)$ can be computed from just the remainder polynomial when $f(x)$ is divided by $h(x) = (x - 2)(x - 1)x = x^3 - 3x^2 + 2x$ and $\left(f(0) \ f(1) \ f(2) \right)^T = V(0, 1, 2)R_h\mathbf{f}$.

Accordingly, the block matrix equation: represents a halfway point in computing the vector of f values because only the **first two**

$$\begin{pmatrix} R_g\mathbf{f} \\ \hline R_h\mathbf{f} \end{pmatrix} = \begin{pmatrix} R_g \\ \hline R_h \end{pmatrix} \mathbf{f}$$

coordinates are required to compute $f(-2)$, $f(-1)$ and only the **last three** coordinates are required to compute $f(0)$, $f(1)$, $f(2)$. To complete the factorization just multiply on the left by a block diagonal matrix of corresponding Vandermonde matrices. Thus, $V(-2,-1,0,1,2) =$

$$
\left(\begin{array}{cc|ccc}
1 & -2 & & & \\
1 & -1 & & & \\
\hline
& & 1\ 0\ 0 & & \\
& & 1\ 1\ 1 & & \\
& & 1\ 2\ 4 & &
\end{array}\right)
\left(\begin{array}{ccccc}
1 & 0 & -2 & 6 & -14 \\
0 & 1 & -3 & 7 & -15 \\
1 & 0 & 0 & 0 & 0 \\
0 & 1 & 0 & -2 & -6 \\
0 & 0 & 1 & 3 & 7
\end{array}\right)
=
\left(\begin{array}{c|c}
V(-2,-1) & \\
\hline
& V(0,1,2)
\end{array}\right)
\left(\begin{array}{c}
R_g \\
\hline
R_h
\end{array}\right).
$$

b. The factorization in part a is really not good enough to have much value because the rightmost factor matrix is not sufficiently sparse. There is a simple way to improve this. It is to reorder the data points so that the factor polynomials have a minimum number of nonzero coefficients. Here is a much improved matrix factorization of $V(0,1,-1,2,-2)$ based on the polynomial factorization $(x)(x^2-1)(x^2-4)$:

$$
\left(\begin{array}{c|cc}
1 & & \\
\hline
& 1\ \ 1 & \\
& 1\ {-1} & \\
\hline
& & 1\ \ 2 \\
& & 1\ {-2}
\end{array}\right)
\left(\begin{array}{ccccc}
1 & 0 & 0 & 0 & 0 \\
1 & 0 & 1 & 0 & 1 \\
0 & 1 & 0 & 1 & 0 \\
1 & 0 & 4 & 0 & 16 \\
0 & 1 & 0 & 4 & 0
\end{array}\right)
=
\left(\begin{array}{c|c}
V(0) & \\
\hline
& V(1,-1) \\
& \quad V(2,-2)
\end{array}\right)
\left(\begin{array}{c}
R_x \\
\hline
R_{x^2-1} \\
R_{x^2-4}
\end{array}\right).
$$

c. Finally, assume the data points are the eighth roots of unity, and let $\omega = \frac{1}{\sqrt{2}}(\cos(\pi/4) + i\sin(\pi/4))$. List the f values at the eighth roots of unity in the order $(f(1)\ f(-1)\ f(i)\ f(-i)\ f(\omega)\ f(-\omega)\ f(i\omega)\ f(-i\omega))$. Then the fast Fourier transform comes from the polynomial factorization

$$
((x^2-1)(x^2+1))\,((x^2-i)(x^2+i)) = (x^4-1)(x^4+1) = x^8 - 1,
$$

and the 8-by-8 Vandermonde matrix $V(1,-1,i,-i,\omega,-\omega,i\omega,-i\omega)$ factors as

$$
\left(\begin{array}{cc|cc|cc|cc}
1 & 1 & & & & & & \\
1 & -1 & & & & & & \\
\hline
& & 1 & i & & & & \\
& & 1 & -i & & & & \\
\hline
& & & & 1 & \omega & & \\
& & & & 1 & -\omega & & \\
\hline
& & & & & & 1 & i\omega \\
& & & & & & 1 & -i\omega
\end{array}\right)
\left(\begin{array}{c|c}
I_2 \quad I_2 & \\
I_2 \ {-I_2} & \\
\hline
& I_2 \quad iI_2 \\
& I_2 \ {-iI_2}
\end{array}\right)
\left(\begin{array}{c|c}
I_4 & I_4 \\
\hline
I_4 & -I_4
\end{array}\right)
$$

where, as usual, I_2 and I_4 are two by two and four by four identity matrices. □

One way to compare the speed of different algorithms is to compare the number of arithmetic operations required. Let's compare the different factorizations in the preceding example in this way. The matrix times vector product

$$V(-2,-1,0,1,2)\mathbf{f} = \begin{pmatrix} 1 & -2 & 4 & 8 & 16 \\ 1 & -1 & 1 & -1 & 1 \\ 1 & 0 & 0 & 0 & 0 \\ 1 & 1 & 1 & 1 & 1 \\ 1 & 2 & 4 & 8 & 16 \end{pmatrix} \begin{pmatrix} f_0 \\ f_1 \\ f_2 \\ f_3 \\ f_4 \end{pmatrix}$$

requires an addition/subtraction (A) for each nonzero entry of the matrix and a multiplication (M) for each matrix entry $\neq 0, \pm 1$. For this method $M=8$ and $A=21$.

The factorization in part a requires $M = 10+3$ and $A = 15+11$, but the factorization in part b requires $M = 5$ and $A = 20$. This shows that matrix factorization can be a good or a bad idea depending on the sparseness of the matrix factors.

The comparison of the Fourier transform in part c is more dramatic. Check that the unfactored form has $M = 32$ and $A = 64$, while the factored form given in part c has $M = 10$ and $A = 48$ — a clear winner — and hence the epithet **fast**.

But if truth be known, this scheme would be even more impressive if 8 were replaced by 16 or 32. In fact, this example is the the third case of a recursive pattern based on the factorization $x^{2n} - 1 = (x^n - 1)(x^n + 1)$, for n is a power of 2. In the n-th case there is a factorization of the 2^n-by-2^n Vandermonde matrix as a product of n factor matrices $M[n, n] \ldots M[n, 1]$. The matrix $M[n + 1, i]$ is obtained from $M[n, i]$ by replacing each numerical entry α with the two by two matrix αI_2, for $i = 1 \ldots n$. The left-most factor matrix $M[n + 1, n + 1]$ is still block diagonal with blocks of size two, but the last half of the blocks are new.

Objective 3

Given two polynomials, $f(x), g(x)$ compute polynomials $a(x), b(x)$ such that
$a(x)f(x) + b(x)g(x)$ equals their greatest common divisor.

If the remainder term $r(x)$ in the division theorem is zero, then $g(x)$ is a **divisor** or **factor** of $f(x)$. When discussing the divisors of a polynomial $f(x)$, we must realize that we could always write a trivial factorization $f(x) = \alpha(1/\alpha)f(x)$ for any nonzero α in $\mathbb{F}$. In order to avoid this kind of silliness, we generally require polynomial divisors to be **monic** and don't even mention factors that are in $\mathbb{F}$. (A polynomial is **monic** if its leading coefficient equals 1.)

If $f(x)$ and $g(x)$ are both multiples of the polynomial $d(x)$, we say $d(x)$ is a **common divisor** of $f(x)$ and $g(x)$. If we were able to factor $f(x)$ and $g(x)$ completely, we could identify their common divisors by identifying the factors that appear in both factorizations. There is a single (monic) **greatest common divisor** $gcd(f, g)$ that itself has every common divisor of $f(x)$ and $g(x)$ as a divisor.

The **Euclidean algorithm** computes the greatest common divisor of two polynomials without factoring anything. Instead, this ancient algorithm works with linear combinations of $f(x)$ and $g(x)$, and relies on the fact that any common divisor of $f(x)$ and of $g(x)$ is also a divisor of any linear combination $a(x)f(x) + b(x)g(x)$. What the Euclidean algorithm really does is compute the least degree monic linear combination of $f(x)$ and $g(x)$.

This algorithm is based on two simple observations. The first is that part 1 of the division theorem could be rewritten as

$$r(x) = f(x) - q(x)g(x),$$

and so $r(x)$ is a linear combination of $f(x)$ and $g(x)$. In other words, the division theorem provides a way to get a nontrivial linear combination, $r(x)$, of the polynomials $f(x)$ and $g(x)$, that has degree **less than** the minimum of their degrees.

The second observation is so obvious you may at first not realize it is worth anything. It is that any linear combination of $f(x)$ and $g(x)$ is also a linear combination of $g(x)$ and $r(x)$ AND vice versa. In particular, the least degree monic linear combination of $f(x)$ and $g(x)$ is the same polynomial (but expressed differently) as the least degree monic linear combination of $g(x)$ and $r(x)$.

The Euclidean algorithm uses the division theorem repeatedly, each time replacing the highest degree polynomial from the last step with the remainder from the current step. It doesn't stop until the remainder is zero. Then the last nonzero remainder is the answer! In order to explain this clearly, we need an **indexing variable** that keeps track of the looping and distinguishes outputs at different stages.

THE EUCLIDEAN ALGORITHM

Suppose $f(x)$ and $g(x)$ are monic polynomials and $deg(f) \geq deg(g)$. Let i be an indexing variable, initially $i = 1$. To emphasize the uniformity of the algorithm, set $r_{-1}(x) = f(x)$ and $r_0(x) = g(x)$ to start.

  a. Apply the division theorem to $r_{i-2}(x)$ and $r_{i-1}(x)$ to obtain $q_i(x), r_i(x)$ in $\mathbb{F}[x]$ such that $r_{i-2}(x) = q_i(x)r_{i-1}(x) + r_i(x)$ and $deg(r_i) < deg(r_{i-1})$.

  b. If $r_i(x) \neq 0$ index i by one and go to step a. Otherwise STOP. The greatest common divisor, $gcd(f, g)$, of $f(x)$ and $g(x)$ is the monic multiple of $r_{i-1}(x)$ (scaled so the leading coefficient is 1).

Notice that when $i = 1$, the conditions of part a could as well have been written:

  1. $f(x) = q_1(x)g(x) + r_1(x)$
  2. $deg(r_1) < deg(g)$.

The algorithm simply builds smaller and smaller r's until they are gone.

Example 2.22 (The Euclidean algorithm.)

a. Suppose $f(x) = x^3 + 2x^2 + x - 2$ and $g(x) = x^2 + 3x + 3$. Begin by setting $r_{-1}(x) = f(x)$ and $r_0(x) = g(x)$. Then long division of $r_{-1}(x) = f(x)$ by $r_0(x) = g(x)$ yields a quotient of $q_1(x) = x - 1$ and remainder $r_1(x) = x + 1$.

$$f(x) = x^3 + 2x^2 + x - 2 = (x-1)(x^2 + 3x + 3) + (x+1) = q_1(x)g(x) + r_1(x).$$

Because $r_1(x) \neq 0$, increment i, so $i = 2$. Now divide $r_0(x) = x^2 + 3x + 3$ by $r_1(x) = x + 1$ obtaining a quotient $q_2(x) = x + 2$ and a remainder $r_2(x) = 1$.

$$r_0(x) = x^2 + 3x + 3 = (x+2)(x+1) + (1) = q_2(x)r_1(x) + r_2(x).$$

Because this remainder is not zero, increment i, so $i = 3$. Now divide $r_1(x) = x + 1$ by $r_2(x) = 1$ obtaining a quotient $q_3(x) = x + 1$ and $r_3(x) = 0$.

$$r_1(x) = x + 1 = (x+1)(1) + (0) = q_2(x)r_2(x) + r_3(x).$$

At last this remainder is zero, and we are done. The greatest common divisor of $f(x)$ and $g(x)$ is $gcd(f,g) = 1$, the monic multiple of the last nonzero remainder. In this case $gcd(f,g) = r_2(x)$.

b. Here are the equations that arise in the Euclidean algorithm computation of the greatest common divisor of $f(x) = x^5 - x^3 + x^2 - 1$ and $g(x) = x^3 + 3x^2 + x - 5$.

$$r_{-1}(x) = x^5 - x^3 + x^2 - 1 = (x^2 - 3x + 7)(x^3 + 3x^2 + x - 5)$$
$$+ (-12x^2 - 22x + 34) = q_1(x)g(x) + r_1(x)$$

$$r_0(x) = x^3 + 3x^2 + x - 5 = \frac{-1}{72}(6x + 7)(-12x^2 - 22x + 34)$$
$$+ \frac{61}{36}(x - 1) = q_2(x)r_1(x) + r_2(x)$$

$$r_1(x) = -12x^2 - 22x + 34 = \frac{-72}{61}(6x+17)\frac{61}{36}(x-1) + 0 = q_3(x)r_2(x) + r_3(x).$$

The last nonzero remainder is r_2. It has monic multiple $gcd(f,g) = x - 1$. This is indeed the greatest common divisor. These polynomials factor as

$$f(x) = (x+1)^2(x^2 - x + 1)(x - 1) \text{ and } g(x) = (x^2 + 4x + 5)(x - 1). \ \square$$

This procedure will certainly always produce a least degree nonzero linear combination of the original polynomials. But to see it is also a common divisor of both $f(x)$ and $g(x)$ is less immediate. To see this, write down all the equations that arise and

$$r_{-1}(x) = q_1(x)r_0(x) + r_1(x)$$
$$r_0(x) = q_2(x)r_1(x) + r_2(x)$$
$$r_1(x) = q_3(x)r_2(x) + r_3(x)$$
$$\cdots$$

$$\cdots$$
$$r_{n-3}(x) = q_{n-1}(x)r_{n-2}(x) + r_{n-1}(x)$$
$$r_{n-2}(x) = q_n(x)r_{n-1}(x) + 0$$

work from bottom up. The output $r_{n-1}(x)$ divides $r_{n-2}(x)$ by the last equation.

As you work upward, at each stage, you already know that $r_{n-1}(x)$ divides the $r's$ on the right-hand side of the equation, so it also divides the left-hand side. Finally, $r_0(x) = g(x), r_{-1}(x) = f(x)$, so it is indeed a common divisor of $f(x)$ and $g(x)$.

I can just hear you —"This is all very interesting, BUT it does not meet this objective! We must find a way to write $gcd(f, g)$ as a linear combination of $f(x)$ and $g(x)$." In fact, there is a nifty way to do it — with MATRICES!

The key fact is that, at each stage, only the last two of the remainder polynomials are used to compute the next remainder polynomial. Just "remember" the necessary data in a **vector**. (Let's again simplify notation, dropping all those "x"s whose only purpose is to remind us that we are working with polynomials.)

$$\mathbf{R}_i = \begin{pmatrix} r_{i-1} \\ r_i \end{pmatrix} \quad \text{also known as} \quad \begin{pmatrix} r_{i-1}(x) \\ r_i(x) \end{pmatrix} \quad \text{with the "x"s present.}$$

This, of course, means that the second coordinate of $\mathbf{R}_{i-1}$ is the first coordinate of $\mathbf{R}_i$, and there is a simple formula relating $\mathbf{R}_{i-1}$ and $\mathbf{R}_i$. To find it, just recall that $r_{i-2} = q_i r_{i-1} + r_i$, and write this equation in terms of the vectors:

$$\mathbf{R}_{i-1} = \begin{pmatrix} r_{i-2} \\ r_{i-1} \end{pmatrix} = \begin{pmatrix} q_i r_{i-1} + r_i \\ r_{i-1} \end{pmatrix} = \begin{pmatrix} q_i & 1 \\ 1 & 0 \end{pmatrix} \begin{pmatrix} r_{i-1} \\ r_i \end{pmatrix} = \begin{pmatrix} q_i & 1 \\ 1 & 0 \end{pmatrix} \mathbf{R}_i.$$

Multiply both sides by the inverse of that last two by two matrix and swap sides:

$$\mathbf{R}_i = \begin{pmatrix} 0 & 1 \\ 1 & -q_i \end{pmatrix} \mathbf{R}_{i-1}, \text{ for } i = 1 \dots n.$$

Let's call the matrix in the last equation Q_i. Combine the equations $\mathbf{R}_1 = Q_1 \mathbf{R}_0$ and $\mathbf{R}_2 = Q_2 \mathbf{R}_1$ to obtain $\mathbf{R}_2 = Q_2(Q_1 \mathbf{R}_0) = (Q_2 Q_1)\mathbf{R}_0$. Continue in this way until you have $\mathbf{R}_n = (Q_n Q_{n-1} \dots Q_2 Q_1)\mathbf{R}_0$. Now look at what we have:

$$\begin{pmatrix} r_{n-1} \\ 0 \end{pmatrix} = \mathbf{R}_n = \begin{pmatrix} 0 & 1 \\ 1 & -q_n \end{pmatrix} \cdots \begin{pmatrix} 0 & 1 \\ 1 & -q_1 \end{pmatrix} \mathbf{R}_0 = \begin{pmatrix} a & b \\ c & d \end{pmatrix} \begin{pmatrix} f \\ g \end{pmatrix},$$

where $Q_n \dots Q_1 = \begin{pmatrix} a & b \\ c & d \end{pmatrix}$.

The last matrix times vector product yields $r_{n-1} = af + bg$ (and $0 = cf + dg$). Finally, as mentioned earlier, $gcd(f, g)$ is required to be monic, so $gcd(f, g) = \frac{r_{n-1}(x)}{L} = \frac{a}{L}f + \frac{b}{L}g$ where L is the leading coefficient of $r_{n-1}(x)$.

***Example* 2.23** (Greatest common divisors as linear combinations.)

a. In part a of the preceding example, we computed the greatest common divisor of $f(x) = x^3 + 2x^2 + x - 2$ and $g(x) = x^2 + 3x + 3$. The sequence of quotients and remainders was:

$$q_1 = x - 1,\ r_1 = x + 1;\quad q_2 = x + 2,\ r_2 = 1;\quad q_3 = x + 1,\ r_3 = 0.$$

All of this is expressed in the matrix equation

$$\mathbf{R}_3 = \begin{pmatrix} 0 & 1 \\ 1 & -q_3 \end{pmatrix} \begin{pmatrix} 0 & 1 \\ 1 & -q_2 \end{pmatrix} \begin{pmatrix} 0 & 1 \\ 1 & -q_1 \end{pmatrix} \mathbf{R}_0,$$

which, with all of the substitutions and simplifications, reads:

$$\begin{pmatrix} 1 \\ 0 \end{pmatrix} = \begin{pmatrix} 0 & 1 \\ 1 & -x-1 \end{pmatrix} \begin{pmatrix} 0 & 1 \\ 1 & -x-2 \end{pmatrix} \begin{pmatrix} 0 & 1 \\ 1 & -x+1 \end{pmatrix} \begin{pmatrix} x^3 + 2x^2 + x - 2 \\ x^2 + 3x + 3 \end{pmatrix}.$$

This simplifies to

$$\begin{pmatrix} 1 \\ 0 \end{pmatrix} = \begin{pmatrix} -x-2 & x^2+x-1 \\ x^2+3x+3 & -x^3-2x^2-x+2 \end{pmatrix} \begin{pmatrix} x^3 + 2x^2 + x - 2 \\ x^2 + 3x + 3 \end{pmatrix}.$$

The desired linear combination is the first entry of the matrix — vector product:

$$1 = -(x+2)(x^3 + 2x^2 + x - 2) + (x^2 + x - 1)(x^2 + 3x + 3).$$

b. In part b of the preceding example, we computed the greatest common divisor of $f(x) = x^5 - x^3 + x^2 - 1$ and $g(x) = x^3 + 3x^2 + x - 5$. The sequence of quotients and remainders was:

$$\begin{aligned}
q_1 &= x^2 - 3x + 7, & r_1 &= -12x^2 - 22x + 34; \\
q_2 &= -(6x + 7)/72, & r_2 &= 61(x - 1)/36; \\
q_3 &= -72(6x + 17)/61, & r_3 &= 0.
\end{aligned}$$

All of this is expressed in the matrix equation

$$\mathbf{R}_3 = \begin{pmatrix} 0 & 1 \\ 1 & -q_3 \end{pmatrix} \begin{pmatrix} 0 & 1 \\ 1 & -q_2 \end{pmatrix} \begin{pmatrix} 0 & 1 \\ 1 & -q_1 \end{pmatrix} \mathbf{R}_0,$$

which, with all of the substitutions and simplification reads:

$$\begin{pmatrix} \frac{61}{36}(x-1) \\ 0 \end{pmatrix} = \begin{pmatrix} \frac{6x+7}{72} & \frac{1}{72}(-6x^3 + 11x^2 - 21x + 23) \\ \frac{36}{61}(x^2 + 4x + 5) & -\frac{36}{61}(x^2 - x + 1)(x + 1)^2 \end{pmatrix} \begin{pmatrix} f \\ g \end{pmatrix}.$$

The desired linear combination is the first entry of the matrix — vector product multiplied by an appropriate scalar:

$$\begin{aligned}
x - 1 = \frac{1}{122} \big(&(6x + 7)(x^5 - x^3 + x^2 - 1) \\
&+ (-6x^3 + 11x^2 - 21x + 23)(x^3 + 3x^2 + x - 5) \big).
\end{aligned}$$ □

SUMMARY

Fix a polynomial $g(x)$ then division by $g(x)$ leads to a remainder matrix R_g.

Block matrices are matrices whose rows and columns are blocked. They multiply and add just like matrices of numbers, as long as the bits and pieces have compatibile shapes.

Given a set of data $\{x_0 \ldots x_n\}$ and a partition of it, there is a factorization of the polynomial $p(x) = (x - x_0) \ldots (x - x_n) = p_1(x)p_2(x)$ that leads to a factorization of the associated Vandermonde matrix where the right factor is stacked remainder matrices R_{p_1} and R_{p_2} and the left factor is block diagonal with blocks the Vandermonde matrices associated with the zeros of $p_1(x)$ and $p_2(x)$.

If the polynomial factors have few nonzero coefficients, then the matrix factors are **sparse**. Sometimes this kind of matrix factorization leads to more efficient matrix times vector computation, as in the **(discrete) fast Fourier transform**.

The **Euclidean algorithm** uses repeated long division to find the greatest common divisor $gcd(f, g)$ of two polynomials $f(x)$ and $g(x)$. Using the quotients that arise in this process, there is a simple way of expressing $gcd(f, g)$ as a linear combination of $f(x)$ and $g(x)$ that use two by two matrices.

Looking further

The topics in this section are "computer algebra."[9]

In case $f(x)$ and $g(x)$ have a greatest common divisor of 1, the last objective can be rewritten

$$\frac{1}{f(x)g(x)} = \frac{a(x)}{g(x)} + \frac{b(x)}{f(x)},$$

which is the basis of "partial fractions" integration in calculus. This objective is achieved by the Maple command gcdex and is sometimes called "Bezout's Lemma." It also has many mathematical generalizations, both to more abstract coefficient number systems and to many variable polynomials.

Cooley and Tukey[10] are credited with the invention of the fast Fourier transform but it was "anticipated" by many, not the least of whom was Gauß. Some enthusiasts have called it "perhaps the single algorithmic discovery that has had the greatest practical impact on history." It is used in optics, acoustics, quantum physics, speech recognition data compression, as well as telecommunications and signal processing.

---

[9]**Modern Computer Algebra** by Joachim von zur Gathen and Jürgend Gerhard, Cambridge University Press, 1999.

[10]J. W. Cooley and J. W. Tukey, "An algorithm for machine computation of complex Fourier series," Math. Comp. **19** (1965), pp. 297-301.

Practice Problems

1. Suppose $g(x)$ is the indicated polynomial and we are working with polynomials of degree n as given. Write the associated remainder matrix R_g.

   a. $x - 2$ $n = 4$ b. $x^3 - 1$ $n = 5$ c. $x^2 + x + 1$ $n = 4$

2. Let ω be the complex number $\cos(\pi/3) + i\sin(\pi/3)$, so $\omega^6 = 1$. Consider the discrete Fourier transform based on the zeros of

$$x^6 - 1 = (x^2 - 1)(x^2 + x + 1)(x^2 - x + 1)$$
$$= ((x - 1)(x + 1))\left((x - \omega^2)(x - \omega^4)\right)\left((x - \omega)(x - \omega^5)\right)$$

   Write the factorization (into the product of two matrices) of the associated Vandermonde matrix.

3. By counting additions /subtractions and multiplications compare the "speed" of the multiplying by the factorization given in the preceding problem to that of just multiplying by the associated Vandermonde matrix.

4. Express the greatest common divisor of the indicated polynomials as a linear combination.

   a. $x^2 + x + 1$ $x + 1$ b. $x^3 - 1$ $x^2 - 1$
   c. $2x^3 - 7x^2 + 7x - 2$ $2x^3 + x^2 + x - 1$

5. Suppose $g(x)$ is a monic polynomial of degree m. Define an "operator" that takes polynomials to polynomials by

$$\star(f(x)) = \text{ the remainder after division of } xf(x) \text{ by } g(x).$$

   The **companion matrix** C_g is obtained by listing as columns the coefficient vectors of the result of applying $\star$ to $1 = x^0, x, x^2...x^{m-1}$. Show that each R_g in Problem 1 is the augmented matrix (as in Section 2.3) associated with C_g and starts with the vector $(1, 0, \ldots 0)^T$.

6. Suppose $g(x)$ is a polynomial of degree m and $\star : \mathbb{F}_{m-1}[x] \to \mathbb{F}_{m-1}[x]$ is the operator defined in the preceding problem.

   a. Suppose $f_1(x), f_2(x) \in \mathbb{F}_{m-1}[x]$ and $\alpha \in \mathbb{F}$. Show that

$$\star(f_1(x) + f_2(x)) = \star(f_1(x)) + \star(f_2(x))$$
$$\star(\alpha f_1(x)) = \alpha \star (f_1(x)).$$

   b. Part a shows that $\star$ is a "linear function" from $\mathbb{F}_{m-1}[x]$ to $\mathbb{F}_{m-1}[x]$. EXPLAIN why C_g is the associated matrix on coefficient vectors.

Chapter 3

Diagonalizable matrices

Suppose a square matrix A is given. What is the simplest matrix D for which there is a matrix P — of largest possible rank — such that $AP = PD$?

This fundamental question, first addressed by Cauchy in 1826 (in substance but not in form), is the subject of this chapter. If there is an invertible matrix, say P, such that $P^{-1}AP = D$, then A is *similar* to D. As A takes all values, the nature of necessary matrices D depends on the number system of the matrix entries. One of the reasons we use $\mathbb{C}$ is that relatively few types of Ds are required. Eventually we will show Frobenius' theorem: every complex matrix is similar to an upper triangular matrix.

The main focus of this chapter is, however, to turn the similarity question on its head and insist that D be diagonal. We say a matrix A is *diagonalizable* if it is similar to a diagonal matrix, and then proceed to decide exactly which matrices are diagonalizable. This in itself is very useful.

Even when A is not diagonalizable, we are interested in largest rectangular matrices P (of full rank) for which $AP = PD$, D diagonal. The columns of P are called *eigenvectors* of A, and the diagonal entries of D are called *eigenvalues* of A.

We begin by using transform plots to understand the geometric meaning of eigenvalues and eigenvectors, and then show that the eigenvectors of a matrix A are easily computed from its eigenvalues. The eigenvalues of A are determined to be exactly the zeros of the *minimal polynomial*. A matrix is diagonalizable if and only if its minimal polynomial has no multiple zeros.

We compute $\mu_A(x)$ by the method of Krylov. It has logical structure analogous to Gaussian elimination but differs in several ways. Elementary row operations are replaced by annihilator polynomial computations, and the working matrix shrinks in rank, not size.

We use digraphs and transfer matrices to solve certain enumeration problems with recurrence relations that are obtained from minimal polynomials. The evolution of the matrix sequence $A, A^2, A^3, \cdots$ is analyzed as a preliminary to the analysis of the three state discrete dynamical systems — Markov chains and Leslie population models. Finally, matrix components are defined. But we can only work effectively with them for diagonalizable matrices and for these we present the basic ideas of principal component analysis.

Because we must find the zeros of polynomials, there are some computational difficulties that must be addressed. In order to minimize these, we generally work with 3-by-3 matrices. We will need to work with complex numbers and the quadratic formula (on page 55). We will also use the calculator to graph cubic polynomials and to find numerical approximations to their real zeros.

3.1   Eigenvectors and eigenvalues

Objective 1

Given a square matrix A and vector $\mathbf{v}$, decide if $\mathbf{v}$ is an eigenvector of A.

Recall the transform plots in Example 1.18 on page 42 (partially reproduced below). This time we wish to study straight lines through the origin under the linear function f_A. Since $A\mathbf{0} = \mathbf{0}$, the origin is always mapped to itself. As shown in Section 1.6, straight lines are always mapped to straight lines by linear functions, so each such line through the origin maps to another line through the origin. Sometimes such a straight line is mapped to itself, even though its points are moved around. When this happens we say the line is f_A-*invariant*.

Example 3.1 (Geometry of eigenvectors.)

a. For $A = \begin{pmatrix} 2 & 0 \\ 0 & 1 \end{pmatrix}$, we have the transform plot:

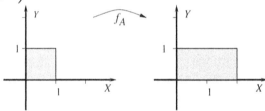

Both the X- and Y-axis are f_A-invariant, while every other line through the origin is mapped to a line with half the slope of the original line.

b. For $A = \frac{1}{4} \begin{pmatrix} 5 & 3 \\ 3 & 5 \end{pmatrix}$, we had the transform plot

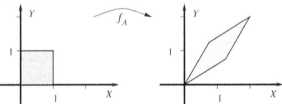

Notice that

$$\frac{1}{4}\begin{pmatrix} 5 & 3 \\ 3 & 5 \end{pmatrix}\begin{pmatrix} 1 \\ 1 \end{pmatrix} = 2\begin{pmatrix} 1 \\ 1 \end{pmatrix} \text{ and } \frac{1}{4}\begin{pmatrix} 5 & 3 \\ 3 & 5 \end{pmatrix}\begin{pmatrix} -1 \\ 1 \end{pmatrix} = \frac{1}{2}\begin{pmatrix} -1 \\ 1 \end{pmatrix}.$$

Therefore, the lines $y = \pm x$ are also invariant under f_A. It turns out that there are no other f_A-invariant lines.

c. Finally, suppose $A = \begin{pmatrix} 0 & 1 \\ -1 & 2 \end{pmatrix}$ and observe that $\begin{pmatrix} 0 & 1 \\ -1 & 2 \end{pmatrix}\begin{pmatrix} 1 \\ 1 \end{pmatrix} = \begin{pmatrix} 1 \\ 1 \end{pmatrix}.$

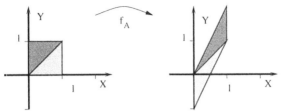

Because f_A shifts up and to the right for points above the line $y = x$ and down and to the left for points below, the only f_A-invariant line is $y = x$. □

An *eigenvector* for the matrix A is a **nonzero** vector $\mathbf{v}$ for which there is a number λ satisfying:
$$A\mathbf{v} = \lambda\mathbf{v}.$$
The number λ is the *eigenvalue*[1] of A (associated with $\mathbf{v}$). Each eigenvector has an f_A-invariant line as its direction. The associated eigenvalue represents a "stretching factor" that f_A affects along the associated f_A-invariant line.

In the above example, the first matrix has eigenvalue $\lambda = 2$ with associated eigenvectors of the form $\mathbf{x}$ and $\lambda = 1$ with associated eigenvectors of the form $\mathbf{y}$. The second matrix has eigenvalues $\lambda = 2$, $\lambda = \frac{1}{2}$ with associated eigenvectors $\mathbf{u}$, $\mathbf{v}$, respectively. The

$$\mathbf{x} = \begin{pmatrix} x \\ 0 \end{pmatrix}; \quad \mathbf{y} = \begin{pmatrix} 0 \\ y \end{pmatrix}; x, y \neq 0$$

$$\mathbf{u} = \begin{pmatrix} x \\ x \end{pmatrix}; \quad \mathbf{v} = \begin{pmatrix} x \\ -x \end{pmatrix}; x \neq 0$$

third matrix has only one eigenvalue, $\lambda = 1$. It has associated eigenvector $\mathbf{u}$.

Recall that it is much easier to check the correctness of a linear system's solutions than it is to find its solutions. Similarly, given a matrix A, it is much easier to verify that a number λ is indeed an eigenvalue of A with associated eigenvector $\mathbf{v}$ than it is to find either λ or $\mathbf{v}$. We begin with these easier situations.

Given A and $\mathbf{v}$, all that is needed is old-fashioned matrix multiplication: $A\mathbf{x} \stackrel{?}{=} \lambda\mathbf{x}$.

Example 3.2 (Identifying eigenvectors.)

a. Let $A = \begin{pmatrix} 1 & 1 \\ 2 & 0 \end{pmatrix}$. Then A has eigenvectors $\mathbf{v}_1 = \begin{pmatrix} -1 \\ 2 \end{pmatrix}$ and $\begin{pmatrix} 1 \\ 1 \end{pmatrix}$. The associated eigenvalues are $\lambda_1 = -1$ and $\lambda_2 = 2$. Here is how to CHECK that this information is correct.

CHECK: $\begin{pmatrix} 1 & 1 \\ 2 & 0 \end{pmatrix} \begin{pmatrix} -1 \\ 2 \end{pmatrix} \stackrel{?}{=} -1 \begin{pmatrix} -1 \\ 2 \end{pmatrix}$ and $\begin{pmatrix} 1 & 1 \\ 2 & 0 \end{pmatrix} \begin{pmatrix} 1 \\ 1 \end{pmatrix} \stackrel{?}{=} 2 \begin{pmatrix} 1 \\ 1 \end{pmatrix}$.

b. Let $\mathbf{e}_1, \mathbf{e}_2, \cdots, \mathbf{e}_n \in \mathbb{R}^n$ be the columns of the identity matrix I_n. Then each $\mathbf{e}_i$ is an eigenvector for any n-by-n diagonal matrix D. The eigenvalue associated with $\mathbf{e}_i$ is the diagonal entry $d_{i,i}$ of D. For example:

$$\begin{pmatrix} a & 0 & 0 \\ 0 & b & 0 \\ 0 & 0 & c \end{pmatrix} \begin{pmatrix} 1 \\ 0 \\ 0 \end{pmatrix} = \begin{pmatrix} a \\ 0 \\ 0 \end{pmatrix} = a \begin{pmatrix} 1 \\ 0 \\ 0 \end{pmatrix}.$$

---

[1]The "Looking further" at the end of this section gives the etymology of this term and some common alternatives.

c. The eigenvalues of even simple-looking matrices can be irrational and complex. The indicated digraph has adjacency matrix A given below. The eigenvalues of A are $\omega, \bar{\omega}, \sigma, \tau$ where

$$x^2 + x + 1 = (x - \omega)(x - \bar{\omega})$$

$$x^2 - x - 1 = (x - \sigma)(x - \tau).$$

The first two are complex, and the second two are irrational. The calculation below shows that the vector $\mathbf{v}$ is an eigenvector of A with the eigenvalue ω.

$$A = \begin{pmatrix} 0 & 1 & 0 & 0 \\ 0 & 0 & 1 & 1 \\ 0 & 0 & 0 & 1 \\ 1 & 1 & 0 & 0 \end{pmatrix} \quad \mathbf{v} = \begin{pmatrix} -\bar{\omega} \\ -1 \\ \bar{\omega} \\ 1 \end{pmatrix} ; \quad A\mathbf{v} = \begin{pmatrix} -1 \\ 1+\bar{\omega} \\ 1 \\ -\bar{\omega}-1 \end{pmatrix} = \begin{pmatrix} -\omega\bar{\omega} \\ -\omega \\ \omega\bar{\omega} \\ \omega \end{pmatrix} = \omega \begin{pmatrix} -\bar{\omega} \\ -1 \\ \bar{\omega} \\ 1 \end{pmatrix}.$$

(The middle step uses $\omega + \bar{\omega} = -1$ and $\omega\bar{\omega} = 1$ which follow from comparing coefficients of $x^2 + x + 1 = (x - \omega)(x - \bar{\omega})$.)

To check that each of the other columns of the matrix P is a eigenvector of A, you will also have to use the equations $\sigma + \tau = 1$ and $\sigma\tau = -1$ which follow from comparing coefficients of $x^2 - x - 1 = (x - \sigma)(x - \tau)$.

$$A = \begin{pmatrix} 0 & 1 & 0 & 0 \\ 0 & 0 & 1 & 1 \\ 0 & 0 & 0 & 1 \\ 1 & 1 & 0 & 0 \end{pmatrix} \quad P = \begin{pmatrix} -\bar{\omega} & -\omega & -\sigma & -\tau \\ -1 & -1 & 1 & 1 \\ \bar{\omega} & \omega & -\sigma & -\tau \\ 1 & 1 & 1 & 1 \end{pmatrix}.$$ □

Objective 2

Given a square matrix A and a number λ, decide if λ is an eigenvalue of A, and if so, find the associated basic eigenvectors.

Now reverse the above process and go from numbers to vectors. The number λ is an *eigenvalue* of A if there is a **nonzero** vector $\mathbf{v}$ such that $A\mathbf{v} = \lambda\mathbf{v}$. This is equivalent to the (homogeneous) **shifted linear system**

$$(A - \lambda I)\mathbf{v} = 0$$

having nontrivial basic solutions. The list of basic solutions to this linear system is called the list of **basic eigenvectors of A associated with** λ, or simply **basic λ-eigenvectors of A**. The **geometric multiplicity** of λ is the number of basic eigenvectors of A that are associated with λ. From our study of linear systems, we know that any solution to the linear system $(A - \lambda I)\mathbf{x} = 0$ is a linear combination of its basic solutions. Thus, any eigenvector of A associated with λ is a linear combination of the basic eigenvectors of A associated with λ.

Example 3.3 (The list of basic eigenvectors.)

a. Test if $\lambda = 1$, 4 or 6 are eigenvalues of the given matrix A, $\quad A = \begin{pmatrix} 3 & 1 \\ 2 & 2 \end{pmatrix}$ and if so, find the associated list of basic eigenvectors.
    We must treat each possible eigenvalue separately. The first case is $\lambda = 1$.

(λ goes here.)

$$A- \overset{\downarrow}{1I} = A - I = \begin{pmatrix} 3 & 1 \\ 2 & 2 \end{pmatrix} - \begin{pmatrix} 1 & 0 \\ 0 & 1 \end{pmatrix} = \begin{pmatrix} 2 & 1 \\ 2 & 1 \end{pmatrix} \overset{RREF}{\longrightarrow} \begin{pmatrix} 1 & \frac{1}{2} \\ 0 & 0 \end{pmatrix}.$$

There is exactly one basic solution to the system $(A - I)\mathbf{x} = \mathbf{0}$, which is $\mathbf{x} = (-\frac{1}{2} \ 1)^T$. Therefore, $\lambda = 1$ is an eigenvalue of A and it has geometric multiplicity 1. Observe that any other eigenvector of A having associated eigenvalue 1 must be a multiple of $\mathbf{x}$.

The case $\lambda = 4$: (λ goes here too.)

$$A- \overset{\downarrow}{4I} = \begin{pmatrix} 3 & 1 \\ 2 & 2 \end{pmatrix} - \begin{pmatrix} 4 & 0 \\ 0 & 4 \end{pmatrix} = \begin{pmatrix} -1 & 1 \\ 2 & -2 \end{pmatrix} \overset{RREF}{\longrightarrow} \begin{pmatrix} 1 & -1 \\ 0 & 0 \end{pmatrix}.$$

There is exactly one basic solution of the system $(A - 4I)\mathbf{x} = \mathbf{0}$, which is $\mathbf{x} = \begin{pmatrix} 1 & 1 \end{pmatrix}^T$. Since this system does have a nonzero solution, we know that 4 is indeed an eigenvalue of A. Since this system has exactly **one** basic solution, the eigenvalue 4 has geometric multiplicity **one**.

The case $\lambda = 6$:

$$A - 6I = \begin{pmatrix} 3 & 1 \\ 2 & 2 \end{pmatrix} - \begin{pmatrix} 6 & 0 \\ 0 & 6 \end{pmatrix} = \begin{pmatrix} -3 & 1 \\ 2 & -4 \end{pmatrix} \longrightarrow \begin{pmatrix} 1 & 3 \\ 0 & -5 \end{pmatrix}$$

There is no point in going further. This matrix has rank 2 and the only solution to $(A - 6I)\mathbf{x} = \mathbf{0}$ is $\mathbf{x} = \mathbf{0}$. Therefore, $\lambda = 6$ is **not** an eigenvalue of A!

b. Check if B has eigenvalues $\lambda = 1$ or 2.

For $\lambda = 1$ the "shifted matrix" is $B - 1I$, and the shifted system $(B - I)\mathbf{x} = \mathbf{0}$ has one basic solution:

$$B = \begin{pmatrix} 0 & 0 & -2 \\ 1 & 2 & 1 \\ 1 & 0 & 3 \end{pmatrix}$$

$$B - I = \begin{pmatrix} -1 & 0 & -2 \\ 1 & 1 & 1 \\ 1 & 0 & 2 \end{pmatrix} \overset{RREF}{\longrightarrow} \begin{pmatrix} 1 & 0 & 2 \\ 0 & 1 & -1 \\ 0 & 0 & 0 \end{pmatrix} \quad \begin{matrix} \text{basic} \\ \text{eigenvector} \\ \text{for } \lambda = 1: \end{matrix} \quad \begin{pmatrix} -2 \\ 1 \\ 1 \end{pmatrix}.$$

But $\lambda = 2$ has geometric multiplicity 2 because $B - 2I =$

$$\begin{pmatrix} -2 & 0 & -2 \\ 1 & 0 & 1 \\ 1 & 0 & 1 \end{pmatrix} \overset{RREF}{\longrightarrow} \begin{pmatrix} 1 & 0 & 1 \\ 0 & 0 & 0 \\ 0 & 0 & 0 \end{pmatrix} \quad \begin{matrix} \text{basic} \\ \text{eigenvectors} \\ \text{for } \lambda = 2: \end{matrix} \quad \begin{pmatrix} 0 \\ 1 \\ 0 \end{pmatrix}, \begin{pmatrix} -1 \\ 0 \\ 1 \end{pmatrix}.$$

c. Check if C has eigenvalues $\lambda = 2, 4$ or 6.

$$C = \begin{pmatrix} 4 & 2 & -2 & 2 \\ 1 & 3 & 1 & -1 \\ 0 & 0 & 2 & 0 \\ 1 & 1 & -3 & 5 \end{pmatrix}$$

The first shifted matrix is $C - 2I =$

$$\begin{pmatrix} 2 & 2 & -2 & 2 \\ 1 & 1 & 1 & -1 \\ 0 & 0 & 0 & 0 \\ 1 & 1 & -3 & 3 \end{pmatrix} \overset{RREF}{\longrightarrow} \begin{pmatrix} 1 & 1 & 0 & 0 \\ 0 & 0 & 1 & -1 \\ 0 & 0 & 0 & 0 \\ 0 & 0 & 0 & 0 \end{pmatrix} \quad \begin{matrix} \text{basic} \\ \text{eigenvectors} \\ \text{for } \lambda = 2: \end{matrix} \quad \begin{pmatrix} -1 \\ 1 \\ 0 \\ 0 \end{pmatrix}, \begin{pmatrix} 0 \\ 1 \\ 0 \\ 1 \end{pmatrix}.$$

Thus, the geometric multiplicity of $\lambda = 2$ as an eigenvalue for C is two.

The geometric multiplicity of $\lambda = 4$ as an eigenvalue of C is one because

$$C - 4I = \begin{pmatrix} 0 & 2 & -2 & 2 \\ 1 & -1 & 1 & -1 \\ 0 & 0 & -2 & 0 \\ 1 & 1 & -3 & 1 \end{pmatrix} \xrightarrow{RREF} \begin{pmatrix} 1 & 0 & 0 & 0 \\ 0 & 1 & 0 & 1 \\ 0 & 0 & 1 & 0 \\ 0 & 0 & 0 & 0 \end{pmatrix} \begin{array}{c} \text{basic} \\ \text{eigenvector} \\ \text{for } \lambda = 4: \end{array} \begin{pmatrix} 0 \\ -1 \\ 0 \\ 1 \end{pmatrix}.$$

The geometric multiplicity of $\lambda = 6$ as an eigenvalue of C is one.

$$C - 6I = \begin{pmatrix} -2 & 2 & -2 & 2 \\ 1 & -3 & 1 & -1 \\ 0 & 0 & -4 & 0 \\ 1 & 1 & -3 & -1 \end{pmatrix} \xrightarrow{RREF} \begin{pmatrix} 1 & 0 & 0 & -1 \\ 0 & 1 & 0 & 0 \\ 0 & 0 & 1 & 0 \\ 0 & 0 & 0 & 0 \end{pmatrix} \begin{array}{c} \text{basic} \\ \text{eigenvector} \\ \text{for } \lambda = 6: \end{array} \begin{pmatrix} 1 \\ 0 \\ 0 \\ 1 \end{pmatrix}. \; \square$$

It can be proven that the sum of the geometric multiplicities of all the eigenvalues of an n-by-n matrix A cannot be more than n. **Moreover, A is diagonalizable exactly when this sum equals n. Each of the matrices in this example is diagonalizable.**

We generally list eigenvalues in increasing numerical order $\lambda_1 \leq \lambda_2 \leq \cdots$. When λ is an eigenvalue with geometric multiplicity greater than one, we generally list the basic eigenvectors according to the position of the associated $RREF(A - \lambda I)$ nonpivotal column. Combining these, we use the term *the standard list of basic eigenvectors* of A to mean the list of basic λ_1-eigenvectors of A, followed by the list of basic λ_2-eigenvectors of A, etc. For example, the standard list of basic eigenvectors for the matrices in parts a and c of the preceding example are

$$A : \begin{pmatrix} -\frac{1}{2} \\ 1 \end{pmatrix}, \begin{pmatrix} 1 \\ 1 \end{pmatrix} \text{ and } C : \begin{pmatrix} -1 \\ 1 \\ 0 \\ 0 \end{pmatrix}, \begin{pmatrix} 0 \\ 0 \\ 1 \\ 1 \end{pmatrix}, \begin{pmatrix} 0 \\ -1 \\ 0 \\ 1 \end{pmatrix}, \begin{pmatrix} 1 \\ 0 \\ 0 \\ 1 \end{pmatrix}$$

in that order, because for A, 1 is less than 4, so the basic eigenvector(s), associated with 1 are listed before those associated with 4 and in that order for C because $2 < 4 < 6$ and the first basic eigenvector of C associated with $\lambda = 2$ is $(-1\ 1\ 0\ 0)^T$.

Sometimes the ordering of eigenvalues and eigenvectors is not important. In these cases we use terms like "the set of basic λ-eigenvectors of A." The distinction is that a **list** is ordered but a **set** is unordered.

SUMMARY

The zero vector is not allowed as an eigenvector — ever! If **v** is a nonzero vector, and $A\mathbf{v} = \lambda\mathbf{v}$ for some number λ, then **v** is an *eigenvector* for the matrix A and λ is the associated *eigenvalue*.

The list of *basic eigenvectors* of A associated with the eigenvalue λ is the list of basic solutions of $(A - \lambda I)\mathbf{x} = \mathbf{0}$.

The *geometric multiplicity* of an eigenvalue is the number of associated basic eigenvectors.

The *standard list of basic eigenvectors* is defined.

Looking further

Sometimes different names for a concept reflect its importance, and other times they reflect historical controversy or national claims of priority. For example, the different notations for the derivative of $y = f(x)$ in calculus, namely $f'(x)$ and $\frac{dy}{dx}$, reflect the English/German dispute about whether Newton/Leibnitz should be given priority.

The etymology of "eigen" is more interesting. Apparently, the earliest publication of the concept we have called eigenvalue is with the name "latent root" by J. Sylvester in 1883 in Phil. Mag. XVI. 267. It seems that the American term "eigenvalue" derives from the German word Eigenwert, which was introduced by D. Hilbert in 1904 (to mean something else). It is well-established folklore that this word came to America with the many academic refugees created by World War II.

In 1942, P. Halmos wrote "Almost every combination of the adjectives proper, latent, characteristic, eigen and secular, with the nouns root, number and value, has been used in the literature for what we call ..."

Sections 3.5 and 3.7 develop another view on how best to order eigenvalues.

$$* \; * \; * \; * \; * \; *$$

Practice Problems

1. For each of the following matrices, determine which of the given vectors are eigenvectors. For each eigenvector, find the corresponding eigenvalue.

   a. $\begin{pmatrix} 6 & 0 \\ 1 & 9 \end{pmatrix}$; $\mathbf{r} = \begin{pmatrix} -3 \\ 1 \end{pmatrix}$, $\mathbf{s} = \begin{pmatrix} -6 \\ 2 \end{pmatrix}$, $\mathbf{t} = \begin{pmatrix} 1 \\ 0 \end{pmatrix}$, $\mathbf{u} = \begin{pmatrix} 0 \\ 0 \end{pmatrix}$.

   b. $\begin{pmatrix} -3 & -4 & 7 \\ 12 & 8 & -20 \\ 3 & -4 & 1 \end{pmatrix}$; $\mathbf{u} = \begin{pmatrix} -1 \\ 8 \\ 5 \end{pmatrix}$ $\mathbf{v} = \begin{pmatrix} 11 \\ 3 \\ 3 \end{pmatrix}$, $\mathbf{w} = \begin{pmatrix} 1 \\ -2 \\ 1 \end{pmatrix}$, $\mathbf{x} = \begin{pmatrix} 1 \\ -8 \\ -5 \end{pmatrix}$.

   c. $\begin{pmatrix} -3 & 6 & 6 & 4 \\ 0 & 0 & 0 & -2 \\ 0 & 0 & 0 & 2 \\ 0 & 0 & 0 & 3 \end{pmatrix}$; $\mathbf{w}_1 = \begin{pmatrix} 1 \\ -2 \\ 2 \\ 3 \end{pmatrix}$, $\mathbf{w}_2 = \begin{pmatrix} 2 \\ -2 \\ 2 \\ 3 \end{pmatrix}$, $\mathbf{w}_3 = \begin{pmatrix} 2 \\ 1 \\ 0 \\ 0 \end{pmatrix}$, $\mathbf{w}_4 = \begin{pmatrix} 2 \\ 0 \\ 1 \\ 0 \end{pmatrix}$.

2. Determine which of the given numbers (if any) are eigenvalues of the matrix.

   a. $\begin{pmatrix} -2 & 2 & -2 \\ -2 & -2 & -6 \\ 2 & 0 & 4 \end{pmatrix}$; $\begin{array}{l} \lambda_1 = -3, \\ \lambda_2 = 0, \\ \lambda_3 = 2. \end{array}$ b. $\begin{pmatrix} 2 & -1 & -2 \\ 52 & -20 & -28 \\ -28 & 11 & 16 \end{pmatrix}$; $\begin{array}{l} \lambda_1 = -3, \\ \lambda_2 = 0, \\ \lambda_3 = 4. \end{array}$

3. For each of the following matrices, all eigenvalues are given. Compute the standard list of eigenvectors.

   a. $\begin{pmatrix} 4 & -2 \\ 1 & 1 \end{pmatrix}$; $\lambda_1 = 2, \lambda_2 = 3$.   b. $\begin{pmatrix} 1 & -8 \\ 1 & -5 \end{pmatrix}$; $\lambda_1 = -3, \lambda_2 = -1$.

   c. $\begin{pmatrix} 1 & 3 & -2 \\ 3 & 1 & 2 \\ -4 & 4 & -2 \end{pmatrix}$; $\begin{array}{l} \lambda_1 = -6, \\ \lambda_2 = 2, \\ \lambda_3 = 4. \end{array}$ d. $\begin{pmatrix} -3 & 3 & 1 \\ 3 & -3 & -1 \\ 2 & -2 & -6 \end{pmatrix}$; $\begin{array}{l} \lambda_1 = -8, \\ \lambda_2 = -4, \\ \lambda_3 = 0. \end{array}$

4. What are the eigenvalues of an idempotent matrix?

5. Suppose A is a symmetric matrix having $\lambda \neq \mu$ as eigenvalues. Let $\mathbf{v}$ be an λ-eigenvector of A and $\mathbf{w}$ be a μ-eigenvector of A. Show that $\mathbf{v}$ and $\mathbf{w}$ are perpendicular. (**HINT:** consider $\mathbf{v}^T A \mathbf{u}$.)

6. The standard list of basic eigenvectors for the matrix A is given. Find the eigenvalue associated with each. Define P to be the matrix whose columns are the given eigenvectors (in order) and let D be a diagonal matrix having $d_{ii} = $ the eigenvalue associated with the i-th column of P. Finally, compute AP and PD. They will be equal. Can you explain why?

a. $A = \begin{pmatrix} 2 & -2 \\ -1 & 3 \end{pmatrix}$; $\mathbf{w}_1 = \begin{pmatrix} 2 \\ 1 \end{pmatrix}$, $\mathbf{w}_2 = \begin{pmatrix} -1 \\ 1 \end{pmatrix}$.

b. $A = \begin{pmatrix} -1 & -1 & 1 \\ 2 & -4 & 2 \\ -1 & 1 & -3 \end{pmatrix}$; $\mathbf{w}_1 = \begin{pmatrix} -1 \\ -2 \\ 1 \end{pmatrix}$, $\mathbf{w}_2 = \begin{pmatrix} -1 \\ 0 \\ 1 \end{pmatrix}$, $\mathbf{w}_3 = \begin{pmatrix} 1 \\ 1 \\ 0 \end{pmatrix}$.

7. Let $\mathbf{v}$ be an λ-eigenvector for A. Suppose $f(x)$ is a polynomial. Show that $\mathbf{v}$ is an eigenvector for $f(A)$ and find the associated eigenvalue.

8. Suppose A is the adjacency matrix of a digraph and regard a numerical labeling of the vertices as a column vector $\mathbf{v}$. Show that $\mathbf{v}$ is a λ-eigenvector of A

 whenever, for every vertex x, the sum $s(x)$ of the labels of the out neighbors of x equals λ times the label of x. Use this to construct eigenvectors of indicated digraph with associated eigenvalues $1, -1, i, -i$.

9. Let λ, μ with $\lambda \neq \mu$ be eigenvalues of A with associated eigenvectors $\mathbf{v}, \mathbf{w}$, respectively. Suppose $\mathbf{u} = a\mathbf{v} + b\mathbf{w}$ is an eigenvector of A. Show that either $a = 0$ or $b = 0$.

10. The ***trace*** $tr(M)$ of a square matrix M is the sum of its diagonal entries. Suppose X is an 1-by-n matrix and Y is an n-by-1 matrix. Show that

$$tr(XY) = tr(YX).$$

Next suppose A is an m-by-n matrix and B is an n-by-m matrix. View A as a stack of row vectors and B as a list of column vectors. Show that $tr(XY) = tr(YX)$ follows. (Compare to problem 9 in Section 2.2)

11. Suppose A is diagonalizable with diagonalizing matrix P (this means $P^{-1}AP = D$ and equivalently $A = PDP^{-1}$).

Show that $tr(A) = tr(D)$. Thus, the trace of A is the sum of its eigenvalues. (**HINT:** Try $X = PD, Y = P^{-1}$.)

3.2   The minimal polynomial algorithm

Objective 1

Given a square matrix A, a vector $\mathbf{v}$ and a polynomial $f(x)$, decide if $f(x)$ annihilates the pair $A, \mathbf{v}$. If not, decide the A annihilator of $\mathbf{v}$.

We determine the eigenvalues of a matrix A in two parts. First find a polynomial, $\mu_A(x)$ called the minimal polynomial of A, whose zeros are exactly the eigenvalues of A. Then factor $\mu_A(x)$.

To compute minimal polynomials, we use a method of А. Н. Крылов (A. N. Krylov)[2]. This method combines two ideas that we have already seen. The first is the shift in emphasis (to treating the **coefficients** of a polynomial function as **variables**) seen with polynomial interpolation. The second idea is the use of auxiliary matrices to compute $f(A)\mathbf{v}$ as in Section 2.3.

We must find polynomials $f(x)$ that have A as a "zero" ($f(A) = 0$). But these zeros are **matrices**, not **numbers**! That is a whole new ball game.

We work up to the new task by using a vector $\mathbf{v}$, chosen in advance, and merely requiring $f(A)\mathbf{v} = \mathbf{0}$. Such a polynomial $f(x)$ is said to ***annihilate the pair*** $A, \mathbf{v}$.

Example 3.4 ($f(x)$ annihilates the pair $A, \mathbf{v}$.)

a. Let A, $\mathbf{v}$ be as given.   Consider $p(A)\mathbf{v}$ $A = \begin{pmatrix} 4 & 1 & -3 \\ -6 & 0 & 4 \\ 2 & 0 & -3 \end{pmatrix}, \mathbf{v} = \begin{pmatrix} -1 \\ 0 \\ 1 \end{pmatrix}$
where $p(x) = x^2 - 2$. Using the second method of Example 2.9 on page 71, we have

$$p(A)\mathbf{v} = (A^2 - 2I)\mathbf{v} = A(A\mathbf{v}) - 2\mathbf{v} = A \begin{pmatrix} -7 \\ 10 \\ -5 \end{pmatrix} - \begin{pmatrix} -2 \\ 0 \\ 2 \end{pmatrix} = \begin{pmatrix} -1 \\ 22 \\ -1 \end{pmatrix} \neq \mathbf{0}.$$

This shows that $p(x)$ does not annihilate the pair $A, \mathbf{v}$.

b. Let $p(x) = x^2 - x$ and $q(x) = x^2 - x - 1$. Also set $A = \begin{pmatrix} 1 & 1 \\ 1 & 0 \end{pmatrix}$.

Then $p(A) = I_2$ and $q(A) = 0$, so $p(x)$ annihilates the pair $A, \mathbf{v}$ only if $\mathbf{v} = \mathbf{0}$, but $q(x)$ annihilates pair $A, \mathbf{r}$ for every $\mathbf{r}$ in $\mathbb{R}^2$.   □

When looking for polynomials $f(x)$ that annihilate a given pair of $A, \mathbf{v}$, we obtain a linear system with the variables being the coefficients $f_0, f_1, \ldots$ of a polynomial

$$y = f_0 + f_1 x + \cdots + f_{n-1} x^{n-1} + f_n x^n.$$

This is just like polynomial interpolation, except that this time the linear system doesn't come from data points, rather it is (TRUMPETS):

---

[2]http://www.math.uu.nl/people/vorst/kryl.html
The Theory of Matrices by F. R. Gantmacher, §7.8 (p. 202), Chelsea, 1959.

$M\mathbf{f} = \mathbf{0}$ where $M = (\mathbf{v}\ A\mathbf{v}\ \cdots\ A^n\mathbf{v})$ is the **auxiliary matrix**.

Now every basic solution to $M\mathbf{f} = \mathbf{0}$ builds back to a polynomial $f(x)$, and every basic solution has a 1 as its last nonzero term. Because the coefficient vector $\mathbf{f}$ lists coefficients of $f(x)$ in order of increasing exponents, that "last nonzero term" is the leading coefficient of $f(x)$

Therefore, **every basic solution of $M\mathbf{f} = \mathbf{0}$ builds back to a monic polynomial**. (Recall, the **leading coefficient** of $f(x)$ is f_n, the coefficient of x^n, for $f(x)$ of degree n. Also recall that a polynomial is **monic** if its leading coefficient is 1.)

The first basic solution of $M\mathbf{f} = \mathbf{0}$ ends with the longest string of zeros. Because the coefficient vector f lists coefficients in order of increasing exponents of A, the first basic solution of this linear system $\mathbf{f}$ has nonzero entries that are coefficients of only low powers of A. **Therefore, the LEAST degree monic polynomial that annihilates the pair $A, \mathbf{v}$ comes from the first basic solution.** The polynomial $f(x)$ with this coefficient vector is called the A **annihilator of** $\mathbf{v}$.

The A annihilator of $\mathbf{v}$ is the least degree monic polynomial that annihilates the pair $A, \mathbf{v}$.

THE A ANNIHILATOR OF $\mathbf{v}$

Suppose a vector $\mathbf{v} \in \mathbb{R}^n$ and an n-by-n matrix A are given.

1. Form the **auxiliary matrix** M with columns (in order):

$$\mathbf{v}, A\mathbf{v}, A^2\mathbf{v}, A^3\mathbf{v}, \ldots, A^n\mathbf{v}.$$

2. Compute $RREF(M)$. (Because M has more columns than rows, $RREF(M)$ must have nonpivotal columns.)

3. As you read the columns of $RREF(M)$ from the left, they look just like those of an identity matrix UNTIL you come to the first non pivotal column: $(-f_0, \ldots, -f_{k-1}, 0\ldots0)^T$.
   The **FIRST** basic solution of $M\mathbf{f} = \mathbf{0}$ is $(f_0, \ldots, f_{k-1}, 1, 0\ldots0)^T$ and the A **annihilator** of $\mathbf{v}$ is

$$f(x) = f_0 + f_1 x + \cdots + f_{k-1}x^{k-1} + x^k.$$

Example 3.5 (The A annihilator of $\mathbf{v}$.)

a. Using the matrix A and vector $\mathbf{v}$, build an auxiliary matrix M whose columns are $\mathbf{v}, A\mathbf{v}, A(A\mathbf{v}), A(A(A\mathbf{v}))$. Building $\quad A = \begin{pmatrix} 1 & 2 & -1 \\ 2 & 2 & -1 \\ -1 & -1 & 4 \end{pmatrix}, \mathbf{v} = \begin{pmatrix} 1 \\ 0 \\ 0 \end{pmatrix}$

M isn't that hard because each column is obtained from its predecessor by multiplication by A. In this way, we compute $A^3\mathbf{v}$ without knowing A^3.

$$M = \begin{pmatrix} 1 & & & \\ 0 & A\mathbf{v} & A^2\mathbf{v} & A^3\mathbf{v} \\ 0 & & & \end{pmatrix} = \begin{pmatrix} 1 & 1 & & \\ 0 & 2 & A(A\mathbf{v}) & A^3\mathbf{v} \\ 0 & -1 & & \end{pmatrix} =$$

$$\begin{pmatrix} 1 & 1 & 6 & \\ 0 & 2 & 7 & A(A^2\mathbf{v}) \\ 0 & -1 & -7 & \end{pmatrix} = \begin{pmatrix} 1 & 1 & 6 & 27 \\ 0 & 2 & 7 & 33 \\ 0 & -1 & -7 & -41 \end{pmatrix} \xrightarrow{\text{RREF}} \begin{pmatrix} 1 & 0 & 0 & -7 \\ 0 & 1 & 0 & -8 \\ 0 & 0 & 1 & 7 \end{pmatrix}$$

Because the **first** nonpivotal column of $RREF(M)$ is $(-7 \ -8 \ 7)^T$, the first (and only) basic solution $M\mathbf{f} = \mathbf{0}$ is $\mathbf{f} = \begin{pmatrix} 7 & 8 & -7 & 1 \end{pmatrix}^T$.

The entries f are the coefficients of the A annihilator of $\mathbf{v}$ because a linear combination of vectors can be expressed as matrix times vector multiplication:

$$\mathbf{0} = M\mathbf{f} = \begin{pmatrix} 1 \\ 0 \\ 0 \end{pmatrix} 7 + \begin{pmatrix} 1 \\ 2 \\ -1 \end{pmatrix} 8 + \begin{pmatrix} 6 \\ 7 \\ -7 \end{pmatrix} (-7) + \begin{pmatrix} 27 \\ 33 \\ -41 \end{pmatrix}$$

$$= 7\mathbf{v} + 8A(\mathbf{v}) - 7A^2(\mathbf{v}) + A^3(\mathbf{v}) = (7 + 8A - 7A^2 + A^3)\mathbf{v}.$$

Thus, the A annihilator of $\mathbf{v}$ is therefore $f(x) = 7 + 8x - 7x^2 + x^3$.

b. Let B and $\mathbf{v}$ be as indicated. Begin by constructing the auxiliary matrix M. When working by hand it is easiest to do this computation by first writing B and then developing M directly to its right as follows: $\qquad B = \begin{pmatrix} 3 & 2 \\ 3 & -2 \end{pmatrix}, \mathbf{v} = \begin{pmatrix} 1 \\ 0 \end{pmatrix}.$

$$B||M = \begin{pmatrix} 3 & 2 \\ 3 & -2 \end{pmatrix} \begin{pmatrix} 1 & 3 & 15 \\ 0 & 3 & 3 \end{pmatrix}.$$

(The $||$ serve no purpose except to separate B and M.)

The next step is to solve the homogeneous linear system $M\mathbf{f} = \mathbf{0}$.

$$\begin{pmatrix} 1 & 3 & 15 \\ 0 & 3 & 3 \end{pmatrix} \xrightarrow{\text{RREF}} \begin{pmatrix} 1 & 0 & 12 \\ 0 & 1 & 1 \end{pmatrix} : \text{ the first basic solution is } \begin{pmatrix} -12 \\ -1 \\ 1 \end{pmatrix}.$$

The B-annihilator of $\mathbf{v}$ is therefore $f(x) = -12 - x + x^2$. Since this polynomial is built from a basic solution, annihilator polynomials are always a **monic**.

NOTICE: also that we would get the same B annihilator of $\mathbf{v}$ if we were overzealous in building M and added a fourth column.

$$\begin{pmatrix} 1 & 3 & 15 & 51 \\ 0 & 3 & 3 & 39 \end{pmatrix} \xrightarrow{\text{RREF}} \begin{pmatrix} 1 & 0 & 12 & 12 \\ 0 & 1 & 1 & 13 \end{pmatrix}.$$

The answer would be unchanged because the basic solutions of the new homogeneous system are: $\begin{pmatrix} -12 & -1 & 1 & 0 \end{pmatrix}^T$ and $\begin{pmatrix} -12 & -13 & 0 & 1 \end{pmatrix}^T$. But the FIRST basic solution STILL leads to $f(x) = -12 - x + x^2 = (x-4)(x+3)$. Observe further that the polynomial associated with the WRONG solution

$$g(x) = -12 - 13x + x^3 = (x-4)(x+3)(x+1)$$

is a multiple of $f(x)$ but has an "extra" zero $x = -1$.

It turns out that we will only be interested in A-annihilators because of their zeros, and to mistake $g(x)$ for $f(x)$ would lead to the incorrect value $x = -1$.

c. Consider the matrix C and vector $\mathbf{v}$. To compute the C-annihilator of $\mathbf{v}$, begin by writing C and developing the auxiliary matrix M.

$$C = \begin{pmatrix} 4 & 2 & -2 & 2 \\ 1 & 3 & 1 & -1 \\ 0 & 0 & 2 & 0 \\ 1 & 1 & -3 & 5 \end{pmatrix}, \quad \mathbf{v} = \begin{pmatrix} 1 \\ 1 \\ 1 \\ 1 \end{pmatrix}$$

$$C \| M = \begin{pmatrix} 4 & 2 & -2 & 2 \\ 1 & 3 & 1 & -1 \\ 0 & 0 & 2 & 0 \\ 1 & 1 & -3 & 5 \end{pmatrix} \begin{pmatrix} 1 & 6 & 36 & 216 & 1296 \\ 1 & 4 & 16 & 64 & 256 \\ 1 & 2 & 4 & 8 & 16 \\ 1 & 4 & 24 & 160 & 1056 \end{pmatrix} \xrightarrow{\text{RREF}} \begin{pmatrix} 1 & 0 & 0 & 48 & 576 \\ 0 & 1 & 0 & -44 & -480 \\ 0 & 0 & 1 & 12 & 100 \\ 0 & 0 & 0 & 0 & 0 \end{pmatrix}.$$

The fourth column is the FIRST nonpivotal column and the associated polynomial is $-48 + 44x - 12x^2 + x^3$. Therefore, the C annihilator of $\mathbf{v}$ is

$$f(x) = -48 + 44x - 12x^2 + x^3 = (x-2)(x-4)(x-6).$$

The C annihilator of $\mathbf{v}$ has zeros at 2, 4, 6. As discussed in part c of Example 3.3 on page 113, these numbers are eigenvalues of C! **This is not mere coincidence — it is the reason for our interest in annihilator polynomials.**

d. This example shows that the same matrix can lead to different annihilator polynomials for different vectors. The auxiliary matrix for Q and $\mathbf{v}$ is:

$$Q = \begin{pmatrix} 0 & 0 & 1 \\ 0 & 0 & 0 \\ 0 & 0 & 0 \end{pmatrix}, \quad \mathbf{v} = \begin{pmatrix} 1 \\ 1 \\ 1 \end{pmatrix}$$

$$\begin{pmatrix} 1 & 1 & 0 \\ 1 & 0 & 0 \\ 1 & 0 & 0 \end{pmatrix} \xrightarrow{\text{RREF}} \begin{pmatrix} 1 & 0 & 0 \\ 0 & 1 & 0 \\ 0 & 0 & 0 \end{pmatrix}, \quad \text{so } \mathbf{f} = \begin{pmatrix} 0 \\ 0 \\ 1 \end{pmatrix}$$

is the first basic solution and the Q-annihilator of $\mathbf{v}$ is $p(x) = x^2$.

On the other hand, for $\mathbf{w} = \begin{pmatrix} 1 & 0 & 0 \end{pmatrix}^T$, the auxiliary matrix has the refreshing property of already being reduced. The vector $\begin{pmatrix} 0 & 1 & 0 \end{pmatrix}^T$ is the basic solution of the associated homogeneous system, and so the Q-annihilator of $\mathbf{w}$ is $f(x) = x$.

Thus, the A-annihilator of $\mathbf{v}$ really does depend on $\mathbf{v}$.   □

Calculators
It is easy to build up an auxiliary matrix using the AUGMENT feature in the matrix math menu of a calculator. Here is a step-by-step method to compute the auxiliary matrix for a square matrix A and vector $\mathbf{v}$:

1. Store the coefficient matrix, in say [A], and store the vector as a matrix with 1 column, say [B].

2. Select and display [B] (so that [B] is the answer).

3. Compute Augment([B],[A]*ans). ("Augment" is the seventh item in the matrix math menu and "ans" is the key sequence 2nd (−).)

4. Repeat the last step until you have a matrix that has more columns than rows (repeatedly pressing the enter key does this automatically).

For the next objective we use random vectors. A ***random*** vector **v** is a numerical vector whose coordinates are chosen randomly. The MATRIX MATH MENU has an item "randM" that will do this for you.

Objective 2

Compute the minimal polynomial of a square matrix.

If $f(A)$ annihilates every vector **v**, then $f(A) = 0$ is the zero matrix and we say that $f(x)$ ***annihilates*** A. The ***minimal polynomial***, $\mu_A(x)$, of A is the least degree monic polynomial $f(x)$ that annihilates A.

In the next section, we show that the eigenvalues of A are exactly the zeros of $\mu_A(x)$. The minimal polynomial $\mu_A(x)$ (together with its factorization) carries very valuable information about a matrix. From this information, it is possible to find all eigenvalues and eigenvectors and to decide if the matrix is diagonalizable.

As already mentioned, the following algorithm uses a random vector **r** to start. By choosing **r** to be the first column of the identity matrix, the algorithm starts faster, but the total process may wind up longer than if a different initial vector were used.

Like the Gaussian elimination algorithm, this algorithm involves choices and there are many paths (sequences of elementary row operations) to a single, final result (RREF(A)). It also has a "working matrix" W that keeps getting smaller and smaller, but this algorithm doesn't stop until the working matrix is 0.

THE MINIMAL POLYNOMIAL OF A

Start with the identity $W = I$ as working matrix.

I. Pick a random vector **r**.
If $\mathbf{v} = W\mathbf{r} = \mathbf{0}$, then repeat this step.

II. Compute the A-annihilator $f(x)$ of **v** and
OUTPUT the FACTOR $f(x)$.

III. Compute the NEW WORKING MATRIX $f(A)W$.
If it is not zero, return to step I, or else go to step IV

IV. $\mu_A(x)$ is the PRODUCT of the factors output from step II
(one for each pass through the algorithm).

Notice that step I involves $\mathbf{v} = W\mathbf{r}$, not just r.
This is CRUCIAL and cannot be omitted!

Note that every time you return to step I, $W \neq 0$. Consequently there is always some vector $\mathbf{r}$ for which $\mathbf{v} = W\mathbf{r} \neq \mathbf{0}$. (In fact, you can arrange that $\mathbf{v}$ is the first nonzero column of W by picking $\mathbf{r}$ to be the corresponding column of an identity matrix.)

In contrast to the Gaussian elimination algorithm, the basic steps (computing A-annihilator) are much more complicated, and the shape of this algorithm's working matrix does not change, rather the **rank** of the working matrix keeps getting smaller. A third difference is that each pass through the steps has, not one, but **two** outputs:

> 1. A polynomial factor from step II
> 2. A new working matrix from step III.

In the end the minimal polynomial $\mu_A(x)$ of A is the PRODUCT of the factors that were output along the way.

We show, at the end of this objective, that successive working matrices have strictly decreasing ranks. Therefore the looping cannot go on indefinitely. We also hint at how **it can be proven that the product of the output factors does not depend on the random vectors chosen or on the number of passes required!**

It is convenient to use subscripts, as in Section 2.7, to distinguish data that changes in different passes through the algorithm. Thus, call the initial working matrix W_1, the second working matrix W_2, etc., and similarly call the first output factor $f_1(x)$, the second output factor $f_2(x)$, etc.

Example 3.6 (The minimal polynomial of A.)

a. Find the minimal polynomial of the matrix $\quad A = \begin{pmatrix} 2 & 1 & 2 \\ 1 & 2 & 3 \\ 1 & 1 & 3 \end{pmatrix}$.

The initial working matrix $W_1 = I_3$

Step I. Pick a random vector r so that $\mathbf{v} = W\mathbf{r} \neq \mathbf{0}$.

Take as first random vector $\mathbf{r}_1 = (1\,0\,0)^T$. Then $\mathbf{v}_1 = W_1\mathbf{r}_1 = (1\,0\,0)^T \neq \mathbf{0}$.

As indicated above, we use a subscript 1 to indicate that this is the first pass output. On the first pass there is no difference between $\mathbf{v}_1$ and $\mathbf{r}_1$, but after that they may well be different!

Step II. Compute the A-annihilator of $\mathbf{v}_1$. Output the (first) factor $f_1(x)$:

$$\begin{pmatrix} 2 & 1 & 2 \\ 1 & 2 & 3 \\ 1 & 1 & 3 \end{pmatrix} \Bigg\| \begin{pmatrix} 1 & 2 & 7 & 33 \\ 0 & 1 & 7 & 39 \\ 0 & 1 & 6 & 32 \end{pmatrix} \overset{RREF}{\longrightarrow} \begin{pmatrix} 1 & 0 & 0 & 4 \\ 0 & 1 & 0 & -10 \\ 0 & 0 & 1 & 7 \end{pmatrix}$$

Save $f_1(x) = -4 + 10x - 7x^2 + x^3$, the A-annihilator of $\mathbf{v}_1$.

Step III. Compute the new working matrix $W_2 = f_1(A)W_1$.
If $W_2 \neq 0$, then return to step I, or else go to step IV

$$W_2 = f_1(A)W_1 = (-4I_3 + 10A - 7A^2 + A^3)I_3 = \cdots = 0.$$

Step IV. $\mu_A(x)$ **is the product of the factors from step III.**

Since $W_2 = 0$, the algorithm has done the job in only one pass. The minimal polynomial of A is: $\mu_A(x) = f_1(x) = -4 + 10x - 7x^2 + x^3$.

b. Let's repeat part a but with a different "random" starting vector. Suppose $\mathbf{r}_1 = \mathbf{v}_1 = (1\,1\,1)^T$ instead.

Step II. Compute the A-annihilator of $\mathbf{v}_1$. Output the (first) factor $f_1(x)$:

$$\begin{pmatrix} 2 & 1 & 2 \\ 1 & 2 & 3 \\ 1 & 1 & 3 \end{pmatrix} \left\|\begin{pmatrix} 1 & 5 & 26 & 136 \\ 1 & 6 & 32 & 168 \\ 1 & 5 & 26 & 136 \end{pmatrix}\right. \xrightarrow{RREF} \begin{pmatrix} 1 & 0 & -4 & -24 \\ 0 & 1 & 6 & 32 \\ 0 & 0 & 0 & 0 \end{pmatrix}$$

Save $f_1(x) = 4 - 6x + x^2$, the A-annihilator of $\mathbf{v}_1$.

Step III. Compute the new working matrix $W_2 = f_1(A)W_1$.
If $W_2 \neq 0$, then return to step I, or else go to step IV:

$$W_2 = f_1(A)W_1 = (4I_3 - 6A + A^2)I_3 = \cdots = \begin{pmatrix} -1 & 0 & 1 \\ 1 & 0 & -1 \\ 0 & 0 & 0 \end{pmatrix} \neq 0.$$

Because this matrix is not zero, we must loop back to step I.

Step I. (2nd pass) Pick a random vector r so that $\mathbf{v} = W\mathbf{r} \neq 0$.

Take $\mathbf{r}_2 = (0\,0\,1)^T$. Then $\mathbf{v}_2 = (1\,-1\,0)^T$ is just the third column of W_2.

Step II. (2nd pass) Compute the A-annihilator of $\mathbf{v}_2$. Output the second factor $f_2(x)$:

$$\begin{pmatrix} 2 & 1 & 2 \\ 1 & 2 & 3 \\ 1 & 1 & 3 \end{pmatrix} \left\|\begin{pmatrix} 1 & 1 & 1 & 1 \\ -1 & -1 & -1 & -1 \\ 0 & 0 & 0 & 0 \end{pmatrix}\right. \xrightarrow{RREF} \begin{pmatrix} 1 & 1 & 1 & 1 \\ 0 & 0 & 0 & 0 \\ 0 & 0 & 0 & 0 \end{pmatrix}$$

Save $f_2(x) = -1 + x$ because it is the A-annihilator of $\mathbf{v}_2$.

Step III. (2nd pass) Compute the new working matrix.
If $W_3 = f_2(A)W_2 \neq 0$, then return to step I, or else go to step IV:

$$W_3 = f_2(A)W_2 = (-I_3 + A)\begin{pmatrix} -1 & 0 & 1 \\ 1 & 0 & -1 \\ 0 & 0 & 0 \end{pmatrix} = \cdots = 0.$$

Step IV. (2nd pass) $\mu_A(x)$ is the product of the factors from step III.

This time we needed two passes and we now output

$$\mu_A(x) = f_1(x)f_2(x) = (4 - 6x + x^2)(-1 + x).$$

DO CHECK that this is just a factored form of same polynomial that we came up with in part a of this example!

c. Here is another example where looping leads to a factored minimal polynomial. There are also a of couple shortcuts indicated. Again, start with $W_1 = I_3$ and the not-so-random $\mathbf{v}_1 = W_1\mathbf{r}_1 = (1\,0\,0)^T$.

$$C = \begin{pmatrix} 2 & -1 & 1 \\ 0 & 1 & 3 \\ 0 & 3 & 1 \end{pmatrix}$$

Step II. Compute the C-annihilator of $\mathbf{v}_1$. Output the (first) factor $f_1(x)$:

$$C||M = \begin{pmatrix} 2 & -1 & 1 \\ 0 & 1 & 3 \\ 0 & 3 & 1 \end{pmatrix} \left|\left| \begin{pmatrix} 1 & 2 \\ 0 & 0 \\ 0 & 0 \end{pmatrix} \right.\right. .$$

Shortcut: You can stop the development of the auxiliary matrix M as soon as you have a matrix that will lead to a nonzero basic solution.

Strictly speaking, we are supposed to compute two more columns of M before going on, but you should recognize that each additional column will have zeros in all but the first entry. Therefore, the matrix M has rank one and the linear system having M as coefficient matrix has **first** basic solution $(-2,1,0)^T$. Thus, $f_1(x) = -2 + x$.

Step III. Compute the new working matrix $W_2 = f_1(A)W_1$.
If $W_2 \neq 0$, then return to step I, or else go to step IV:

$$W_2 = f(C)W_1 = (-2I + C)I = \begin{pmatrix} 0 & -1 & 1 \\ 0 & -1 & 3 \\ 0 & 3 & -1 \end{pmatrix} \neq 0.$$

Step I. (2nd pass) Pick a random vector r so that $\mathbf{v} = W\mathbf{r} \neq 0$.

Let's take $\mathbf{r} = (0\,1\,0)^T$. Then $0 \neq \mathbf{v}_2 = W_2\mathbf{r}_2 = (-1\,-1\,3)^T$.

Step II. (2nd pass) Compute the A-annihilator of $\mathbf{v}_2$. Output the (2nd) factor $f_2(x)$:

$$C||M = \begin{pmatrix} 2 & -1 & 1 \\ 0 & 1 & 3 \\ 0 & 3 & 1 \end{pmatrix} \left|\left| \begin{pmatrix} -1 & 2 & -4 & 8 \\ -1 & 8 & 8 & 80 \\ 3 & 0 & 24 & 48 \end{pmatrix} \right.\right. \overset{RREF}{\longrightarrow} \begin{pmatrix} 1 & 0 & 8 & 16 \\ 0 & 1 & 2 & 12 \\ 0 & 0 & 0 & 0 \end{pmatrix}.$$

Therefore, $f_2(x) = -8 - 2x + x^2$.

Step III. (2nd pass) Compute the new working matrix.
If $W_3 = f_2(A)W_2 \neq 0$, then return to step I, or else go to step IV:

$$W_3 = f_2(A)W_2 = (-8I_2 - 2A + A^2) \begin{pmatrix} 0 & -1 & 1 \\ 0 & -1 & 3 \\ 0 & 3 & -1 \end{pmatrix} = \cdots = 0.$$

Shortcut: You can skip Step III if the sum of the degrees of the output factors equals the size of the matrix. (This is implied by the Cayley Hamilton theorem — see page 210.) Because the product $f_2(x)f_1(x)$ has degree 3, the size of the matrix A, theory predicts that $\mu_A(x) = f_1(x)f_2(x)$, and the check in step III is automatic.

Step IV. (2nd pass) $\mu_A(x)$ **is the product of the factors from step III.** Again two passes lead to a factored polynomial!

$$\mu_A(x) = f_1(x)f_2(x) = (x-2)(-8 - 2x + x^2).$$

d. When the algorithm is applied to a diagonal matrix, the only way you will fail to get the minimal polynomial in one pass is if the first vector r_1 that you choose has some coordinate 0. In fact, the second working matrix $\quad D = \begin{pmatrix} 1 & 0 & 0 \\ 0 & 2 & 0 \\ 0 & 0 & 3 \end{pmatrix}$

will have nonzero columns exactly in the coordinates where r_1 has zeros. For example, if $r_1 = (1\ 1\ 0)^T$, then

$$f_1(x) = x^2 - 3x + 2 \text{ and } W_2 = \begin{pmatrix} 0 & 0 & 0 \\ 0 & 0 & 0 \\ 0 & 0 & 6 \end{pmatrix}.$$

The only thing that matters about r_2 is that it must have nonzero 3rd coordinate. Then $f_2(x) = (x-3)$ and $W_3 = 0$, so $\mu_D(x) = (x^2 - 3x + 2)(x-3)$.

Do check that $x^3 - 6x^2 + 11x - 6 = (x^2 - 3x + 2)(x - 3)$ as promised. ☐

Here is a simplified version of the minimal polynomial algorithm:

THE GAMBLER'S DELIGHT

Pick a random vector $r \neq 0$. Compute the A-annihilator $f(x)$ of r.
If $f(A) \neq 0$, you lose. Start over.
If $f(A) = 0$, you win and $\mu_A(x) = f(x)$.

Unlike the lottery, it can actually be shown to be a winning strategy, provided your initial vectors are truly random (whatever that means). The proof of this fact is hinted at in part d of the preceding example, but requires much more machinery than we can now muster.

Proof of termination
Let's compare two successive working matrices W_i and $W_{i+1} = f(A)W_i$ that arise in this algorithm. Recall that $f(x)$ is the A-annihilator of v_i and that $v_i = W_i r_i \neq 0$. This implies that r_i is not a solution to the linear system $W_i x = 0$ (since $v_i = W_i r_i \neq 0$) but is a solution to the linear system $W_{i+1} x = 0$. Indeed, $W_{i+1} r_i = f(A)(W_i r_i) = f(A)v_i = 0$ by construction of $f(x)$.

Since every solution to the first system is a solution to the second (after all, $W_{i+1}\mathbf{x} = f(A)(W_i\mathbf{x}) = f(A)\mathbf{0} = \mathbf{0}$), this can only happen if $RREF(W_{i+1})$ has more non-pivotal columns than $RREF(W_i)$. Thus, the rank of W_{i+1} is strictly less than the rank of W_i. It follows that the algorithm must terminate after at most n passes where n is the size of A.

Independence of random choices

This algorithm always produces a polynomial that annihilates A. It is conceivable, however, that the output polynomial $f(x) = f_1(x) \ldots f_t(x)$ overshot and (by Problems 4 and 5 of this section) is a proper multiple of $\mu_A(x)$.

This cannot occur for the gambler's delight because it produces only one polynomial that is the A-annhilator of some vector and that annhilates A.

In case $\mu_A(x)$ is a power of an irreducible polynomial, the random vector $\mathbf{r}$ in (step I) of the last pass actually has the output product polynomial as its A-annihilator. Therefore overshooting cannot occur in this case either.

Problem 9 of Section 3.4 hints at how the general case can be reduced to this special case.

SUMMARY

The A *annihilator of* $\mathbf{v}$ is the least degree monic polynomial $f(x)$ such that $f(A)\mathbf{v} = \mathbf{0}$. Its coefficient vector $\mathbf{f}$ is the first basic solution to $M\mathbf{f} = \mathbf{0}$ where M is the auxiliary matrix $M = (\mathbf{v}\ A\mathbf{v}\,A^2\mathbf{v}\cdots A^n\mathbf{v})$.

The *minimal polynomial* algorithm is:

  Start with the identity $W = I$ as working matrix.

 I. Pick a random vector $\mathbf{r}$ so that $\mathbf{v} = W\mathbf{r} \neq 0$.

 II. Compute the A-annihilator $f(x)$ of $\mathbf{v}$ and output the factor $f(x)$.

 III. Compute the new working matrix $f(A)W$. If it is not zero, return to step I, or else go to step IV.

 IV. $\mu_A(x)$ is the PRODUCT of the factors output from step II (one for each pass through the algorithm).

The gambler's delight is a less systematic, but still winning, strategy to compute minimal polynomials.

Looking further

Although the material in this section is not now standard for courses at this level, it is this book's raison d'être. Further explanation of this departure from common practice appears in a pedagogical postscript.

Practice Problems

1. Find the A-annihilator of the indicated vectors, A the given matrix.

a. $\begin{pmatrix} -2 & 4 \\ -3 & 6 \end{pmatrix}$; $\begin{pmatrix} 2 \\ 3 \end{pmatrix}$ $\begin{pmatrix} 3 \\ 4 \end{pmatrix}$

b. $\begin{pmatrix} 2 & -1 \\ 5 & -2 \end{pmatrix}$; $\begin{pmatrix} 1 \\ 2 \end{pmatrix}$ $\begin{pmatrix} 1 \\ 2-i \end{pmatrix}$

c. $\begin{pmatrix} 1 & -3 & -3 \\ 1 & 2 & 0 \\ 0 & 1 & 2 \end{pmatrix}$; $\begin{pmatrix} 0 \\ 1 \\ 0 \end{pmatrix}$ $\begin{pmatrix} -1 \\ 2 \\ 0 \end{pmatrix}$

d. $\begin{pmatrix} 2 & -1 & 1 \\ -1 & 4 & 1 \\ 1 & 1 & 4 \end{pmatrix}$; $\begin{pmatrix} 1 \\ 1 \\ -1 \end{pmatrix}$ $\begin{pmatrix} 1 \\ -1 \\ 2 \end{pmatrix}$

e. $\begin{pmatrix} 2 & -1 & -2 \\ 52 & -20 & -28 \\ -28 & 11 & 16 \end{pmatrix}$; $\begin{pmatrix} 1 \\ 0 \\ 1 \end{pmatrix}$ $\begin{pmatrix} 1 \\ 1 \\ 1 \end{pmatrix}$

f. $\begin{pmatrix} 1 & -4 & 1 & 1 \\ 2 & -18 & 6 & 10 \\ 4 & -12 & 6 & 8 \\ 1 & -24 & 7 & 13 \end{pmatrix}$; $\begin{pmatrix} 0 \\ 1 \\ 1 \\ 1 \end{pmatrix}$ $\begin{pmatrix} 1 \\ 0 \\ 0 \\ 0 \end{pmatrix}$

2. For each part in Problem 1, suppose that the minimal polynomial algorithm was executed starting with the first vector given above. Then your work in Problem 1 is essentially step II in the first pass of the algorithm. Finish the algorithm from this point giving the output polynomials $f_i(x)$ for each pass required through the steps.

3. Find the minimal polynomial for each matrix by either method.

a. $\begin{pmatrix} 3 & 2 \\ 1 & 2 \end{pmatrix}$

b. $\begin{pmatrix} 3 & -1 \\ -1 & 3 \end{pmatrix}$

c. $\begin{pmatrix} 1 & 1 \\ -2 & 4 \end{pmatrix}$

d. $\begin{pmatrix} 3 & 1 & 0 \\ 1 & 3 & 0 \\ 0 & 0 & 2 \end{pmatrix}$

e. $\begin{pmatrix} -2 & 1 & -2 \\ 0 & -1 & -6 \\ 0 & 1 & 4 \end{pmatrix}$

f. $\begin{pmatrix} 1 & 1 & 1 \\ 2 & 1 & -1 \\ -3 & 2 & 4 \end{pmatrix}$

4. Suppose two monic polynomials $f(x) \neq g(x)$ of the same degree d annihilate A. Show that there is a monic polynomial of smaller degree that also annihilates A. Thus the minimal polynomial is unique. (**HINT:** What is the degree of $f(x) - g(x)$?)

5. Use Problem 4 to show that $\mu_A(x)$ is a factor of any polynomial $f(x)$ that annihilates A.

6. Show that any monic polynomial that annihilates A also annihilates its transpose A^T. Conclude that $\mu_A(x) = \mu_{A^T}(x)$.

7. Suppose that A and B are similar (similar is defined on page 107 to mean $A = P^{-1}BP$ for some invertible matrix P). Show that any monic polynomial that annihilates A also annihilates B. Conclude that $\mu_A(x) = \mu_B(x)$.

3.3   Linear recurrence relations

Objective 1

Given a digraph $\mathcal{D}$ and a specified walk type, let w_n be the number of walks of length n in $\mathcal{D}$ and of the specified type. Compute the initial values and a recurrence relation satisfied by the sequence w_n.

We interrupt the development of "eigenlife" to give a nifty application of the minimal polynomial. This section also introduces recursion, a basic tool in Chapter 4.

One of the basic branches of combinatorics is enumeration. Although ancient in origin, it is an important tool in the design and analysis of both algorithms and large discrete models (e.g., in quantum mechanics and with large integrated circuits).

In a typical enumeration problem, we are given a family $S_1, S_2, \ldots$ of sets of increasing size that are specified in a uniform, but perhaps intricate, way. One objective is to give a simple formula for the size f_k of S_k.

The nature of this formula may vary. Sometimes the formulas are so complicated that they just change one hard problem into another hard problem. The easier such a formula is to evaluate, the better it is. A common type of formula is called a **recurrence relation**. In a recurrence relation the formula for f_k is allowed to involve earlier values in the **sequence** $\{f_k\}$.

Example 3.7 (Recursion.)
  a. There is a famous formula for the sum f_k of the first k odd integers.
$$f_k = k^2.$$
Please check that this formula really works for $k = 1, 2, 3$. This is a "closed form" formula. It requires only k to compute f_k. Everyone agrees that this is the best kind of formula — provided it is reasonably simple.

To express the numbers f_k in terms of a recurrence relation, just notice that the $k + 1$-th odd integer is $2k + 1$. Therefore,
$$f_{k+1} = f_k + (2k + 1).$$
This formula is a "first order recurrence relation" because it requires k AND the previous value in the sequence $\{f_n\}$ to compute the next term.

Recurrence relations are proved by **recursion** (also called mathematical induction). The idea is to **first** verify initial cases of the formula. Then **second** show the $(k+1)$-th instance of the formula **presuming the formula to be correct for all earlier instances**.

In this example, this second proof step means that one must show $f_{k+1} = (k+1)^2$, BUT one may use the formula $f_k = k^2$ in the proof process. Because of the recurrence relation, this is a truly simple matter:

$$f_{k+1} = f_k + (2k + 1) = k^2 + (2k + 1) = (k + 1)^2.$$

Many people have trouble with this really simple idea because it employs sophisticated mathematical notation.

Think of a "staircase to the sky." On the k-th step is the formula $f_k = \ldots$. Your job is to climb the staircase and find the first step where the formula is wrong. I claim that your job is hopeless if the following two things happen.

FIRST The formula on the first step is right.

NEXT Never are there two successive steps with the lower
formula correct and the upper formula incorrect.

The sophisticated mathematical notation uses one variable k to say:
"your job is hopeless" $\Leftrightarrow$ f_k is correct for every k.

AND another variable with, for the sake of confusion, the same name to say:

"NEXT" $\Leftrightarrow$ Whenever the k-th step is on my side, the $k + 1$-th step is too.

The "real" mathematician is not confused by this. She realizes these instances of k are separately quantified. It is just a mathematical version of using the same word "jerk" to refer to several different guys in the same paragraph — what's wrong with that?

b. Here is an ancient enumeration problem that was posed by Fibonacci in 1202: Each month, beginning after she is two months old, the female of a pair of rabbits gives birth to a pair of rabbits (of different sexes). Assuming no mortality, how many rabbit pairs will there be at the end of the n-th month?

The Fibonacci numbers F_n count the number of rabbit pairs **after** n months. By inspection, $F_0 = 0, F_1 = 1, F_2 = 1, F_3 = 2, F_4 = 3 \ldots$. Notice that in the n-th month, each rabbit pair is either immature and born the previous month, or mature. The number of immature pairs is F_{n-2} and the number of mature pairs is F_{n-1}. Therefore, the Fibonacci numbers satisfy the recurrence relation

$$F_n = F_{n-1} + F_{n-2}.$$

This is a **second order** recurrence relation because it requires the previous **two** values in the sequence $\{F_n\}$ to compute "the next" term.

Problem 6 outlines a proof of the closed form formula for the Fibonacci numbers

$$F_n = \frac{(1 + \sqrt{5})^n - (1 - \sqrt{5})^n}{2^n \sqrt{5}}.$$

Now really, would you rather work out the Fibonacci numbers using this formula or the recurrence relation? □

A recurrence relation on the sequence $\{w_k\}$ that gives w_k as a linear combination of the preceding n terms is called a **n-*th* order *linear*** recurrence relation. Because F_k is given as a linear combination of the preceding two values, the Fibonacci numbers satisfy a second order linear recurrence relation.

It is important to realize that a recurrence relation alone **does not** determine the sequence $\{w_k\}$. The first few values of the sequence $w_1, \ldots, w_{n-1}$ are also required. In practice these initial values are found by inspection of the first few cases of the problem or are given in advance.

Because of our work with digraph adjacency matrices in Section 2.2, there is a type of enumeration problem that we already know how to solve.

> The i, j-entry of the k-th power of the adjacency matrix $A(\mathcal{D})_{ij}^k$ counts the number of walks, w_k, in $\mathcal{D}$ of length k from i to j.

A recurrence relation on the sequence $\{w_k\}$ can be obtained from the **minimal polynomial** of $A(\mathcal{D})$. The idea is to express the k-th power $A(\mathcal{D})^k$ of the adjacency matrix as a linear combination of lower powers. Suppose $A(\mathcal{D})$ has minimal polynomial $\mu(x) = x^n + f_{n-1}x^{n-1} + \cdots + f_0$. Then the equation $\mu(A) = 0$ can be rewritten:

$$A^n = -f_{n-1}A^{n-1} - \cdots - f_1 A + -f_0.$$

Now multiply both sides by A^{k-n} and simplify

$$A^{k-n}A^n = -f_{n-1}A^{k-n}A^{n-1} - \cdots - f_1 A^{k-n}A + -f_0 A^{k-n}$$
$$A^k = -f_{n-1}A^{k-1} - \cdots - f_1 A^{k-n+1} - f_0 A^{k-n}.$$

Finally take the i, j-entry of both sides and remember that $w_k = A(\mathcal{D})_{ij}^k$ to obtain the n-th order linear recurrence relation:

$$w_k = -f_{n-1}w_{k-1} - \cdots f_1 w_{k-n+1} - f_0 w_{k-n}.$$

Example 3.8 (Recurrence relations on digraph walks.)

a. Let w_n be the number of closed walks starting and ending at vertex 2 in the indicated digraph. The adjacency matrix of this digraph is the matrix A. As shown in Section 2.2, w_k is the $2, 2$ entry of A^k. It is a simple matter to compute the minimal polynomial $\mu_A(x) = x^2 - x - 1$ using the methods of the last section.

$$A = \begin{pmatrix} 0 & 1 \\ 1 & 1 \end{pmatrix}$$

Therefore $A^2 - A - I = 0$, or equivalently, $A^2 = A + I$ and just as above,

$$A^k = A^{k-2}A^2 = A^{k-2}(A + I) = A^{k-1} + A^{k-2}.$$

(The middle step uses $A^2 = A + I$!) By taking the 22 entry of both sides of this matrix equation, we obtain the recurrence relation:

$$w_k = w_{k-1} + w_{k-2}, \; k \geq 3.$$

Notice also that none of the above computations has used the fact that $\{w_n\}$ counts closed walks starting (and ending) at vertex 2. It would have been equally appropriate for counting the total number of walks of length n. We are not finished until we determine initial values of $\{w_n\}$ to "prime the pump" and get the recurrence relation working.

We generally do this by "inspection." Because there is a single loop at vertex 2, $w_1 = 1$. There are two ways to build a walk of length two from 2 to 2 ($2 \to 1 \to 2$ or looping $2 \to 2 \to 2$). Therefore $w_2 = 2$. Now the pump is primed and the recursion relation tells us the rest: $w_3 = w_2 + w_1 = 2 + 1 = 3$, $w_4 = w_3 + w_3 = 3 + 2 = 5$, etc.

This sequence is exactly the Fibonacci sequence!

b Let w_n be the number of walks starting at vertex i and ending at vertex j in the digraph that appears in part c of Example 3.2 on page 112. In view of that example, we know that the digraph's adjacency matrix,

$$A = \begin{pmatrix} 0 & 1 & 0 & 0 \\ 0 & 0 & 1 & 1 \\ 0 & 0 & 0 & 1 \\ 1 & 1 & 0 & 0 \end{pmatrix}$$

has four eigenvalues and has minimal polynomial

$$\mu_A(x) = (x^2 + x + 1)(x^2 - x - 1) = x^4 - x^2 - 2x - 1,$$

so the numbers w_k satisfy the fourth-order linear recurrence relation

$$w_k = w_{k-2} + 2w_{k-3} + w_{k-4}, \text{ for } k \geq 5$$

independent of the starting and ending vertices i, j; BUT different i and j correspond to different initial values. For example,

if $i = 1$ and $j = 2$ then

$w_1 = 1$ $(1 \to 2)$,

$w_2 = 0$,

$w_3 = 1$ $(1 \to 2 \to 4 \to 2)$,

$w_4 = 1$ $(1 \to 2 \to 3 \to 4 \to 2)$;

but if $i = 2 = j$ then, by inspection

$w_1 = 0$ (there is no loop at vertex 2),

$w_2 = 1$ $(2 \to 4 \to 2)$,

$w_3 = 1$ $(2 \to 3 \to 4 \to 2)$,

$w_4 = 2$ $(2 \to 3 \to 4 \to 1 \to 2$ or $2 \to 4 \to 2 \to 4 \to 2)$.

There are lots of variations, all with the same recurrence relation but with different initial conditions. If we wanted all possible starts and all possible endings, we would simply add up all entries in A^k to compute the initial values. If we wanted to count closed walks, we would just count diagonal entries of A^k for the initial values. □

When a counting problem is "modeled" as walks in a digraph in this way, the adjacency matrix is called a ***transfer matrix***. To illustrate the power of this method, we take up a more complicated class of counting problems.

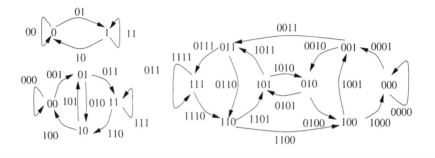

Objective 2

Let b_n be the number of binary sequences of length n having no sublists in a given list L and starting and ending in specified ways. Determine the initial values of b_n and a recurrence relation satisfied by $\{b_n\}$ using a transfer matrix.

A ***binary sequence of length*** n is a list of length n all of whose members are 0's and 1's. There are eight binary sequences of length 3:

$$000 \quad 001 \quad 010 \quad 011 \quad 100 \quad 101 \quad 110 \quad 111.$$

This is not quite the same as counting in binary. For example, binary 1 is not usually written with the string 001 . A ***sublist*** is a sequence of **successive** terms. For example, the sequence 11 is not a sublist of the of the 6th binary sequence (101) listed above, but it does appear as a sublist of the 4th (011), 7th (110) and 8th (111).

Imagine a binary sequence moving across a video screen (like the stock market ticker tape on CNN). Each time a bit appears on the right, one falls out of view on the left. The binary sequence 0001011100 could be built up with the screen shot sequence

$$00 \quad 00 \quad 01 \quad 10 \quad 01 \quad 11 \quad 11 \quad 10 \quad 00.$$

The k-th ***DeBruijn graph*** is a digraph having as vertices the binary sequences of length k, and arcs labeled by binary sequences of length $k+1$. The left-most k terms

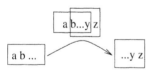

of an arc-label match its tail vertex and the right-most k terms of the arc-label match its tip vertex as in the adjacent figure. The first three DeBruijn graphs are at the top of this page. Notice that every vertex has two arcs entering and two arcs leaving.

If k bits are visible on the video screen, then the sequence of "screen shots" corresponds to a sequence of binary sequences of length k AND **successive screen shots label adjacent vertices in the k-th DeBruijn graph**. Put another way, the whole ticker-tape sequence corresponds to a walk in the k-th DeBruijn graph from the vertex labeling the first screen shot to the vertex labeling the last screen shot. Use the above sequence 8 of two bit lists to map out a walk in the second DeBruijn graph. Notice that it is an optimal scenic tour — traversing each edge exactly once.

In this objective we must count the number of binary sequences of length n that **DO NOT** have any sublists from a specified list L. The key idea is:

> To insure that the forbidden sublists do not appear in the sequence, we work with a DeBruijn graph from which the ARCS labeled by lists containing the forbidden sublists have been removed.

For example, suppose the binary sequences we are to count must start with 1 and end with 1 and never have two consecutive 0's . We simply take the first DeBruijn graph and remove the loop labeled 00. The resulting graph looks remarkably like the graph appearing in Part 1 of Example 3.10. In fact, these specifications correspond exactly to the counting problem worked out in that example.

The second part of Example 3.8 on page 131 also corresponds to a binary sequence counting problem. It is essentially a second DeBruijn graph that has had its loops removed. For this reason, it "counts" binary sequences that have no three like symbols in a row (i.e, have no 000 or 111 sublists).

Example 3.9 (Counting binary sequences.)

a. Let b_n be the number of binary sequences of length n that start 10... and end ...01 and have no sublists of the form 101 or 010. Give initial values and a recurrence relation for $\{b_n\}$.

Let's begin with initial values. We have to do this by inspection. The first sequences that actually meet the restrictions are 1001, then 10001 and 100001. Therefore,
$$b_2 = 0, b_3 = 0, b_4 = 1, b_5 = 1 \text{ and } b_6 = 1.$$

Both of the forbidden sublists have length 3, so we will use the (second) DeBruijn graph because its arcs are labeled by 3 bits. However, a walk of length 1 in this graph corresponds to a binary sequence of length 3, **and shorter binary sequences do not correspond to anything in it.** This is the reason we must find initial values by inspection.

To build the transfer matrix **all we do is remove the arcs in it that correspond to the forbidden sublists**. The resulting digraph is:

The matrix A is adjacency matrix of the digraph. This matrix has minimal polynomial $\mu_A(x) = x^4 - 2x^3 + x^2 - 1$. Therefore, the sequence satisfies the fourth-order recurrence relation:
$$b_k = 2b_{k-1} - b_{k-2} + b_{k-4}.$$

Because we have the initial values, the "pump is primed" and we have

$$A = \begin{pmatrix} 1 & 1 & 0 & 0 \\ 0 & 0 & 0 & 1 \\ 1 & 0 & 0 & 0 \\ 0 & 0 & 1 & 1 \end{pmatrix}$$

$$b_7 = 2b_6 - b_5 + b_3 = 2 - 1 + 0 = 1.$$

b. Let c_n be the number of binary sequences of length n that start and end with the same 3 bits and have no sublists of the form 101 or 0000. Give initial values and a recurrence relation for $\{c_n\}$. n01

The forbidden sublists have lengths 3 and 4. Make a new list L' that consists of all lists of length 4 that contain any of the original list.

$L' = \{1010, 1011, 0101, 1101, 0000\}$. Now work with the **third** DeBruijn graph with these arcs removed: The resulting digraph and adjacency matrix are

$$A = \begin{pmatrix} 0 & 1 & 0 & 0 & 0 & 0 & 0 & 0 \\ 0 & 0 & 1 & 1 & 0 & 0 & 0 & 0 \\ 0 & 0 & 0 & 0 & 1 & 0 & 0 & 0 \\ 0 & 0 & 0 & 0 & 0 & 0 & 1 & 1 \\ 1 & 1 & 0 & 0 & 0 & 0 & 0 & 0 \\ 0 & 0 & 0 & 0 & 0 & 0 & 0 & 0 \\ 0 & 0 & 0 & 0 & 1 & 0 & 0 & 0 \\ 0 & 0 & 0 & 0 & 0 & 0 & 1 & 1 \end{pmatrix}.$$

This matrix has minimal polynomial $\mu_A(x) = x^6 - x^5 - x^3 - x^2$. Therefore, the sequence $\{c_n\}$ satisfies the recurrence relation:

$$c_k = c_{k-1} + c_{k-3} + c_{k-4},$$

and six initial values must be specified because $\mu_A(x)$ has degree 6.

The first few sequences of length at least 4 that meet these restrictions are 1111, 11111, and 111111, 001001, 010010, 100100, so $c_3 = 7, c_4 = 1, c_5 = 1,$ $c_6 = 4$. But pretty soon it becomes easier to count closed walks in the digraph $c_7 = 9, c_8 = 11, c_9 = 16....$ □

The order of the recurrence relation satisfied by a sequence is a natural measure of the complexity of a counting problem. This measure is bounded by the degree of the minimal polynomial of an appropriate associated transfer matrix.

SUMMARY

Recurrence relations solve *enumeration* problems recursively by giving a formula for an instance of the problem in terms of earlier instances. A k-th *order linear* recurrence relation for the sequence $f_1 f_2 \ldots$ expresses f_{n+1} as a linear combination of $f_n \ldots f_{n-k}$.

If an enumeration problem is modeled by a *transfer matrix* A, then the enumerated sequence satisfies a linear recurrence relation of order equal to the degree of $\mu_A(x)$. Thus, the degree of the minimal polynomial of A is a natural indicator of the complexity of the situation under study, because it is the order of the recurrence relation that arises.

The k-th *DeBruijn* graph is a digraph with vertices labeled by k-bit binary sequences and edges labeled by $k + 1$-bit binary sequences. We build transfer matrix models for counting binary sequences without certain forbidden *sublists* by selectively removing edges from DeBruijn graphs.

Looking further

The key insight of DeBruijn graphs is that you can build "memory" into a transfer matrix model by making the underlying digraph more complicated. The role of 2-by-2 matrices in the final objective of chapter 2 was somewhat analogous. This is a very powerful idea that appears in the study of finite state machines and many other places.

* * * * * * *

Practice Problems

1. Write the adjacency matrix for the indicated digraph and give a recurrence relation, together with initial conditions, for the sequence $\{w_n\}$ of closed walks of length n. Determine $w_i, i = 1, 2, 3, 4, 5$ by inspection, then compute w_6 and w_7 from the recurrence relation.

2. Let b_n be the number of binary sequences of length n that start with 00 and end with 00 but have no sublists in the indicated list L. Write the transfer matrix model for $\{b_n\}$. Find a recurrence relation for $\{b_n\}$ and give the first 4 values b_2, b_3, b_4, b_5. Use the recurrence relation to compute the next two values in the sequence.

   a. $L = \{010, 101\}$ b. $L = \{011, 110\}$

3. Same as 2, but this time no restrictions on walk starting or ending points.

   (**Hint:** sum all entries in the matrix.)

   a. $L = \{000, 111\}$ b. $L = \{011, 110\}$

4. Prove by recursion that $1 + \cdots + n^3 = (n(n+1)/2)^2$.

5. Let A be the adjacency matrix in Example 3.8 a.

   Show that $tr(A^k) = F_{k+1} + F_{k-1}, k \geq 2$ where F_k is the k-th Fibonacci number defined in Example 3.7 b.

   Hint: count closed walks starting and ending at vertex 1.

6. Let A be the adjacency matrix in Example 3.8 as above. Show that the eigenvalues λ_1, λ_2 of A are $\frac{1 \pm \sqrt{5}}{2}$. Then use Problem 11 of Section 3.1 to show that $tr(A^k) = \lambda_1^k + \lambda_2^k$. Finally, use the preceding problem to give a recursive proof of the formula for F_k that appears in Example 3.7.

7. The figure hints at a geometric proof of the result in part a of Example 3.7. Can you fill in the details? Can you find a geometric proof of the result in problem 4 above?

3.4   Properties of the minimal polynomial

Objective 1

Given an n-by-n matrix $(n \leq 3)$, write $AP = PD$, where D is diagonal and P has columns the standard list of basic eigenvectors of A.

The main purpose of this section is to support our most important claims about minimal polynomials. These arguments lead naturally to what we call the **components** of a square matrix. Components provide a way to "approximate" a matrix that is the basis of many powerful applications.

This objective refers to a simple, but very important, matrix equation that summarizes the properties of eigenvalues and eigenvectors (and appeared already in the Problem 6 of Section 3.1). It is a little more than a summary of earlier objectives. First, compute the minimal polynomial of A. Next, search for zeros of the minimal polynomial. Third, find the basic eigenvectors associated with each eigenvalue. The final step is to put it all together in THE equation — which we now describe.

Recall that geometric multiplicity of an eigenvalue is the number of associated basic eigenvectors. Therefore, the total number of basic eigenvectors, t, is the sum of the eigenvalue geometric multiplicities. Make two lists the first of basic eigenvectors and the second of eigenvalues (repeated according to geometric multiplicity):

$$\text{eigenvectors:} \qquad \mathbf{p}_1 \quad \mathbf{p}_2 \quad \cdots \quad \mathbf{p}_t$$
$$\text{eigenvalues:} \qquad \lambda_1 \leq \lambda_2 \leq \cdots \leq \lambda_t.$$

Then form a new matrix P, called the **basic diagonalizing matrix** for A, whose columns are $\mathbf{p}_1 \ \mathbf{p}_2 \ldots \mathbf{p}_t$ (in order) and a second new t-by-t matrix D that is diagonal with λ_i in the (i, i) position. Now compute AP using the fact that the columns of P are eigenvectors:

$$AP = A \left(\mathbf{p}_1 \ \mathbf{p}_2 \ \cdots \ \mathbf{p}_t \right) = \left(\lambda_1 \mathbf{p}_1 \ \lambda_2 \mathbf{p}_2 \ \cdots \ \lambda_t \mathbf{p}_t \right).$$

This result is just the same as PD! There you have THE equation $AP = PD$.

If $t = n$ (the number of basic eigenvectors for A equals the size of A), then P is square and P is invertible. In this case A is **diagonalizable**.

It is important to realize that THE equation $AP = PD$ still works if P is rectangular $(t < n)$! It just doesn't tell us as much about A.

WARNING: We generally use the matrix P whose columns are the list of basic eigenvectors of A, BUT there are other matrices of the same shape and rank as P for which $AP = PD$ with the same D. Investigating this non-uniqueness is just the kind of thing we are unprepared for, and that occurs in more advanced classes. **Read carefully when asked to find such a P to see what is required.**

Example 3.10 (The whole enchilada.)

a. **The first step is to find the minimal polynomial.** Let $A = \begin{pmatrix} 1 & 4 \\ 4 & 1 \end{pmatrix}$ and work through the minimal polynomial algorithm, taking the first "random vector" to be $\mathbf{r} = (1\ 0)^T$ and, as always, $W_1 = I_3$. The first auxiliary matrix is

$$\begin{pmatrix} 1 & 4 \\ 4 & 1 \end{pmatrix} \Bigg|\Bigg| \begin{pmatrix} 1 & 1 & 17 \\ 0 & 4 & 8 \end{pmatrix} \rightarrow \begin{pmatrix} 1 & 0 & 15 \\ 0 & 1 & 2 \end{pmatrix},$$

so the A-annihilator of $\mathbf{v}_1 = (1\ 0)^T$ is $f_1(x) = -15-2x + x^2$.
Now $W_2 = f_1(A)W_1 = \cdots = 0$, so only one pass was necessary, and we conclude $\mu_A(x) = f_1(x) = -15-2x + x^2 = (x + 3)(x-5)$.

The second step is to find the eigenvalues of A.
The eigenvalues of A are the zeros of $\mu_A(x)$, namely $-3, 5$.

The third step is to find the associated basic eigenvectors.

$$\lambda = -3: \quad A + 3I = \begin{pmatrix} 4 & 4 \\ 4 & 4 \end{pmatrix} \rightarrow \begin{pmatrix} 1 & 1 \\ 0 & 0 \end{pmatrix}, \text{ basic eigenvector: } \begin{pmatrix} -1 \\ 1 \end{pmatrix}.$$

$$\lambda = 5: \quad A-5I = \begin{pmatrix} -4 & 4 \\ 4 & -4 \end{pmatrix} \rightarrow \begin{pmatrix} 1 & -1 \\ 0 & 0 \end{pmatrix}, \text{ basic eigenvector: } \begin{pmatrix} 1 \\ 1 \end{pmatrix}.$$

The final step is to build the matrix P and check $AP = PD$.
The columns of P are the basic eigenvectors of A — **in order of increasing eigenvalues**.

$$AP = \begin{pmatrix} 1 & 4 \\ 4 & 1 \end{pmatrix}\begin{pmatrix} -1 & 1 \\ 1 & 1 \end{pmatrix} = \begin{pmatrix} 3 & 5 \\ -3 & 5 \end{pmatrix} = \begin{pmatrix} -1 & 1 \\ 1 & 1 \end{pmatrix}\begin{pmatrix} -3 & 0 \\ 0 & 5 \end{pmatrix} = PD.$$

b. Let B be the indicated matrix.

The first step is to find the minimal polynomial. $\quad B = \begin{pmatrix} 0 & -2 & -2 \\ 0 & 1 & -2 \\ 1 & 2 & 5 \end{pmatrix}$
If the initial "random" vector $\mathbf{r} = (1\ 0\ 0)^T$,
then the first auxiliary matrix is:

$$\begin{pmatrix} 1 & 0 & -2 & -6 \\ 0 & 0 & -2 & -12 \\ 0 & 1 & 5 & 19 \end{pmatrix} \rightarrow \begin{pmatrix} 1 & 0 & 0 & 6 \\ 0 & 1 & 0 & -11 \\ 0 & 0 & 1 & 6 \end{pmatrix}.$$

The B-annihilator of $(1\ 0\ 0)^T$ is $f_1(x) = -6 + 11x-6x^2 + x^3$.
Now $W_2 = f_1(A)I_3 = \cdots = 0$, so we conclude that

$$\mu_B(x) = f_1(x) = x^3-6x^2 + 11x-6 = (x-1)(x-2)(x-3).$$

The second step is to find the eigenvalues of B.
The eigenvalues of B are $\lambda_1 = 1, \lambda_2 = 2$ and $\lambda_3 = 3$.

The third step is to find the associated basic eigenvectors.

Diagonalizable matrices.

$$\lambda = 1: \quad B\text{-}I = \begin{pmatrix} -1 & -2 & -2 \\ 0 & 0 & -2 \\ 1 & 2 & 4 \end{pmatrix} \rightarrow \begin{pmatrix} 1 & 2 & 0 \\ 0 & 0 & 1 \\ 0 & 0 & 0 \end{pmatrix} \qquad \begin{pmatrix} -2 \\ 1 \\ 0 \end{pmatrix}.$$

$$\lambda = 2: \quad B\text{-}2I = \begin{pmatrix} -2 & -2 & -2 \\ 0 & -1 & -2 \\ 1 & 2 & 3 \end{pmatrix} \rightarrow \begin{pmatrix} 1 & 0 & -1 \\ 0 & 1 & 2 \\ 0 & 0 & 0 \end{pmatrix} \qquad \begin{pmatrix} 1 \\ -2 \\ 1 \end{pmatrix}.$$

$$\lambda = 3: \quad B\text{-}3I = \begin{pmatrix} -3 & -2 & -2 \\ 0 & -2 & -2 \\ 1 & 2 & 2 \end{pmatrix} \rightarrow \begin{pmatrix} 1 & 0 & 0 \\ 0 & 1 & 1 \\ 0 & 0 & 0 \end{pmatrix} \qquad \begin{pmatrix} 0 \\ -1 \\ 1 \end{pmatrix}.$$

These vectors form the columns of the matrix P and

$$BP = \begin{pmatrix} 0 & -2 & -2 \\ 0 & 1 & -2 \\ 1 & 2 & 5 \end{pmatrix} \begin{pmatrix} -2 & 1 & 0 \\ 1 & -2 & -1 \\ 0 & 1 & 1 \end{pmatrix} = \begin{pmatrix} -2 & 1 & 0 \\ 1 & -2 & -1 \\ 0 & 1 & 1 \end{pmatrix} \begin{pmatrix} 1 & 0 & 0 \\ 0 & 2 & 0 \\ 0 & 0 & 3 \end{pmatrix} = PD.$$

c. Let C be the indicated matrix.

The first step is to find the minimal polynomial. $\quad C = \begin{pmatrix} -9 & -8 & -3 \\ 17 & 17 & 7 \\ -26 & -28 & -12 \end{pmatrix}$
If the initial "random" vector $\mathbf{r} = (1\ 0\ 0)^T$, then
the first auxiliary matrix is:

$$\begin{pmatrix} 1 & -9 & 23 & -49 \\ 0 & 17 & -46 & 99 \\ 0 & -26 & 70 & -150 \end{pmatrix} \rightarrow \begin{pmatrix} 1 & 0 & 0 & -2 \\ 0 & 1 & 0 & -5 \\ 0 & 0 & 0 & -4 \end{pmatrix}.$$

The C-annihilator of $(1\ 0\ 0)^T$ is $f_1(x) = 2 + 5x + 4x^2 + x^3$.

Because this has the same degree as the size of C, we take the shortcut and
conclude that $\mu_C(x) = f_1(x) = 2 + 5x + 4x^2 + x^3 = (x+1)^2(x+2)$.

**Because $\mu_C(x)$ has a multiple zero, we know that C is NOT diagonaliz-
able. Nonetheless we can still write $AP = PD$.**

The second step is to find the eigenvalues of C.
Since $\mu_C(x) = (x+1)^2(x+2)$, we know that C has two eigenvalues $\lambda_1 = -2$
and $\lambda_2 = -1$ but is **NOT DIAGONALIZABLE.**

The third step is to find the associated basic eigenvectors.

$$\lambda = -2: \quad C + 2I = \begin{pmatrix} -7 & -8 & -3 \\ 17 & 19 & 7 \\ -26 & -28 & -10 \end{pmatrix} \rightarrow \begin{pmatrix} 1 & 0 & -1/3 \\ 0 & 1 & 2/3 \\ 0 & 0 & 0 \end{pmatrix} \qquad \begin{pmatrix} 1/3 \\ -2/3 \\ 1 \end{pmatrix}.$$

$$\lambda = -1: \quad C + I = \begin{pmatrix} -8 & -8 & -3 \\ 17 & 18 & 7 \\ -26 & -28 & -11 \end{pmatrix} \rightarrow \begin{pmatrix} 1 & 0 & -1/4 \\ 0 & 1 & 5/8 \\ 0 & 0 & 0 \end{pmatrix} \qquad \begin{pmatrix} 1/4 \\ -5/8 \\ 1 \end{pmatrix}.$$

Thus,

$$CP = \begin{pmatrix} -9 & -8 & -3 \\ 17 & 17 & 7 \\ -26 & -28 & -12 \end{pmatrix} \begin{pmatrix} 1/3 & 1/4 \\ -2/3 & -5/8 \\ 1 & 1 \end{pmatrix} = \begin{pmatrix} 1/3 & 1/4 \\ -2/3 & -5/8 \\ 1 & 1 \end{pmatrix} \begin{pmatrix} -2 & 0 \\ 0 & -1 \end{pmatrix} = PD. \quad \square$$

Matrices that are not diagonalizable are sometimes called **defective.** The matrix A that appears in part c of Example 3.1 on page 110 is defective because it has only one f_A-invariant line and has only one basic eigenvector. The matrix C in the preceding example is also defective. These matrices illustrate the second fundamental property of the minimal polynomial — that a matrix is diagonalizable if and only if its minimal polynomial has no multiple zeros. Check that $\mu_A(x) = (x - 1)^2$.

Frobenius' Theorem

We have shown that any matrix is equivalent to a unique RREF matrix. It is natural to look for an analogous set of "canonical form" matrices, where every matrix is similar to a unique one in the list. We leave this fundamental question for a more advanced course.[3] We are content to show:

THEOREM Let A be a square matrix with complex entries. Then there is an upper triangular matrix T that is similar to A.

Although this part is an introduction to proof by recursion (also known as induction), the actual result is only used in Chapter 4 to prove the Cayley Hamilton theorem. This part may be omitted on first reading.

Two preliminary results are required.

> **Suppose a nonzero vector v is given. Then there is an invertible matrix P whose first column is v** (see Problem 9 of Section 2.3 on page 76).
> Since $\mathbf{v} \neq 0$ there is some coordinate $v_i \neq 0$. Take the remaining columns of P to be all but the i-th column of the identity matrix.

The second preliminary result is really the heart of the argument.

> A square matrix A is similar to a matrix, we shall call C, that has its first column all zeros except perhaps for the first entry.

> Since the complex numbers are algebraically closed (see Section 2.1), the minimal polynomial $\mu_A(x)$ must have at least one, possibly complex, zero λ. This implies that λ is an eigenvalue of A. Let $\mathbf{v} \neq 0$ be a λ-eigenvector of A. By the first preliminary result, there is an invertible matrix $P = (\mathbf{v}|P')$ whose first column is $\mathbf{v}$. (And whose last $n - 1$ columns will be called P'.)

> Now define $\mathbf{c} = (\lambda\ 0 \cdots\ 0)^T$ to have length n and note that $P\mathbf{c} = (\mathbf{v}|P')(\lambda\ 0 \cdots\ 0)^T = \lambda\mathbf{v}$ because of all those zeros at the end of $\mathbf{c}$.

$$AP = (A\mathbf{v}|AP') = (\lambda\mathbf{v}|AP') = (P\mathbf{c}|PP^{-1}AP') = P(\mathbf{c}|P^{-1}AP').$$

> Left multiply by P^{-1} to get $P^{-1}AP = (\mathbf{c}|P^{-1}AP') = C$, as advertised.

The proof of Frobenius' theorem is by recursion (or induction) on the size of A. There is nothing to show if A is 1-by-1, so the initial case is true. Notice that when A is a 2-by-2 matrix, the matrix C produced by the second preliminary result is upper triangular, and so it alone implies Frobenius' theorem for 2-by-2 matrices.

---

[3]"Looking further," at the end of this section, gives one possible answer.

The general case requires more elaborate notation than we have. The usual response to this dilemma is to omit many details and rely on the reader to fill in the blanks. Because this is our first induction argument, we will instead work through the next case, then sketch the general case changing only the bare minimum of text.

Here is how to show that Frobenius' theorem holds for 3-by-3 matrices, using that it holds for all 2-by-2 matrices.

Suppose A is 3-by-3. By the second preliminary result there is an invertible matrix P such that:

$$P^{-1}AP = \begin{pmatrix} \lambda_1 & c_{12} & c_{13} \\ 0 & c_{22} & c_{23} \\ 0 & c_{32} & c_{33} \end{pmatrix}.$$

Because the submatrices have smaller size than A we know that there is some number * (whose exact value is unimportant) and a matrix Q such that:

$$Q^{-1}\begin{pmatrix} c_{22} & c_{23} \\ c_{32} & c_{33} \end{pmatrix}Q = \begin{pmatrix} \lambda_2 & * \\ 0 & \lambda_3 \end{pmatrix}.$$

Now define R as indicated and notice that R is invertible with its inverse having the same structure but with bottom right corner Q^{-1}.

$$R = \begin{pmatrix} 1 & 0 & 0 \\ 0 & c_{22} & c_{23} \\ 0 & c_{32} & c_{33} \end{pmatrix}.$$

Finally consider $(PR)^{-1}A(PR) = R^{-1}(P^{-1}AP)R$. Using the above formulas for $P^{-1}AP$ and R, this matrix can be written and actually multiplied out to be:

$$\begin{pmatrix} 1 & 0 & 0 \\ 0 & q_{11} & q_{12} \\ 0 & q_{21} & q_{22} \end{pmatrix}^{-1}\begin{pmatrix} \lambda_1 & c_{12} & c_{13} \\ 0 & c_{22} & c_{23} \\ 0 & c_{32} & c_{33} \end{pmatrix}\begin{pmatrix} 1 & 0 & 0 \\ 0 & q_{11} & q_{12} \\ 0 & q_{21} & q_{22} \end{pmatrix} = \begin{pmatrix} \lambda_1 & * & * \\ 0 & & \\ 0 & Q^{-1}\begin{pmatrix} c_{22} & c_{23} \\ c_{32} & c_{33} \end{pmatrix}Q \end{pmatrix}.$$

Here we have again used * for some (possibly new) number whose exact values are unimportant.

Combining the last equation with the formula for $Q^{-1}\begin{pmatrix} c_{22} & c_{23} \\ c_{32} & c_{33} \end{pmatrix}Q$ gives the desired result:

$$(PR)^{-1}A(PR) = \begin{pmatrix} \lambda_1 & * & * \\ 0 & \lambda_2 & * \\ 0 & 0 & \lambda_3 \end{pmatrix}.$$

Here is the general case:
Suppose A is an n-by-n matrix. By the second preliminary result there is an invertible matrix P such that:

$$P^{-1}AP = \begin{pmatrix} \lambda_1 & * \cdots * \\ \vdots & C \\ 0 & \end{pmatrix}.$$

Because the submatrices have smaller size than A, we know that Frobenius' theorem holds for C and there is a matrix Q such that:

$$Q^{-1}CQ = \begin{pmatrix} \lambda_2 & * & * \\ \vdots & \ddots & * \\ 0 & \cdots & \lambda_n \end{pmatrix}.$$

Now define R as indicated and notice that R is invertible with its inverse having the same structure but with bottom right corner Q^{-1}.

$$R = \begin{pmatrix} 1 & 0 \ldots & 0 \\ \vdots & Q & \\ 0 & & \end{pmatrix}.$$

Finally consider $(PR)^{-1}A(PR) = R^{-1}(P^{-1}AP)R$. Using the above formulas for $P^{-1}AP$ and R, this matrix can be written and actually multiplied out to be:

$$\begin{pmatrix} 1 & 0\cdots0 \\ 0 & \\ \vdots & Q^{-1} \\ 0 & \end{pmatrix} \begin{pmatrix} \lambda_1 & *\cdots* \\ 0 & \\ \vdots & C \\ 0 & \end{pmatrix} \begin{pmatrix} 1 & 0\cdots0 \\ 0 & \\ \vdots & Q \\ 0 & \end{pmatrix} = \begin{pmatrix} \lambda_1 & *\cdots* \\ 0 & \\ \vdots & Q^{-1}CC \\ 0 & \end{pmatrix}.$$

Combining the last two equations gives the desired result:

$$(PR)^{-1}A(PR) = \begin{pmatrix} \lambda_1 & * & * & * \\ 0 & \lambda_2 & * & * \\ \vdots & 0 & \ddots & * \\ 0 & \cdots & 0 & \lambda_n \end{pmatrix}.$$

Objective 2

Given a diagonalizable matrix A and its minimal polynomial,
write A as a sum of its components.

We now work with polynomials in the matrix A (linear combinations of powers of A). There are both practical and theoretical reasons for this study. Suppose A has minimal polynomial,

$$\mu_A(x) = (x - \lambda_1)^{m_1} \ldots (x - \lambda_t)^{m_t}, \; \lambda_i \neq \lambda_j \text{ for } i \neq j.$$

Then there exist polynomials in A, $A_{\lambda_1} \ldots A_{\lambda_t}$, such that

$$(A - \lambda_i I)^{m_i} A_{\lambda_i} = 0 \text{ for each } i \text{ and } A = A_{\lambda_1} + \cdots + A_{\lambda_t}.$$

The matrices A_λ are called **components** of A. The actual polynomials that give rise to the components of A are not obtained by a simple formula when some zero λ_i of $\mu_A(x)$ has multiplicity $m_i > 1$. These polynomials, however, can be obtained from the Euclidean algorithm discussed in Section 2.7 as shown in Problem 8.

We shall see, at the bottom of page 142, that $\mu_A(x)$ has multiple zeros only if A is not diagonalizable. Since this chapter is primarily about diagonalizable matrices, we focus on the case when $\mu_A(x)$ has no multiple zeros. In this case the fundamental Lagrange interpolants from Section 2.3 are just what the doctor ordered.

Components

Suppose the matrix A has minimal polynomial $\mu_A(x) = (x - \lambda_1)\ldots(x - \lambda_t)$ (with no multiple zeros). Write $\mu_A(x) = (x - \lambda_1)g_1(x)$ and consider the Lagrange interpolant (see page 74),

$$\Lambda_1(x) = \frac{g_1(x)}{g_1(\lambda_1)} = \frac{(x - \lambda_2)\ldots(x - \lambda_t)}{(\lambda_1 - \lambda_2)\ldots(\lambda_1 - \lambda_t)}.$$

Just as in Section 2.3, $\Lambda_1(x)$ takes the value 1 when $x = \lambda_1$ and $\Lambda_1(\lambda_i) = 0$ for $i \neq 1$. Also observe that it has degree $t - 1$ which is less than the degree t of $\mu_A(x)$. Therefore, as above, $\Lambda_1(A) \neq 0$ but

$$A\Lambda_1(A) - \lambda_1 \Lambda_1(A) = (A - \lambda_1 I)\Lambda_1(A) = (A - \lambda_1 I)g_1(A)/g_1(\lambda_1)$$
$$= \mu_A(A)/g_1(\lambda_1) = 0.$$

This last equation implies that $A\Lambda_1(A) = \lambda_1 \Lambda_1(A)$.

Similarly, let $\Lambda_j(x)$ be the j fundamental Lagrange interpolant for $\lambda_1 \ldots \lambda_t$, and consider

$$f(x) = \Lambda_1(x) + \cdots + \lambda_t(x) - 1.$$

It has degree at most $t-1$, because each $\lambda_i(x)$ has degree $t-1$; BUT it takes the value 0 at EACH of the t distinct eigenvalues λ_i, $= 1 \ldots t$. Therefore, $f(x) = a(x-\lambda_1) \ldots (x-\lambda_t)$, for some polynomial a. The only way such an expression can have degree less than t is for $a = 0$! Therefore $f(x) = \Lambda(x) - 1 = 0$ and $\Lambda(x) = 1$ is constant. (Compare to Problem 11 of Section 2.3.)

It follows that

$$I = \Lambda(A) = \Lambda_1(A) + \Lambda_2(A) + \cdots + \Lambda_t(A).$$

Now multiply the last equation by A and use the formula $A\Lambda_i(A) = \lambda_i\Lambda_i(A)$ (established above for only $i = 1$) to obtain:

$$A = A\Lambda_1(A) + A\Lambda_2(A) + \ldots A\Lambda_t(A) = \lambda_1\Lambda_1(A) + \cdots + \lambda_t\Lambda_t(A).$$

We have obtained the formula for the components of A:

$$A_\lambda = \frac{\lambda g(A)}{g(\lambda)} \quad \text{where } \mu_A(x) = (x-\lambda)g(x).$$

Notice that any nonzero column of A_λ is a λ-eigenvector of A, by the last equation on the preceding page. This gives an **alternative method to compute eigenvectors** that does not require solving a linear system! This method does not necessarily give basic eigenvectors, and works best when λ has geometric multiplicity 1.

Example 3.11 (Matrix components.)

    a. Recall part d of Example 3.6 on page 125, in which we showed that the diagonal matrix D has $\mu_D(x) = (x-1)(x-2)(x-3)$. The components of D are

$$D = \begin{pmatrix} 1 & 0 & 0 \\ 0 & 2 & 0 \\ 0 & 0 & 3 \end{pmatrix} \qquad D_1 = \frac{1(D-2I)(D-3I)}{(1-2)(1-3)} = \begin{pmatrix} 1 & 0 & 0 \\ 0 & 0 & 0 \\ 0 & 0 & 0 \end{pmatrix},$$

$$D_2 = \frac{2(D-1I)(D-3I)}{(2-1)(2-3)} = \begin{pmatrix} 0 & 0 & 0 \\ 0 & 2 & 0 \\ 0 & 0 & 0 \end{pmatrix}, \qquad D_3 = \frac{3(D-1I)(D-2I)}{(3-1)(3-2)} = \begin{pmatrix} 0 & 0 & 0 \\ 0 & 0 & 0 \\ 0 & 0 & 3 \end{pmatrix}.$$

    b. The matrix in part b of Example 3.10 on page 137 was shown to be diagonalizable and similar to D in part a. Because D and B have the same minimal polynomial, their components are given by the same formulas:

$$B = \begin{pmatrix} 0 & -2 & -2 \\ 0 & 1 & -2 \\ 1 & 2 & 5 \end{pmatrix} \qquad B_1 = \frac{1(B-2I)(B-3I)}{(1-2)(1-3)} = \begin{pmatrix} 2 & 2 & 2 \\ -1 & -1 & -1 \\ 0 & 0 & 0 \end{pmatrix},$$

$$B_2 = \frac{2(B-1I)(B-3I)}{(2-1)(2-3)} = \begin{pmatrix} -2 & -4 & -4 \\ 4 & 8 & 8 \\ -2 & -4 & -4 \end{pmatrix}, \qquad B_3 = \frac{3(B-1I)(B-2I)}{(3-1)(3-2)} = \begin{pmatrix} 0 & 0 & 0 \\ -3 & -6 & -9 \\ 3 & 6 & 9 \end{pmatrix}.$$

c. Defective matrices have components too. The matrix C is not diagonalizable because its minimal polynomial is $\mu_C(x) = (x+2)(x+1)^2$.

$$C = \begin{pmatrix} -9 & -8 & -3 \\ 17 & 17 & 7 \\ -26 & -28 & -12 \end{pmatrix} .$$

Nonetheless, the factors of $\mu_A(x)$ are relatively prime, so as in Example 2.23

$$1 = (-x)(x+2) + (1)(x+1)^2.$$

Using this fact, we can obtain formulas for the components of C by a procedure similar to above. Indeed $C = C[(-C)(C+2) + (C+1)^2]$ and has components

$$C_{-1} = -C^2(C+2) = \begin{pmatrix} 3 & 0 & 1 \\ -7 & 1 & 3 \\ 10 & -4 & -6 \end{pmatrix}, \quad C_{-2} = (C+1)^2 = \begin{pmatrix} -12 & -8 & -2 \\ 24 & 16 & 4 \\ -36 & -24 & -6 \end{pmatrix}.$$

Given polynomials, like $x+2$ and $(x+1)^2$, that have no common zeros there is a systematic way of finding a linear combination of them that equals 1 as above. This process is discussed in detail in Objective 3 of Section 2.7. □

We are now ready to prove the fundamental theorem. As always, you may wish to skip the proofs on first reading.

THEOREM Let A be a square matrix. Then:

1. The zeros of $\mu_A(x)$ are exactly the eigenvalues of A.
2. A is diagonalizable if and only if $\mu_A(x)$ has degree equal to the number of distinct eigenvalues of A.

Theorem, part 1.
Suppose that $\mu_A(x) = f(x)g(x)$ is factored as the product of two nonconstant polynomials (so each of $f(x)$ and $g(x)$ have degree greater than 1 and less than the degree of $\mu_A(x)$). Then
$$f(A)g(A) = g(A)f(A) = \mu_A(A) = 0,$$
BUT the matrices $f(A) \neq 0$ and $g(A) \neq 0$ because $\mu_A(x)$ is the LEAST degree monic polynomial annihilating A.

Now suppose $\mu_A(r) = 0$, i.e., that r is a zero of $\mu_A(x)$. Then $\mu_A(x) = g(x)(x - r)$ for some polynomial $g(x)$. As in the preceding paragraph, $(A - rI)g(A) = \mu_A(A) = 0$ and $g(A) \neq 0$. Let $\mathbf{v}$ be a nonzero column of $g(A)$. Then $(A - rI)\mathbf{v} = 0$, $\mathbf{v} \neq 0$, so, by definition, $\mathbf{v}$ is an eigenvector of A and the associated eigenvalue is r. This shows that every zero of $\mu_A(x)$ is an eigenvalue of A.

Suppose $\mathbf{v}$ is a λ-eigenvector of A and let $\mathcal{F}$ be the collection of all polynomials $f(x)$ which satisfy the equation $f(A)\mathbf{v} = f(\lambda)\mathbf{v}$. We claim that $\mathcal{F}$ contains every polynomial (compare to Problem 7 from Section 3.1). Although we could pound this out directly, we will instead use it to illustrate a less pedestrian proof method. However, for the sake of continuity, let us for the moment accept that $f(A)\mathbf{v} = f(\lambda)\mathbf{v}$ for every polynomial $f(x)$ and finish this proof.

Surely then, the minimal polynomial $\mu_A(x)$ of A is in $\mathcal{F}$, and we have:

$$\mu_A(\lambda)\mathbf{v} = \mu_A(A)\mathbf{v} = 0\mathbf{v} = \mathbf{0}.$$

But, being an eigenvector, $\mathbf{v} \neq 0$. It follows that $\mu_A(\lambda) = 0$. Therefore, every eigenvalue of A is a zero of the minimal polynomial $\mu_A(x)$. This completes the proof of the first part of the fundamental theorem.

Now let's return to that less pedestrian proof that every polynomial is in $\mathcal{F}$, the collection of all polynomials that satisfy the equation $f(A)\mathbf{v} = f(\lambda)\mathbf{v}$ for $\mathbf{v}$ a λ-eigenvector of A. Notice that both $f_0(x) = 1$ and $f_1(x) = x$ are in $\mathcal{F}$

$$f_0(A)\mathbf{v} = I\mathbf{v} = 1\mathbf{v} = f_0(\lambda)\mathbf{v} \text{ and } f_1(A)\mathbf{v} = A\mathbf{v} = \lambda\mathbf{v} = f_1(\lambda)\mathbf{v}.$$

Because ANY polynomial can be obtained from these by the operations: scalar multiplication, addition and multiplication, it is enough to check that $\mathcal{F}$ is closed under these operations. Each of these checks comes down to a routine application of matrix "rules of arithmetic." Suppose $c \in R$ and $f(x) \in \mathcal{F}$. Then

$$((cf)(A))\mathbf{v} = c(f(A)\mathbf{v}) = c(f(\lambda)\mathbf{v}) = (cf(\lambda)\mathbf{v}.$$

That middle step uses $f(x) \in \mathcal{F}$, i.e., $f(A)\mathbf{v} = f(\lambda)\mathbf{v}$. Suppose $f(x), g(x) \in \mathcal{F}$. Then $f(A)=f(\lambda)\mathbf{v}$ and $g(A)=g(\lambda)\mathbf{v}$. These facts are used right in the middle of:

$$((f + g)(A))\mathbf{v} = (f(A) + g(A))\mathbf{v} = f(A)\mathbf{v} + g(A)\mathbf{v}$$
$$= f(\lambda)\mathbf{v} + g(\lambda)\mathbf{v} = (f(\lambda) + g(\lambda))\mathbf{v} = ((f + g)(\lambda))\mathbf{v}$$

and
$$((fg)(A))\mathbf{v} = (f(A)g(A))(\mathbf{v}) = f(A)(g(A)\mathbf{v}) = f(A)(g(\lambda)\mathbf{v})$$
$$= f(\lambda)(g(\lambda)(\mathbf{v})) = (f(\lambda)g(\lambda))\mathbf{v} = ((fg)(\lambda))\mathbf{v}.$$

These equations are hard to parse because the meaning of the operations keeps changing and the only clue is the different parentheses groupings. For example, + in the first sequence refers (in order) to polynomials, matrices, vectors, vectors, numbers" and finally polynomials. In a way, that is kind of amazing!

In the next proof another fact about matrix polynomials is required. Suppose A is similar to B and a matrix P such that $P^{-1}AP = B$ is given. Let $\mathcal{G}$ be the collection of all polynomials $g(x)$ for which $g(B) = P^{-1}g(A)P$. Then $\mathcal{G}$ contains every polynomial This result can be proven by exactly the same method as above. A special case is worked out in Example 3.12 on page 147.

Theorem, part 2. (Diagonalizable if and only if no multiple zeros.)
Suppose A is diagonalizable. Then there is a matrix P such that $D = P^{-1}AP$ is diagonal. Since $\mathcal{G}$ consists of all polynomials, it follows that any polynomial annihilating A also annihilates D and vise versa, and therefore $\mu_A(x) = \mu_D(x)$.

It is a simple matter (see part b of Example 3.2 on page 110) to see that the minimal polynomial of a diagonal matrix D is the product of factors $(x - d_{11}) \cdots (x - d_{nn})$ with one term for each **distinct** diagonal entry. Therefore, the minimal polynomial of a diagonalizable matrix has no multiple roots.

Suppose that A is a matrix whose minimal polynomial factors into distinct linear factors. This is exactly the context in which fundamental Lagrange interpol-ants lead to components. As above, write $\mu_A(x) = (x - \lambda_i)g_i(x)$ and define

$$\Lambda_i(x) = \frac{g_i(x)}{g_i(\lambda_i)}, \text{where } i = 1 \dots t. \text{ Then } I = \Lambda_1(A) + \Lambda_2(A) + \cdots + \Lambda_t(A),$$

where every nonzero column of $f_i(A)$ is a λ_i-eigenvector of A. Since every λ_i-eigenvector of A is a linear combination of the basic λ_i-eigenvectors of A, it follows that every column of the identity matrix is a linear combination of basic eigenvectors of A. This means that the matrix P constructed from the standard list of basic eigenvectors of A is an invertible matrix. In other words, A is diagonalizable.

SUMMARY

The list of basic eigenvectors of A form the columns of the **diagonalizing matrix** P, and $AP = PD$ where D is diagonal and has the i-th eigenvalue of A in position (i, i).
A is **diagonalizable** when P is invertible. Otherwise A is **defective**.
Any square matrix is similar to an upper triangular matrix.
The eigenvalues of A really are the zeros of $\mu_A(x)$.
$\mu_A(x)$ has no multiple roots if and only if A is diagonalizable. If A is di-agonalizable and $\mu_A(x) = (x - \lambda)g(x)$, then the associated **component** is obtained from the associated fundamental Lagrange interpolant:

$$A_\lambda = \frac{\lambda g(A)}{g(\lambda)} \text{ satisfies } AA_\lambda = \lambda A_\lambda.$$

A is the sum of its components and each nonzero column of A_λ is a λ-eigenvector of A.

Looking further
Frobenius' theorem can be refined in several ways. One way is that the "triangular-izing" matrix P can be forced to satisfy $\overline{P}P^T = I$ (Schur's theorem). A second way is that the only nonzero entries in T are forced either to be on the diagonal or on the "super diagonal" (positions $t_{i\ i+1}$). The extreme case when all $n - 1$ super diagonal entries are 1's, and all diagonal entries are equal λ, is called a **Jordan block**. You can check that an n-by-n Jordan block has minimal polynomial $(x - \lambda)^n$ and has only one basic eigenvector.

* * * * * * *

Practice Problems

1. For each of the following matrices A, find: (i) the minimal polynomial $\mu(x)$ of A, (ii) the eigenvalues of A, (iii) the standard list of basic eigenvectors of A, (iv) the diagonalizing matrix P. Show $AP = PD$ for a diagonal matrix D and give D. Indicate if the matrix A is defective.

a. $A = \begin{pmatrix} 1 & 1 \\ 4 & -2 \end{pmatrix}$ b. $A = \begin{pmatrix} -3 & 3 & 2 \\ 3 & -3 & -2 \\ 1 & -1 & -6 \end{pmatrix}$ c. $A = \begin{pmatrix} 3 & 1 & 2 \\ 1 & 3 & 2 \\ -4 & 4 & -2 \end{pmatrix}$

d. $A = \begin{pmatrix} 1 & 1 \\ -8 & -5 \end{pmatrix}$ e. $A = \begin{pmatrix} -3 & -2 & 2 \\ -4 & -5 & 4 \\ -10 & -10 & 9 \end{pmatrix}$ f. $A = \begin{pmatrix} 1 & 2 & 4 & 1 \\ -4 & -18 & -12 & -24 \\ 1 & 6 & 6 & 7 \\ 1 & 10 & 8 & 13 \end{pmatrix}$

2. Return to Problems 1 a, b, c and d. Compute the components A_λ for each eigenvalue λ of A.

3. Describe the components of a symmetric projector P_A (as in Section 2.4).

4. Suppose A is diagonalizable with $AP = PD$. Show that $A^T Q = QD$ where $P^T = Q^{-1}$. Use this to show that A and A^T have the same eigenvalues and the same geometric multiplicities. (Compare to Problem 8 of Section 2.3, Problem 6 of Section 3.2.)

5. Suppose A is diagonalizable and has λ as an eigenvalue. Show that the rank of A_λ is the geometric multiplicity of λ as an eigenvalue of A. (Hint: Suppose $AP = PD$. Compute $P^{-1}A_\lambda P$.) It follows that every basic λ-eigenvector of A is a linear combination of the columns of A_λ. (Compare to Problem 7 of Section 3.1.)

6. Let M be a matrix of rank r. Suppose the j-th column, M_j, is the first nonzero column of M and let R be the row of M that contains the first nonzero entry $m_{ij} \neq 0$ of M_j. Show that the matrix $M_j R$ has the same shape as M but has rank 1. Show that the matrix $M - M_j R / m_{i,j}$ has j-th column zero and rank less than r. **HINT:** First work out an explicit 3-by-3 example.

7. Give a recursive argument based on the Preceding problem to show that a matrix of rank r can always be written as a sum of at most r rank one matrices.

8. This problem shows how to define components for defective matrices. Suppose $\mu_A(x) = (x - \lambda)^m g(x)$ where $g(\lambda) \neq 0$. Then as shown in Section 2.7, there exist polynomials $a(x)$ and $b(x)$ such that
$$1 = a(x)(x - \lambda)^m + b(x)g(x).$$
Set $A_\lambda = Ab(A)g(A)$ and $A_g = Aa(A)(A - \lambda I)^m$. Then show that $A = A_\lambda + A_g$ and $(A - \lambda I)^m A_\lambda = 0$ and $g(A)A_g = 0$.

9. Suppose that λ is an eigenvalue of A and let $E = b(A)g(A)$, where $b(x), g(x)$ are as in the preceding problem. Consider an implementation of the minimal polynomial algorithm from Section 3.2 for the matrix A taking m steps. As in that section let $\mathbf{r}_i$, $f_i(x)$, $i = 1 \ldots m$ be the random vector and annihilator polynomials from the i-th pass. Track the minimal polynomial algorithm with A replaced by A_λ and $\mathbf{r}_i$ replaced by $E\mathbf{r}_i$. Show that this change results in $f_i(x)$ being replaced with the highest power of $(x - \lambda)$ dividing $f_i(x)$.

3.5 The sequence $\{A^k\}$

Objective 1

Given an n-by-n matrix A, $n \leq 3$, decide if it is diagonalizable and if so,
find its spectral radius and a formula for an arbitrary power of A.

Suppose a is any (**complex**) number. Then the behavior of the sequence $a, a^2, \ldots$ is exhibited by the graph of the exponential function $y = a^x$ and depends on the value of a. The sequence converges to zero if a has absolute value less than 1, and diverges if a has absolute value greater than 1. If $|a| = 1$ then the sequence is either constant $(a = 1)$ or oscillates $(a = -1)$.

The most important factor in determining the behavior of the matrix sequence $\{A^k\}$ $= A, A^2, A^3, \ldots$ is the largest of the absolute values of the eigenvalues of A. This number is called the *spectral radius* of A. Any eigenvalue having largest absolute value is called a *dominant eigenvalue.*

The behavior of $\{A^k\}$ is determined by the spectral radius of A when A is diagonalizable. This is not so in general, as shown in Problem 9. Fortunately, we can test if A is diagonalizable because A **is diagonalizable if and only if** $\mu_A(x)$ **has no multiple zeros.** This is guaranteed if there are as many eigenvalues of A as its size (why?).

Example 3.12 (Spectral radius and powers of a diagonalizable matrix.)

a. You can check that A has eigenvalues $\lambda_1 = \frac{1}{2}$, $\lambda_2 = 1$ and the associated basic eigenvectors are $\mathbf{v}_1, \mathbf{v}_2$, respectively. $A = \begin{pmatrix} \frac{1}{2} & \frac{1}{2} \\ 0 & 1 \end{pmatrix} \mathbf{v}_1 = \begin{pmatrix} 1 \\ 0 \end{pmatrix} \mathbf{v}_2 = \begin{pmatrix} 1 \\ 1 \end{pmatrix}$

The spectral radius a of A is the largest of the numbers $|1/2|, |1|$. Therefore, $a = 1$ and λ_2 is the (only) dominant eigenvalue. We have the matrix equation:

$$AP = \begin{pmatrix} \frac{1}{2} & \frac{1}{2} \\ 0 & 1 \end{pmatrix} \begin{pmatrix} 1 & 1 \\ 0 & 1 \end{pmatrix} = \begin{pmatrix} \frac{1}{2} & 1 \\ 0 & 1 \end{pmatrix} = \begin{pmatrix} 1 & 1 \\ 0 & 1 \end{pmatrix} \begin{pmatrix} \frac{1}{2} & 0 \\ 0 & 1 \end{pmatrix} = PD.$$

Repeated use of the equation $AP = PD$ gives

$$A^k P = A^{k-1}(AP) = A^{k-1}(PD)$$
$$= A^{k-2}(AP)D = A^{k-2}(PD)D = \cdots = PD^k.$$

(Every time you see an AP, replace it with PD and don't stop until they are all gone.) Since P is invertible, this can be rewritten as:

$$A^k = PD^k P^{-1} = \begin{pmatrix} 1 & 1 \\ 0 & 1 \end{pmatrix} \begin{pmatrix} (\frac{1}{2})^k & 0 \\ 0 & 1^k \end{pmatrix} \begin{pmatrix} 1 & -1 \\ 0 & 1 \end{pmatrix} = \begin{pmatrix} (\frac{1}{2})^k & 1 - (\frac{1}{2})^k \\ 0 & 1 \end{pmatrix}.$$

The alert reader will recall that this is a special case of a result used in the proof of the second part of the fundamental theorem on minimal polynomials in the preceding section.

b. You can check that A has eigen-values $\lambda_1 = 1+i$, $\lambda_2 = 1-i$ and the associated basic eigenvectors $\quad A = \begin{pmatrix} 1 & -1 \\ 1 & 1 \end{pmatrix} \quad \mathbf{v}_1 = \begin{pmatrix} i \\ 1 \end{pmatrix} \quad \mathbf{v}_2 = \begin{pmatrix} -i \\ 1 \end{pmatrix}$

are $\mathbf{v}_1, \mathbf{v}_2$, respectively. The spectral radius a of A is the largest of the numbers $|1+i|, |1-i|$. Therefore, $a = \sqrt{2}$ and both λ_1 and λ_2 are dominant. We have the matrix equation:

$$AP = \begin{pmatrix} 1 & -1 \\ 1 & 1 \end{pmatrix}\begin{pmatrix} i & -i \\ 1 & 1 \end{pmatrix} = \begin{pmatrix} i & -i \\ 1 & 1 \end{pmatrix}\begin{pmatrix} 1+i & 0 \\ 0 & 1-i \end{pmatrix} = PD.$$

Since P is invertible, as above we have,

$$A^k = PD^kP^{-1} = \begin{pmatrix} i & -i \\ 1 & 1 \end{pmatrix}\begin{pmatrix} (1+i)^k & 0 \\ 0 & (1-i)^k \end{pmatrix}\frac{1}{2}\begin{pmatrix} -i & 1 \\ i & 1 \end{pmatrix}$$

$$= \frac{1}{2}\begin{pmatrix} (1+i)^k + (1-i)^k & i(1+i)^k - i(1-i)^k \\ -i(1+i)^k + i(1-i)^k & (1+i)^k + (1-i)^k \end{pmatrix},$$

just in case you were wondering ... (Compare to part c of Example 2.2 on page 58 and Problem 7 of Section 2.2.)   □

Objective 2

Given an n-by-n matrix A, $n \le 3$, describe the evolution of $\{A^k\}$.

We use the term **evolution** and the symbol "$\lim\limits_{k\to\infty}$" to be an abbreviation for "as k grows arbitrarily large." As k grows arbitrarily large, a numerical sequence $\{f_k\}$ can do three things: **converge, oscillate** or **diverge.** The technical meaning of these terms is a major theoretical foundation of both the real number system and calculus. We cannot now deal with their full subtlety.

Roughly speaking, the numerical sequence $\{f_k\}$ converges to a number L if for any "neighborhood of L," no matter how small, eventually the sequence $\{f_k\}$ "moves in to L to stay." The sequence $\{f_k\}$ diverges if it is **unbounded**, having terms with absolute value larger than any number you can write down. A sequence that oscillates neither converges nor diverges. This means it neither "moves in to stay" any place nor "gets too far from home," it just sort of "hangs around."

The situation is essentially the same for matrix sequences $\{A^k\}$, A diagonalizable. Suppose the matrix A is diagonalizable with eigenvalues $\lambda_1 \, \lambda_2 \cdots \lambda_n$. As repeatedly shown, if $A = PDP^{-1}$, then $A^k = PD^kP^{-1}$, for every integer k. Consequently, "continuity" of matrix multiplication implies that

$$\lim_{k\to\infty} A^k = \lim_{k\to\infty} PD^kP^{-1} = P\begin{pmatrix} \lim\limits_{k\to\infty}\lambda_1^k & 0 & 0 \\ 0 & \ddots & 0 \\ 0 & 0 & \lim\limits_{k\to\infty}\lambda_n^k \end{pmatrix}P^{-1}.$$

Once the eigenvalues of A are known, the evolution of the matrix sequence $\{A^k\}$ is completely determined by the numerical limits

$$\lim_{k\to\infty} \lambda_1^k \ldots \lim_{k\to\infty} \lambda_n^k.$$

All we need to do is compare these separate limits. If any of these limits diverges, the spectral radius of A is greater than 1 and the matrix sequence also diverges. Similarly, if all of these limits converge to zero, the spectral radius of A is less than 1 and the matrix sequence also converges to 0.

There remains the possibility that A has spectral radius 1. This means that all the dominant eigenvalues lie on the unit circle in the geometric model of $\mathbf{C}$. If even one of these dominant eigenvalues is not equal to 1, then that part of the matrix D will forever oscillate, forcing A to also oscillate.

THE EVOLUTION OF $\{A^k\}$, A diagonalizable

Suppose A has spectral radius a.

1. If $a > 1$, the matrix sequence $\{A^k\}$ is unbounded and diverges.

2. If $a = 1$ and there is a dominant eigenvalue not equal to 1 ($\lambda \neq 1 = |\lambda|$), the matrix sequence $\{A^k\}$ oscillates.

3. If $a = 1$ and 1 is the **only** dominant eigenvalue, the matrix sequence $\{A^k\}$ converges to a matrix all of whose columns are eigenvectors of A associated with the eigenvalue 1.

4. If $a < 1$, the matrix sequence $\{A^k\}$ converges to the zero matrix.

Because the eigenvalues of A are the zeros of $\mu_A(x)$, this objective involves computing the minimal polynomial AND estimating the absolute value of its zeros. If $\mu_A(x)$ is of degree 2 or less, this is an easy task requiring nothing more than the quadratic formula. In case $\mu_A(x)$ has degree 3, something more is needed.

Suppose $\mu_A(x)$ has degree 3. As in Problem 3 of Section 2.1, you can numerically approximate a real zero r of a $\mu_A(x)$ (see calculators below). Then divide $\mu_A(x)/(x - r)$ to obtain a quadratic polynomial.

In actual practice, the complex zeros are the zeros of a quadratic factor of $\mu_A(x)$:

$$\mu_A(x) = (x - r)(x^2 + c_1 x + c_0) = (x - r)(x - (p + qi))(x - (p - qi)).$$

The quadratic formula gives a formula for p and q in terms of c_1 and c_0 (but it is not necessary for our purposes). When $x = 0$ the above equation reads

$$\mu_A(0) = -r(c_0) = -r(p + qi)(p - qi) = -r|p + qi|^2 = -r|p - qi|^2.$$

Therefore, when $\mu_A(x)$ has degree 3 and **only** one real zero r, $\mu_A(x)$ has complex zeros $p \pm iq$ **both** of absolute value

$$\sqrt{-\mu_A(0)/r}.$$

Example 3.13 (The absolute value of a complex eigenvalue.)
  The indicated matrix A has minimal polynomial $\mu_A(x)$

$$A = \begin{pmatrix} 7 & -8 & 4 \\ 10 & -11 & 6 \\ 10 & -14 & 9 \end{pmatrix}, \qquad \begin{aligned} \mu_A(x) &= x^3 - 5x^2 + 11x - 15 \\ &= (x-3)(x^2 - x + 5) \end{aligned}$$

and the quadratic term has no real factors. Therefore, A has one real eigenvalue $\lambda = 3$ and two complex eigenvalues; BUT the absolute value of (either) complex eigenvalue is $\sqrt{-(-15)/3} = \sqrt{5} \doteq 2.236$, so the spectral radius $a = 3$ of A is the largest of $\{3, \sqrt{5}\} = 3$.   □

Example 3.14 (Evolution of $\{A^k\}$.)

  a. The matrix A has minimal polynomial $\mu_A(x)$

$$A = \begin{pmatrix} 0.10 & 0.96 & 0.18 \\ 0.25 & 0.00 & 0.00 \\ 0.00 & 0.80 & 0.00 \end{pmatrix}, \qquad \mu_A(x) = x^3 - 0.1x^2 - 0.24x - 0.036.$$

  This polynomial has three real zeros: $-.300$, $-.200$ and $.600$. Therefore, A has spectral radius $a = .6 < 1$ and the matrix sequence $\{A^k\}$ converges to 0.

  b. The matrix A has minimal polynomial $\mu_A(x)$

$$A = \begin{pmatrix} 0 & 0 & 2 \\ 1 & 0 & 0 \\ 0 & .5 & 0 \end{pmatrix}, \qquad \begin{aligned} \mu_A(x) &= x^3 - 1 \\ &= (x-1)(x^2 + x + 1). \end{aligned}$$

  This polynomial has only one real zero, namely $r = 1$. Its two complex zeros can be computed from the quadratic formula ($\frac{1 \pm i\sqrt{3}}{2}$), BUT this is unnecessary because they have absolute value $\sqrt{-\mu_A(0)/r} = 1$. Therefore, A has spectral radius $a = 1$ and has three dominant eigenvalues, only one of which is 1. Therefore the sequence $\{A^k\}$ oscillates.   □

Calculators
If you have an explicit numerical matrix A, you can check which of the possibilities *describes*$\{A^k\}$ on a calculator, by computing the sequence $\{A^{2^k}\}$ and seeing what happens. To do this, edit $[A]$ then select $[A]$ from the MATRIX NAMES menu. Next press $\boxed{\text{enter}}$ so that the matrix $[A]$ is also "ans," then press the $\boxed{x^2}$ button. Follow this by repeatedly pressing $\boxed{\text{enter}}$. Unfortunately, in some extreme cases the calculator may take a very long time, and its many approximate computations may lead to inaccurate results. For this reason, test questions will not simply ask for the "evolution of a matrix sequence." Rather they will ask for more detailed information. The graphing feature of the TI-83 calculator is described in Chapter 3 of its handbook. To graph functions, you may need to reset the fourth item in the $\boxed{\text{MODE}}$ menu to display Func.

Here is how to use the calculator to numerically estimate the absolute values of the zeros of a monic cubic polynomial $y = f(x)$.

1. Press the $\boxed{Y =}$ key on the left just below the display, and enter the formula for $y = f(x)$. (Don't even think about $\boxed{\text{alpha}}$ $\boxed{(}$ $\boxed{\cos}$ $\boxed{)}$ ... . You must have the coefficients of $f(x)$ explicitly!)

2. Press the $\boxed{\text{GRAPH}}$ just below the display. The calculator may take some time to display the graph. If the final display does not show the graph clearly, you may want to press the $\boxed{\text{WINDOW}}$ key below the display and adjust the values of Xmin, Xmax, Ymin, Ymax. Then graph again.

3. Once you have the graph clearly displayed, you must estimate the values of each of its real zeros. This can be done in a number of ways using the $\boxed{\text{MATH}}$ $\boxed{\text{SOLVER}}$ feature described in Chapter 2 of the handbook or by using the $\boxed{\text{TRACE}}$ and $\boxed{\text{ZOOM}}$ features.

4. If the graph has only one real zero, r, then you must also find the y-intercept $f(0)$ of the the graph (the constant term in the formula for $f(x)$). In this case it has two complex zeros each of absolute value $\sqrt{-f(0)/r}$.

Shortcut
The minimal polynomial is not the only polynomial that has the eigenvalues of A as its zeros. The ***characteristic polynomial*** $p_A(x)$ is always a multiple of the minimal polynomial and is already "known" to your calculator. This polynomial is defined in terms of determinants in Chapter 4, but we mention it here because we are now letting the calculator do all the work, and it already "knows" how to work with (small) determinants. You can actually "trick" a TI-83 calculator into telling you the eigenvalues of a real 3-by-3 matrix without actually knowing either the minimal or the characteristic polynomial!

Let's suppose you have entered the matrix as $\boxed{\text{MATRIX}}$ $[A]$.

1. Press the $\boxed{Y =}$ key. Then press the $\boxed{\text{MATRIX}}$ $\boxed{\rightarrow}$. The first menu item is
$$det($$
Select it by pressing $\boxed{\text{ENTER}}$.

2. Next enter $\boxed{\text{x, n,t,}\theta}$ (next to the $\boxed{\text{ALPHA}}$ key) followed by the multiplication key $\boxed{\times}$. Then press $\boxed{\text{MATRIX}}$ $\boxed{\rightarrow}$ and select the 5th menu item
$$identity(\quad .$$
This is the calculator's version of an identity matrix; all you need to do is tell it the correct size. Since A is 3-by-3, you enter $\boxed{3}$ $\boxed{)}$.

3. Finally, to enter the matrix and complete the formula, press the subtraction key $\boxed{-}$ followed by the
$$\boxed{\text{MATRIX}} \boxed{\text{ENTER}} \boxed{)} \boxed{\text{ENTER}}.$$

4. By the way, don't forget to graph it. $\boxed{\text{GRAPH}}$ on the right below the display.

Example 3.15 (Eigenvalues by graphing calculator and matrix sequences.)

1. The graph of the characteristic polynomial $p_A(x) = det(xI - A)$ of the matrix A in part a of Example 3.14 on page 150 must be ZOOMed before you can actually see that it has three real zeros. Because largest absolute value is 0.6, the matrix sequence converges to zero. Here are two views of that function.

2. The graph of the characteristic polynomial $p_A(x) = det(xI - A)$ of the matrix A in part b of Example 3.14 on page 150 is easier. Because the graph crosses the x-axis only once, we know it has but one real eigenvalue. Since its y-intercept is -1, we conclude that it has two complex eigenvalues of absolute value $\sqrt{\frac{-(-1)}{1}} = 1$. A has spectral radius 1 and all three of its eigenvalues are dominant. Therefore the matrix sequence $\{A^k\}$ oscillates.

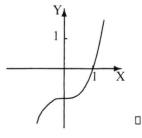

□

Objective 3

Given a 3-by-3 numerical matrix with three distinct real eigenvalues, compute the dominant eigenvalue by iteration and give the deflated matrix.

When confronted with a large numerical matrix A the problem of computing and factoring polynomials ($\mu_A(x)$ or $p_A(x)$) is more difficult than more direct numerical methods of computing eigenvalues. Most such numerical methods for computing eigenvalues and eigenvectors take advantage of special properties of the matrices being studied and require much more background to understand.

However, one basic **iterative method** uses the ideas of this section. As described above, the convergence of $\{A^k\}$ is determined by the spectral radius a of A. The iterative method takes this a step further and uses $\{A^k\}$ to find a. As a bonus, it does not directly involve $\{A^k\}$ and actually produces an eigenvector too!

The key idea is to work with vector sequences rather than matrix sequences. Pick an initial random vector v_1 and define a vector sequence $\{v_k\}$ recursively by

$$v_{k+1} = \frac{1}{|v_k|} A v_k.$$

Then "almost always" $\{v_k\}$ converges to an eigenvector associated with the dominant eigenvalue of A.

This may not work well when A is not diagonalizable and will definitely fail if it has several dominant eigenvalues. Another problem is that the initial "random vector" might be an eigenvector for some nondominant eigenvalue. There are technical ways around all of these issues, but our goal here is to give the basic idea and not to cover every case.

Dominant eigenvalues by iteration

Suppose A is n-by-n, with n eigenvalues having distinct absolute values. Specify an error tolerance $\epsilon > 0$ and pick a random vector $\mathbf{v} \neq \mathbf{0}$.

  a. Compute $\mathbf{w} = \frac{1}{|v|}A\mathbf{v}$.

  b. For each nonzero coordinate v_i of $\mathbf{v}$, compute the ratio of the coordinates $q_i = \mathbf{w}_i/\mathbf{v}_i$, $i = 1, \ldots, n$.

  c. If $|q_i - q_j| \geq \epsilon$ for any i, j, then replace $\mathbf{v}$ with $\mathbf{w}$ and go back to step a. Otherwise STOP.

The vector $\mathbf{v}$ is (approximately) an eigenvector of A and the associated eigenvalue λ is obtained by comparing $A\mathbf{v}$ and $\mathbf{v}$.

Of course finding one eigenvalue should not be compared to finding **all** eigenvalues. However, when combined with **deflation**, iteration forms a recursive process that can find all eigenvalues of a suitable matrix. Deflation replaces A with a matrix of smaller **rank** that has all the same eigenvalues and eigenvectors of A except that the found eigenvector for A has eigenvalue 0 for the deflated matrix.

If the dominant eigenvalue λ has geometric multiplicity 1, then the deflated matrix is just $A - A_\lambda$, where A_λ is the λ-component of A as defined in the preceding section. But, of course, it is computed in a different way.

Deflation

Suppose A is n-by-n, with n eigenvalues having distinct absolute values.

  a. Use iteration to find the dominant eigenvalue a of A and an associated eigenvector is $\mathbf{w}$.

  b. Repeat the process for A^T. It will have the same dominant eigenvalue (by Problem 6 of Section 3.2) but a possibly different eigenvector $\mathbf{x}$.

The component of A is $A_\lambda = \lambda(\mathbf{x}^T\mathbf{w})^{-1}\mathbf{w}\mathbf{x}^T$ and the deflation of A by λ is $A - A_\lambda$.

Here is the idea. By Problem 6 of Section 3.4, A_λ has rank 1 and can be "factored" as a column vector times a row vector. As pointed out just before Example 3.11 on page 142, the columns of A_λ **are** eigenvectors of A and one can show $A_\lambda{}^T = A^T{}_\lambda$. This is actually enough to determine A_λ. More organized support for this algorithm is given at the end of this section.

Example 3.16 (Iterative eigenvectors and deflation.)

a. Consider the diagonal matrix D and initial vector $\mathbf{v}$. Let's also say we are content to have three decimal places of accuracy.

$$D = \begin{pmatrix} 3 & 0 & 0 \\ 0 & 2 & 0 \\ 0 & 0 & 1 \end{pmatrix} \qquad \mathbf{v} = \begin{pmatrix} 1 \\ 1 \\ 1 \end{pmatrix}$$

Before you begin, it is a good idea to set the display to float 3 so things don't get to ugly. Do this with the MODE ↓ sequence.

Begin by entring the matrix D, as say [D], and the initial vector $\mathbf{v}$, as say [C], in the calculator memory. Then select [C] and ENTER so that the initial vector is stored in "Ans" too.

1. Now enter the following sequence MATH → ↓ ENTER . That selects the rounding feature of the calculator so it will round the answer to the specified precision (which comes in step 4).

2. Next enter [D]* Ans*. Then press the 2nd x^2 to obtain $\sqrt{(}$ followed by MATRIX → ENTER to display the determinant function. (The determinant is only used here because your author doesn't know how to extract a matrix entry in the middle of a one line command with this calculator.) The display should now look like:

$$round([D] * Ans * \sqrt{(det(}$$

3. Continue by entering Ans and then press the MATRIX → ↓ ENTER to select the transpose operator. Follow this by *Ans) and x^{-1}). (Note that this construction — from the $\sqrt{(}$ on — is simply dividing by the length of the *Ans* vector.)

4. End with ,5) to specify the calculator rounding precision because it needs to be that than that which we require of the answer. The display should now look like:

$$round([D] * Ans * \sqrt{(det(Ans^T * Ans)^{-1})}, 5)$$

5. Keep pressing ENTER until the displayed vector does not change (up to three decimal places).

The 18th vector is $(3.000 \ 4.000E-4 \ 0.000)^T$ which rounds to three decimal places to the correct answer $(3\ 0\ 0)^T$. (You can check that if the calculator rounding precision is set to three decimal places, it doesn't get there right answer. The rounding error becomes significant and it only gets to $(3.00\ 0.01\ 0.00)^T$ at the 20th vector, then repeats.) The dominant eigenvalue of D is 3 and an associated eigenvector is $\mathbf{w} = (3\ 0\ 0)^T$.

The deflation of A also requires the dominant eigenvector of D^T. Since D is symmetric, we don't have to look far. $\mathbf{x} = \mathbf{w}$ and

$$D_3 = \frac{3}{9}\mathbf{w}\mathbf{x}^T = \begin{pmatrix} 3 & 0 & 0 \\ 0 & 0 & 0 \\ 0 & 0 & 0 \end{pmatrix} \text{ and } D - D_3 = \begin{pmatrix} 0 & 0 & 0 \\ 0 & 2 & 0 \\ 0 & 0 & 1 \end{pmatrix}$$

is the deflated matrix.

b. Consider the symmetric matrix A and initial vector $\mathbf{v}$. Let's say we are content to have three decimal places of accuracy. The iteration algorithm vector sequence stabilizes at the 6th vector $\mathbf{w} = (2.977\ 3.689\ 2.094)^T$ and

$$D = \begin{pmatrix} 2 & 2 & 1 \\ 2 & 3 & 1 \\ 1 & 1 & 2 \end{pmatrix} \qquad \mathbf{v} = \begin{pmatrix} 1 \\ 0 \\ 0 \end{pmatrix}$$

the associated eigenvalue is $\lambda_1 = 5.182$. Since A is symmetric, $\mathbf{x} = \mathbf{w}$ is the dominant eigenvector for A^T. The associated component is

$$A_{\lambda_1} = \frac{\lambda}{\mathbf{x}^T\mathbf{w}}\mathbf{w}\mathbf{x}^T = \begin{pmatrix} 1.71 & 2.12 & 1.20 \\ 2.12 & 2.63 & 1.49 \\ 1.20 & 1.49 & .847 \end{pmatrix} ; \ A_1 = A - A_{\lambda_1} = \begin{pmatrix} 0.290 & -.119 & -.203 \\ -.119 & .375 & -.491 \\ -.203 & -.491 & 1.15 \end{pmatrix}$$

is the first deflated matrix. Again take $\mathbf{v}_1 = (1\ 0\ 0)^T$. The 10th vector in the above defined sequence is $\mathbf{v} = (.169\ .584\ -1.27)^T$ and the associated eigenvalue is $\lambda_2 = 1.41$. The associated component is

$$A_{\lambda_2} = \begin{pmatrix} .020 & .070 & -.152 \\ .070 & .242 & -.527 \\ -.152 & -.527 & 1.14 \end{pmatrix} ; \ A_2 = A_1 - A_{\lambda_2} = \begin{pmatrix} .270 & -.189 & -.0510 \\ -.189 & .132 & .0357 \\ -.0510 & .0357 & .00965 \end{pmatrix}$$

is the second deflated matrix. The 10th vector in the above defined sequence is $\mathbf{v} = (.333\ -.233\ .0630)^T$, and the associated eigenvalue is $\lambda_2 = 0.412$. You can check that $A_2 = A_{\lambda_3}$ as you would expect. □

Proofs

The argument that this iterative method converges uses the fact that if a vector sequence $\{\mathbf{x}_k\}$ converges to a vector $\mathbf{e}$, then the vector sequence $\{P\mathbf{x}_k\}$ converges to $P\mathbf{e}$. This is a consequence of the continuity of the linear function associated with P and will not be proven here.

Suppose that D is a diagonal matrix with dominant eigenvalue $\lambda \neq 0$ in the $(1,1)$ position. As described at the beginning of this section, the matrix sequence $\{(\frac{1}{\lambda}D)^k\}$ converges to the matrix E of all zeros, except $e_{11} = 1$.

Now assume the initial random vector $\mathbf{x}$ has nonzero first coordinate, and work through the algorithm for D, as illustrated in part a of the preceding example. Label the k-th vector $\mathbf{v}$ of the algorithm by $\mathbf{x}_k$ (the reason for this awkward switch from $\mathbf{v}$ to $\mathbf{x}$ will soon be clear). Then $\mathbf{x}_k$ is a scalar multiple of $D^k\mathbf{x}$, and so $\{\mathbf{x}_k\}$ converges to a scalar multiple of the first column, $\mathbf{e}$ of E.

Now suppose A is diagonalizable and we have found P, D where D is diagonal as above, and (as always) $AP = PD$. How does the algorithm applied to A compare with that applied to D?

Let $\mathbf{v}_k$ be the k-th vector when the algorithm is applied to A with initial random vector $\mathbf{v} = P\mathbf{x}$. Then (as in part a of Example 3.12)

$$v_2 = \frac{1}{|v_1|}Av_1 = \frac{1}{|Px_1|}APx_1 = \frac{1}{|Px_1|}PDx_1 = \frac{1}{|Px_1|}Px_2,$$

$$v_3 = \frac{1}{|v_2|}Av_2 = \frac{|Px_1|}{|Px_2|}A\frac{1}{|Px_1|}Px_2 = \cdots = \frac{1}{|Px_2|}Px_3,$$

and by recursion $v_{k+1} = \frac{1}{|Px_k|}Px_{k+1}$.

Thus, each vector in the sequence $\{v_k\}$ is a scalar multiple of the corresponding vector in the sequence $\{Px_k\}$. By continuity of the linear function associated with P, the vector sequence $\{v_k\}$ converges to a scalar multiple of Pe, a dominant eigenvector of A.

To verify the deflation formula, take λ, E, A, P and D as above. Then the first column p_1 of P is a λ-eigenvector of A and since λ has geometric multiplicity 1, the eigenvector w obtained by iteration satisfies $p_1 = sw$ for some nonzero number s. By Problem 4 of Section 3.4, $Q^{-1}A^T Q = D$ where $Q^T = P^{-1}$, and λ has geometric multiplicity 1 for A^T. Consequently, the A^T eigenvector x obtained by iteration must satisfy $q_1 = tx$ for some nonzero number t, and where q_1 is the first column of Q.

Observe that q_1^T is the first ROW of P^{-1}, since $Q^T = P^{-1}$. Note also that the $1,1$ entry of $I = P^{-1}P$ is $1 = q_1^T p_1$. Consequently, $x^T w = (tq_1^T)(sp_1) = st$.

But the actual $A_\lambda = PEP^{-1} = \lambda(PE)(EP^{-1})$ and, lo and behold, PE (respectively EP^{-1}) is square matrix with first column p_1 and all other columns zero (respectively first row q_1^T and all other rows zero). It follows that

$$A_\lambda = \lambda p_1 q_1^T = \lambda(x^T w)^{-1} w x^T.$$

SUMMARY

The **evolution** of the sequence $\{A^k\}$ of matrices depends on the **spectral radius** of A. When A is diagonalizable, the sequence may **diverge, oscillate** or **converge**. When the spectral radius equals 1, the matrix sequence can either oscillate or converge to a nonzero matrix. In order to decide between these, you must decide if there are any **dominant eigenvalues** other than 1.

When a matrix has a unique dominant eigenvalue λ, λ can be numerically computed by **iteration**. When combined with **deflation**, iteration gives a method to compute all eigenvalues of a diagonalizable n-by-n matrix that has n eigenvalues with distinct absolute values.

Looking further

Computations involving very large numerical matrices are necessary in many applications but are only feasible with computers. Here is a general reference for appropriate methods and software.[4]

---

[4]*Numerical linear algebra*, by James W. Demmel, SIAM, 1997; ISBN 0-89871-389-7.

Practice Problems

1. Find the minimal polynomial and spectral radius of the indicated matrices.

   a. $\begin{pmatrix} 2 & 2 \\ 2 & 2 \end{pmatrix}$
   b. $\begin{pmatrix} -19 & 8 \\ -40 & 17 \end{pmatrix}$
   c. $\begin{pmatrix} -3 & 5 & -1 \\ -3 & 5 & -1 \\ -1 & 1 & 1 \end{pmatrix}$

2. For each matrix in the preceding problem, diagonalize the matrix and write a formula for the closed form formula for its n-th power as in Example 3.12.

3. Each of the following is the characteristic polynomial of a 3-by-3 matrix. Use the calculator to graph the polynomial and determine the spectral radius of the matrix.

   a. $x^2 - 1.5x + 0.5$
   b. $x^3 - 1.55x - 0.3$
   c. $x^3 - 0.2x^2 - 0.43x - 0.1$
   d. $x^3 - 1$
   e. $x^3 + 3x^2 + 6x + 4$
   f. $x^3 + 1.8x^2 + 1.8x + 0.8$

4. Determine the spectral radius and describe the evolution of $\{A^k\}$ for each of the following matrices.

   a. $\begin{pmatrix} 3 & -2 \\ 4 & -3 \end{pmatrix}$
   b. $\begin{pmatrix} 2 & 1 & 2 \\ 1 & 2 & 2 \\ 1 & 1 & 3 \end{pmatrix}$
   c. $\begin{pmatrix} 0.3 & 1.0 & 0.4 \\ 0.6 & 0.0 & 0.0 \\ 0.0 & 0.8 & 0.0 \end{pmatrix}$
   d. $\begin{pmatrix} 0.3 & 1.5 & 0.3 \\ 0.4 & 0.0 & 0.0 \\ 0.0 & 0.8 & 0.0 \end{pmatrix}$

5. Use the iterative eigenvalue method to find the dominant eigenvalue and an associated eigenvector of each of the following matrices

   a. $\begin{pmatrix} 7 & -4 \\ 8 & -5 \end{pmatrix}$
   b. $\begin{pmatrix} 3 & 2 & 1 \\ 2 & 3 & 1 \\ 1 & 1 & 0 \end{pmatrix}$

6. For each part of the preceding problem, compute the component associated with the found eigenvector and find the deflated matrix.

7. Find **four** (complex) solutions to the matrix equation $X^2 = I_2$. Show that there are infinitely many other solutions. (Compare to Problem 5 of Section 2.4.)

8. Find **four** solutions to the matrix equation $X^2 = \begin{pmatrix} 0 & -2 \\ 2 & 0 \end{pmatrix}$ (Compare to part c of Example 2.2.)

   Can you prove that there are no more solutions?

9. Let $A = \begin{pmatrix} 1 & 1 \\ 0 & 1 \end{pmatrix}$. Show by recursion that $A^n = \begin{pmatrix} 1 & n \\ 0 & 1 \end{pmatrix}$.

   Describe the evolution of the matrix sequence $\{A^n\}$. Reconcile this fact with the discussion on page 149.

3.6   Discrete dynamical systems

Objective 1

Given a three-state linear discrete dynamical system, write its transition matrix A. Compute the spectral radius of A and determine the system's long-term behavior. Identify any stable steady states.

A discrete **dynamical system** is a mathematical model that is organized somewhat like a motion picture. It records the development of a complicated situation as time goes by with a sequence of "mathematical snapshots." But the "record" is not a picture. Rather, there are a number of critical variables, say x_1, x_2, ..., x_n, each of which varies with time. The record is a sequence of **state vectors** $\{\mathbf{x}(t)\}$.

In contrast to a calculus-based model, where each $x_i(t)$ is a continuous function of a real variable t, the independent variable t of a **discrete** dynamical system takes values in a (possibly infinite) list t_1, t_2, ...

The number, n, of **states** is the number of coordinates in the state vector of the discrete dynamical system. Although realistic models may require hundreds of states, we study only three-state discrete dynamical systems, because they are computationally manageable while rich enough to show the full variety of possible behaviors. When using these models, we can imagine ourselves watching a group of individuals and tracking some particular attribute that can take three possible values for each individual. As time goes by, we seek to identify patterns in behavior.

Like all mathematical models, a discrete dynamical system is based on mathematical assumptions. It is the task of the experimental scientist to match raw data against such models as mentioned at the close of Section 2.5 in "looking further."

The most basic type of discrete dynamical system is a **linear** discrete dynamical system. In a linear discrete dynamical system, the mathematical assumption is that the changing of an individual's attributes is "a fact of nature" and doesn't really vary between individuals or as time goes by. If you skip ahead to the first part of Examples 3.17 a and 3.18 you will have an idea of what makes up one of these models.

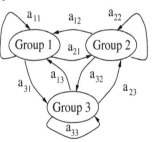

There is a numerical matrix A called the **transition matrix** that relates successive values of $\mathbf{x}$ as follows:

$$\mathbf{x}(t_n) = A\mathbf{x}(t_{n-1}) \text{ where}$$

$$a_{ij} = \begin{cases} \text{the average contribution to the next group} \\ i \text{ that is caused by one in today's group } j. \end{cases}$$

This implies that the **columns** of A might be labeled "from" and **rows** labeled "to."

WARNING: This is OPPOSITE of what you would expect from our digraph notation! Sadly, each subject is well established and it is to late to reverse either world.

In this objective we show how eigenvalues and eigenvectors play a pivotal role in linear discrete dynamical system models. Contrary to intuition, the long-term behavior of these models is related to the eigenvalues, eigenvectors and especially the spectral radius a of the transition matrix A, rather than to the initial state $x(t_0)$.

Let's simplify notation writing $\mathbf{x}_k$ for $\mathbf{x}(t_k)$. The state vectors for a linear discrete dynamical system has the form:

$$\mathbf{x}_k = A\mathbf{x}_{k-1} = A(A\mathbf{x}_{k-2}) = A^2\mathbf{x}_{k-2} = \cdots = A^k\mathbf{x}_0.$$

In the last section we focused on the matrix sequence $\{A^k\}$ but used a vector sequence $\{\mathbf{x}_k\}$ to find dominant eigenvalues and associated eigenvectors of the matrix A. Now we have no control over the initial state vector and, it turns out that, the evolution of the matrix sequence $\{A^k\}$ limits, but does not determine, the possible evolution of the vector sequence $\{\mathbf{x}_k\}$.

One issue is that it is technically possible that the vector sequence to converge to an "unstable equilibrium" associated with an eigenvector of A whose associated eigenvalue is not dominant. A more important issue is that there may be several dominant eigenvalues. This possibility is critical only when the spectral radius is $a = 1$, and then the question whether the state vector sequence oscillates or converges to a steady state really does depend on the initial state vector.

The **stability** of a convergent sequence of vectors measures how much the final result would have changed if the initial state were "perturbed" slightly. **Stable equilibrium occurs only if 1 is the ONLY dominant eigenvalue.** In this case, **any** eigenvector associated to $\lambda = 1$ is a "stable steady state." We will not dwell on the technical subtleties of unstable equilibria beyond acknowledging their existence.

Assuming stability, the spectral radius a of the transition matrix A is also the **asymptotic growth rate** of the state vector sequence $\{\mathbf{x}_k\}$.

We study two important but rather different types of mathematical models based on linear discrete dynamical systems. In practice, the major difference between the models is whether it allows for global growth or decay, and the extent to which certain state transitions are forbidden.

Markov chains

In a **Markov chain** the studied population is measured in probabilities or percentages. The (i, j) entry of the transition matrix a_{ij} measures the **percentage** of the individuals in group j that move to group i between observations. Because only percentages are recorded and the entries $a_{i,j}$ in the j-th column reflect "where percentages of group j go,"

$$1 = a_{1,j} + a_{2,j} + a_{3,j} = 100\%.$$

Moreover, there are no negative entries (because movement in the opposite direction is recorded in the transposed position), so $0 \leq a_{ij} \leq 1$ for entries of the transition matrix of a Markov chain. Population change is NOT considered in a Markov chain model.

Example 3.17 (Markov chains.)

a. Three regions that we will label as America, Europe and Asia, are in economic
    competition. Each commands a percentage of international trade from year
    to year. America begins with 70% of the market, Europe with 20% and Asia
    with 10%. Because of the nature of their products, the regions' market share
    changes from year to year. The following data is available.

    > Of American customers, 20% are satisfied and stay loyal, but 40%
    > switch to European and 40% switch to Asian products from one year to
    > the next. Of the European customers, 20% are satisfied and stay loyal,
    > but 40% switch to American and 40% switch to Asian products. Finally,
    > 60% of the Asian customers are satisfied and stay loyal, and only 20%
    > switch to American and 20% switch to European products.

    The problem is to project the range of possible long-term market share distri-
    butions and decide what determines these possibilities.

    The state vector has as its entries the "market share percentages." Let's list
    these in order: America, Europe, Asia. The initial state vector and transition
    matrix are

$$\mathbf{x}_0 = \begin{pmatrix} 0.7 \\ 0.2 \\ 0.1 \end{pmatrix} \text{ and } A = \begin{pmatrix} 0.2 & 0.4 & 0.2 \\ 0.4 & 0.2 & 0.2 \\ 0.4 & 0.4 & 0.6 \end{pmatrix}.$$

    The minimal polynomial of A is

$$\mu_A(x) = p_A(x) = x^3 - x^2 - 0.04x + 0.04 = (x-1)(x-.2)(x+.2).$$

    Therefore, the spectral radius is $a = 1$, and because the eigenvalue $\lambda = 1$ has
    multiplicity 1 there is a UNIQUE stable steady state. Case 3 of "the evolution
    of the matrix sequence" on page 149 holds. The stable steady state arises as an
    eigenvector of A associated with $\lambda = 1$ BUT is not the basic eigenvector.

    The basic eigenvector $(1/2, 1/2, 1)^T$ must be "normalized" so its coordinates
    add up to 1 (=100%). Thus the projected stable steady state is America 25%,
    European 25% and Asia 50%. (The exact process of this normalization is
    explained in part b below)

    Note that this distribution is almost achieved in only three years.

$$\mathbf{x}_3 = A^3\mathbf{x}_0 = \begin{pmatrix} 0.248 & 0.256 & 0.248 \\ 0.256 & 0.248 & 0.248 \\ 0.496 & 0.496 & 0.504 \end{pmatrix} \begin{pmatrix} 0.7 \\ 0.2 \\ 0.1 \end{pmatrix} = \begin{pmatrix} 0.2496 \\ 0.2536 \\ 0.4968 \end{pmatrix}$$

    Observe also that the initial vector $\mathbf{x}_0$ plays no role in the analysis and is only
    used to decide how fast the stable steady-state is achieved.

b. Professor Relbeil's ski area, Aftermath, has (only) three snow reports.

    > If today's report is champagne powder, the probabilities of tomorrow's
    > report are 50% champagne powder, 10% sunny and 40% packed powder.
    > When today's report is sunny, the probabilities of tomorrow's report are

30% champagne powder, 40% sunny and 30% packed powder. If today's report is packed powder, the probabilities of tomorrow's report
are 10% champagne powder, 10% sunny and 80% packed powder.
Find the transition matrix and identify a stable steady-state vector if one exists.

List the snow reports in the order given.
The transition matrix is :
The minimal polynomial of A is

$$A = \begin{pmatrix} 0.5 & 0.3 & 0.1 \\ 0.1 & 0.4 & 0.1 \\ 0.4 & 0.3 & 0.8 \end{pmatrix}$$

$$\mu_A(x) = (x-1)(x-2/5)(x-3/10),$$

which has three real zeros 1, 0.4, 0.3. Therefore, the spectral radius of A is
$a = 1$. The basic eigenvector associated with $\lambda = 1$ is $\mathbf{v} = (1/3, 2/9, 1)^T$.

But this is **NOT** the steady-state distribution because it is not expressed in terms
of percentages. To do this we multiply $\mathbf{v}$ by reciprocal of the sum of the entries
of $\mathbf{v}$ $\frac{1}{1/3+2/9+1} = 9/14$ to obtain

$$(3/14, 2/14, 9/14)^T = (21.4\%, 14.3\%, 64.3\%)^T.$$

Observe that $\mathbf{x} = (-1, 0, 1)^T$ is also an eigenvector of A and that if somehow
$\mathbf{x}$ were the initial state vector then the discrete dynamical system would not
deviate from this distribution. This would be an example of an unstable steady
state. Of course this whole scenario is completely unrealistic because the entries
in the state vector of a Markov chain are percentages and must be non-negative.

c. On graduation night there are three parties in Fort Fun. The
highly excited new graduates are anxious to celebrate ev-
erywhere, so they move from party to party every half hour.
What do you expect to happen if the transition matrix is A?

$$A = \begin{pmatrix} 0 & 1 & 0 \\ 0 & 0 & 1 \\ 1 & 0 & 0 \end{pmatrix}$$

This transition matrix has minimal polynomial

$$\mu_A(x) = x^3 - 1 = (x-1)(x^2 + x + 1).$$

This polynomial has one real zero , $\lambda_1 = 1$, and two complex zeros. The
complex zeros have absolute value $\sqrt{-\mu_A(0)/\lambda_1} = 1$. Since all three zeros
of this polynomial have absolute value 1, A has spectral radius $a = 1$ and case
2 of "The evolution of the matrix sequence" on page 149 holds. This system
oscillates and won't predictably settle down to a stable steady-state.
The graduates just keep moving from party to party, never stopping — what a
surprise! □

Suppose that A is the transition matrix of a Markov chain. Because the entries in each
column sum to 100%, the transposed matrix A^T has $(1, 1, 1)^T$ as an eigenvector with
eigenvalue 1. In view of Problem 4 of Section 3.4, this implies that A has $\lambda = 1$ as
an eigenvalue — always.

There is also a wonderful theory of matrices having all entries non-negative due
to Perron and Frobenius. It predicts that every Markov chain transition matrix has
spectral radius 1 and that there are no **feasible** unstable steady states — for the same
reason as given in part b of the above example, an eigenvector with a nondominant

eigenvalue must have negative coordinates and so cannot represent a state distribution. **Therefore, every Markov chain model either converges to a stable steady state or oscillates.**(An even more careful analysis shows that the only way a three-state Markov chain can oscillate is if its transition matrix has at least four zeros.)

If it does not oscillate AND if $\lambda = 1$ has geometric multiplicity equal to one, then ANY initial state vector **converges to THE UNIQUE stable steady state.** But if the geometric multiplicity of the eigenvalue 1 is greater than one, then there are infinitely many stable steady states and ANY associated eigenvector, when normalized, represents a stable steady state of the system. (This exceptional case would occur if the transition matrix were the identity matrix.)

Incidently, there do exist Markov chains with transition matrices that are not diagonalizable (try the transition matrix in part b of Example 3.17 with $a_{22} = .5$ and $a_{32} = .2$), but the Perron-Forbenius theory still prevails and they cause no problems.

Leslie population models

Leslie models are basic tools in population ecology. They are used to project species extinction and survival. Game and fish officials also use them to set hunting and fishing license quotas.

The individuals of the studied population are grouped by "maturity" (for us: juvenile, mature, old). This model's state vector is called the *population distribution* and it records the **number** of individuals in each age bracket.

$$A = \begin{pmatrix} f_1 & f_2 & f_3 \\ s_1 & 0 & 0 \\ 0 & s_2 & 0 \end{pmatrix}$$

The transition matrix A has two types of nonzero entries. The entries $s_1 = a_{12}$, $s_2 = a_{23}$ record the survival rates from juvenile to mature and mature to old. The entries $f_1 = a_{11}$, $f_2 = a_{12}$, $f_3 = a_{13}$ record fertility rates (the average number of offspring per sample period created by members) of each of the three age groups

We also make a number of unreasonable simplifying assumptions. We assume that all individuals have the same birthday and that the time required to move through a maturity phase is the same for all individuals and all phases. Finally, we assume that "there is no surviving old age," that is, the life-span is limited to the three sample periods. Sounds grim, huh?

These models can have the full variety of long-term behaviors seen in the evolution of the matrix sequence $\{A^k\}$. For this reason, there are no shortcuts. Only the full analysis is adequate.

Example 3.18 (Leslie population models.)[5]

    a. Serena studies squirrels. She observes that squirrels $0 - 1$ year old produce 0.2 offspring per year on average and 50% survive 1 year. Squirrels $1-2$ years old

---

[5]The numerical data and situations described in this and other examples/problems of this type are simplified and are not scientifically accurate.

produce an average of 1.2 offspring and 80% survive 2 years. Squirrels over 2 years old produce an average of .6 offspring. Serena assumes that squirrels do not live more than 3 years.

What does this model predict about the squirrel population, long term.

We find the transition matrix A and compute its spectral radius.

The transition matrix of this population model is:
The minimal polynomial of A is

$$A = \begin{pmatrix} 0.2 & 1.2 & 0.6 \\ 0.50 & 0.0 & 0.0 \\ 0.0 & 0.8 & 0.0.0 \end{pmatrix}$$

$$\mu_A(x) = -0.24 - 0.6x - 0.2x^2 + x^3$$
$$= (x - 1.019)(x^2 + 0.8194x + 0.2354).$$

Because this polynomial has only one real zero, we must find the absolute value of its complex eigenvalues to determine the spectral radius a of A. By the formula in the preceding section

$$|z| = \sqrt{-(-0.24)/1.019} \doteq 0.485.$$

Therefore, the real eigenvalue 1.019 is the only dominant eigenvalue and the spectral radius of A is $a = 1.019$. Thus, the **asymptotic growth rate** is 1.019. This means that the model projects that next year's squirrel population will be 101.9% of this year's or about 2% growth per year.

b. Greenpeace is studying the humpback whale population. The whale population is surveyed every decade. The surveyors have broken the whale population into groups according to their age. The researchers compare the results of the first two surveys and find that:

whales 0–10 years old average 0.1 offspring, and 25% live to ten.
whales 10–20 years old average 0.96 offspring, and 80% live to twenty.
whales over 20 years old average 0.18 offspring, and none live to thirty.

(To do this they had make some assumptions about the "mom" of each "pup.") They also ASSUME that survival rates and birth rates do not change and that no whales live more than 30 years. We are asked:

What will happen to the whales if current trends continue?

Of course, we determine the transition matrix A and compute its spectral radius.

The transition matrix of this population model is:
The minimal polynomial of A is

$$A = \begin{pmatrix} 0.1 & 0.96 & 0.18 \\ 0.25 & 0 & 0 \\ 0 & 0.8 & 0.0 \end{pmatrix}$$

$$\mu_A(x) = x^3 - 0.1x^2 - 0.24x - 0.036.$$

This polynomial has three real zeros, namely $-.300, -.200$ .600 and so A has spectral radius $a = .6 < 1$. Therefore, case 1 of the "Evolution of the matrix sequence" on page 149 holds, and the system converges to 0. This model predicts the rapid extinction of the humpback whale!

The researchers are not happy. They decide to play "Whatif."

> "Suppose somehow the infant survival rate
> for offspring of young mothers could be increased."

We go back and make the matrix entry $a_{2,1}$ a variable and solve for spectral radius 1 — hmm! Well anyway, it turns out that a 90% survival rate would be required before the unhappy fate of the whales would be reversed by this change alone. In this case, The transition matrix of this population model is A and its minimal polynomial is

$$A = \begin{pmatrix} 0.1 & 0.96 & 0.18 \\ 0.9 & 0 & 0 \\ 0 & 0.8 & 0.0 \end{pmatrix}$$

$$\mu_A(x) = x^3 - 0.1x^2 - 0.864x - .129.$$

It also has three real zeros, namely $-.787$, $-.157$, 1.045, and so its spectral radius is $a = 1.045$. This means that the whale population in this model is projected to grow, increasing at a rate of 4.5% per decade.

> Okay, so what about parenting classes for whales?

The point here is that the real value of such models is that they allow "management" of resources to achieve maximal effect.

c. The department of wildlife is studying big horned sheep sampling every 5 years. They observe that big horned sheep 0–5 years old produce 0.2 offspring and 50% survive the sample period. Big horned sheep 5–10 years old produce 1.0 offspring per sample period on average and 100% survive the sample period. Big horned sheep over 10 years produce an average of 0.6 offspring per sample period and none survive the sample period.

According to this model, what will happen to the big horned sheep population long-term?

We determine the transition matrix A and compute its spectral radius.

The transition matrix of this population model is:
The minimal polynomial of A is

$$A = \begin{pmatrix} 0.2 & 1 & 0.6 \\ 0.5 & 0 & 0 \\ 0 & 1 & 0 \end{pmatrix}$$

$$\mu_A(x) = x^3 - 0.3x^2 - 0.5x - 0.2.$$

This polynomial has one real zero, namely 1.00, and two complex conjugate zeros. By the formula on page 149, their absolute value is $\sqrt{\frac{-(-0.2)}{1}} < 1$, so A has spectral radius $a = 1$.

Therefore, case 2 of the "Evolution of the matrix sequence" holds, and the system converges to 0. Therefore, this model predicts a stable steady-state in the big horn sheep population.

As with Markov chains, the stable steady-state is the eigenvector of A associated with the eigenvalue 1 and whose coordinates sum to 1. The predicted population distribution is 50% infants and 25% mature and old sheep. □

SUMMARY

The *evolution* of a *discrete dynamical system* having *transition* matrix A is determined by the *spectral radius* a of A and is also the *asymptotic growth rate* of the model. Just as with matrix sequences, a discrete dynamical system may *diverge, converge to a steady state, oscillate* or *converge to 0*.

Markov chains are discrete dynamical systems whose transition matrices have non-negative entries and whose columns sum to 1. They either converge to a steady state or oscillate.

Leslie population models depend on age grouped fertility and survival rates. These discrete dynamical systems exhibit all possibilities — overpopulation, extinction, oscillation and sustained stability.

Looking further

Discrete dynamical systems form an important class of mathematical models that are widely used in government, business and science. Here is a fine reference for population models.[6]

* * * * * * *

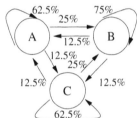

Practice Problems

1. The figure describes a linear discrete dynamical system. Write the transition matrix and compute its spectral radius. Does this model predict a stable steady-state? If so, what is it?

2. Yucca Mountain now has three grocery stores: Jose's, Bob's and Carla's. Last month, two new stores opened at the same time, Jose's and Carla's. People in Yucca Mountain now have a choice. It seems that from week-to-week 10% of Jose's customers are loyal, 80% switch to Bob's and 10% switch to Carla's. From week-to-week 40% of Bob's customers switch to Jose's, 20% are loyal and 40% switch to Carla's. Also from week-to-week 10% of Carla's customers switch to Jose's, 80% switch to Bob's and 10% are loyal.

   Write the transition matrix and compute its spectral radius. Find the steady-state distribution of customers. How many weeks after the opening of the new stores will it take to get to within 10% of this steady state?

3. The men of medieval Camelot are nobility, guildsmen and peasants. Eighty percent of the sons of nobility become nobility, 10% guildsmen and 10% peasants. Twenty percent of the sons of guildsmen become nobility, 70% guildsmen and 10% peasants. Twenty percent of the sons of peasants become nobility, 10% guildsmen and 70% peasants.

---

[6]*Matrix Population Models*, Hal Caswell, Sinauer Press, Sunderland, MA, 2001.

Write the transition matrix and compute its spectral radius. Does this model predict a steady-state distribution of the male population? If so, what is it?

4. Westslope is a town with many apple orchards. The annual harvests are fabulous, average or terrible. If this year's is fabulous, next year has probabilities 0.2 (thats 20%) of being fabulous, 0.4 of average and 0.4 of terrible. If this year's is average, then next year's harvest has probabilities 0.2 of being fabulous, 0.6 of average and 0.2 of terrible. If this year's is terrible, then next year's harvest has probabilities 0.4 of being fabulous, 0.4 of average and 0.2 of terrible. Write the transition matrix and compute its spectral radius. Does this model predict that the distribution of apple harvests has a steady state? If so, what is the probability that the harvest will be average in the steady state?

5. The Department of Fish and Wildlife is modeling the elk population. They estimate that elk 0–3 years old average 0 offspring, and 100% live three more years. Elk 3–6 years old average .8125 offspring, and 25% live three more years. Elk over 6 years old average 0.75 offspring. No elk survive 9 years. Write the transition matrix and compute its spectral radius. What is the asymptotic growth rate, and what long-term behavior does this model predict?

6. A biology class is studying a fruit fly population. They found that flies 0–2 days old average 0 offspring, and 100% live two more days. Flies 2–4 days old average 1.55 offspring, and 50% live two more days. Flies over 4 days old average 0.6 offspring. No fruit flies survive 6 days. Write the transition matrix and compute its spectral radius. What is the asymptotic growth rate and what long-term behavior does this model predict?

7. Greenpeace is studying the killer whale population. The researchers believe that whales 0–10 years old average 0.4 offspring, and 80% live ten more years. Whales 10–20 years old average 0.65 offspring, and 50% live ten more years. Whales over 20 years old average 0.2 offspring. No whales survive 30 years. Write the transition matrix and compute its spectral radius. What is the asymptotic growth rate and what long-term behavior does this model predict?

8. The Department of Fish and Wildlife is modeling the deer population. They estimate that 0–2 year old deer average 0.2 fawns, and 50% live two more years. Deer 2–4 years old average 0.86 fawns, and 80% live two more years. Deer over 4 years old average 0.25 fawns. No deer survive 6 years. Write the transition matrix and compute its spectral radius. What is the asymptotic growth rate, and what long-term behavior does this model predict?

9. Consider a three state Leslie model with incremental survival rates s_1, s_2 and fertility rates f_1, f_2, f_3. Show that the resulting transition matrix A is diagonalizable unless $s_1 = 0$ or some fertility rate is zero.
   HINT: Show that $\mu_A(x) = x^3 - f_1 x^2 - s_1 f_2 x - s_1 s_2 f_3$. Then compute the greatest common divisor of $\mu_A(x)$ and its derivative. Software recommended!

3.7   Matrix compression with components

Objective 1

Given $r \leq 2$ and a 4 by 4 matrix with distinct eigenvalues, compute its components and write an optimal rank r approximation to A.

In this section matrices are assumed to have real entries and be diagonalizable.

It is natural to approximate a matrix with a simpler matrix, if possible. The basic idea of this section is to do this using a matrix of smaller rank. Problem 6 of Section 3.4 shows that an n-by-n matrix of rank r can be reconstructed from at most $2nr$ numbers. When r is small the dramatic difference between $2nr$ and n^2 justifies the term **compression**.

But how best to choose a low rank approximation to A? This is a more subtle problem than that discussed in Section 2.5 with least squares approximations for two reasons. The first is that the set of rank r matrices is a much more complicated set than the set of solutions of a linear system. The second reason is more practical: How you measure closeness between matrices depends on how you are using them! If the matrix sequence $\{A^k\}$ is also to be approximated, then Section 3.5 makes it clear that matching the eigenvalues (and associated eigenvectors) of largest absolute value first is essential for success.

But what if you are just using the matrix as a tabular array of numbers as, for example, with frames of video images from outer space and different numbers distinguish brightness in the various pixel positions? Then you may wish to find an approximation $\tilde{A}$ of low rank to A by an analog to geometric projection in the sample space as done in Section 2.5. In this case the sum of the squares of the differences of the matrix entries:

$$(A_{11} - \tilde{A}_{11})^2 + \cdots + (A_{nn} - \tilde{A}_{nn})^2$$

is to be minimized. Exactly this kind of "data compression" is used transmit more pictures when limited power is available.

Remarkably, these two different "closeness" criteria lead to the SAME answer: components, as appearing in Section 3.4; but the proofs are quite beyond our means.[7]

Recall that each component for A is really just $f(A)$ for a very carefully chosen polynomial $f(x)$. Also recall that the equations relating (diagonalizable) A and its components are:

$$AA_{\lambda_i} = \lambda_i A_{\lambda_i} \text{ for each } i \text{ and } A = A_{\lambda_1} + \cdots + A_{\lambda_t}.$$

---

[7]Hotelling, H, "Analysis of a complex of statistical variables into principal components." J. Educ. Psychol. (1933);
"Eckhart C. and G. Young The approximation of one matrix by another of lower rank." Psycometrika, (1936).

Now take this a step further and define a sequence of approximations to A. First list the eigenvalues of A so that $|\lambda_1| \geq |\lambda_2| \geq \cdots |\lambda_t|$ and let m_i be the geometric multiplicity of λ_i. (Since A is assumed to be diagonalizable, we know that the sum of the m_is is the size n of A.) Define the s-th **component approximation**, $s \leq t$ $C_s(A)$ to A by

$$C_s(A) = A_{\lambda_1} + \cdots + A_{\lambda_s}.$$

WARNING: This definition depends on the order in which you listed eigenvalues of the same absolute value. Also notice that this is similar to, but different from, the deflation of a matrix with respect to a dominant eigenvalue. There we were removing dominant eigenvalue components. Here we are **keeping** dominant eigenvalue components and removing the rest.

Observe that $C_s(A)$ is unique when $|\lambda_s| > |\lambda_{s+1}|$. For us the key fact is that $C_s(A)$ is the UNIQUE BEST approximation of A among all matrices of rank less than or equal to $m_1 + \cdots + m_s$ with respect to either of the above two closeness criteria. Other authors define component approximations for all ranks, but uniqueness issues arise and we avoid the complication.

Example 3.19 (Matrix compression with components.)

a. Imagine the problem of approximating a matrix A by a rank one matrix M so that the sum of the squares of the entries of $A - M$ is minimized. This example gives a visual interpretation for two by two matrices. Let A be the indicated matrix. Because M has rank one it must be of the indicated form

$$A = \begin{pmatrix} 1 & a \\ a & 1 \end{pmatrix} \quad M = \begin{pmatrix} x & mx \\ y & my \end{pmatrix} \quad \text{for some numbers } x, y, m.$$

This is represented in the figure by the two points $\begin{pmatrix} x \\ y \end{pmatrix}$ and $\begin{pmatrix} mx \\ my \end{pmatrix}$,

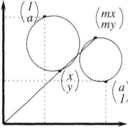

with coordinates the columns of M, being on the same line through the origin. The other labeled points are the columns of the matrix A. The two circles have as diameters the line segments

$$\overrightarrow{\begin{pmatrix} x \\ y \end{pmatrix} \begin{pmatrix} 1 \\ a \end{pmatrix}} \quad \text{and} \quad \overrightarrow{\begin{pmatrix} mx \\ my \end{pmatrix} \begin{pmatrix} a \\ 1 \end{pmatrix}}$$

and these directed line segments represent the geometric vectors that are the columns of $A - M$. The total area of the two circles is proportional to the sum of the squares of the lengths of these line segments which, in turn, equals the sum of squares of the entries of $A - M$.

The problem of finding the optimal matrix M is therefore equivalent to finding a figure as indicated so that the sum of the areas of the circles is minimal.

Check that A has eigenvalues $\lambda_1 = a+1$ and $\lambda_2 = a-1$ and when $a \geq 0$, as in the figure, $|\lambda_1| \geq |\lambda_2|$. The principal component of A is $A_{\lambda_1} = \frac{a+1}{2}\begin{pmatrix} 1 & 1 \\ 1 & 1 \end{pmatrix}$.

b. Let C be the indicated matrix. The eigenvalues of C can be computed via the minimal polynomial, the characteristic polynomial or the iterative method.

$$C = \begin{pmatrix} 1.0 & 0.47645 & 0.65292 \\ 0.47646 & 1.0 & 0.77038 \\ 0.65292 & 0.77038 & 1.0 \end{pmatrix} \qquad \begin{aligned} \lambda_1 &= 2.274 \\ \lambda_2 &= .5348 \\ \lambda_3 &= .1912. \end{aligned}$$

The first two matrix components can be computed from the factorization of $\mu_A(x)$ or by iteration. If we measure "length" of matrices as in part a, then C has "length" 5.494 and C_{λ_1} has length 5.171 or 94% ($= 5.171/5.494$) of the "length" of A. Thus, C can be approximated with respect to this measure by, the rank 1 matrix, C_{λ_1} with an error of only 6%. Check that, the rank 2 matrix, $C_2(A) = C_{\lambda_1} + C_{\lambda_2}$ approximates C with an error of only 1.2%.

$$C_{\lambda_1} = \begin{pmatrix} 0.64633 & 0.69682 & 0.75261 \\ 0.69682 & 0.75127 & 0.81142 \\ 0.75261 & 0.81142 & 0.87632 \end{pmatrix} \qquad C_{\lambda_2} = \begin{pmatrix} 0.3374 & -.2518 & -.0567 \\ -.2518 & 0.18786 & 0.04229 \\ -.0567 & 0.04228 & 0.00952 \end{pmatrix}$$

The theorem of Eckert and Young asserts that no matrix of rank 1 can do better than C_{λ_1}, and that $C_2(C) = C_{\lambda_1} + C_{\lambda_2}$ is the (only) "closest" matrix of rank 2 to C. □

In practice, component approximations are most useful when the size of the matrix n is large, and the rank r of the approximating matrix $\tilde{A}$ is small. This kind of matrix compression is used mostly when the cost of computation is low compared to cost of transmission and storage, as in for example, "remote sensing."

Objective 2

Given data involving three attributes, standardize it and compute the associated correlation matrix. Determine the first principal component and determine what percentage of the data variance it explains.

Principal component analysis is an extremely important type of matrix compression. It was first introduced by K. Pearson in 1901, but has been rediscovered many times in many different contexts. This fundamental "multivariate" statistical technique finds "linear" patterns relating many variables in large data sets (or shows that no such significant pattern exists).

A typical application involves numerical data measuring a **attributes** (also called objects or variables) in s **samples** (also called variables or subjects). The data may be summarized by an a-by-s numerical matrix M whose (i, j) entry is the measure of the i-th attribute in the j-th sample.

There are two ways to organize the data depending on whether the focus of attention is the attributes or the samples. Mathematically, the difference between these views is whether one works with M or M^T. We, of course, need to do both.

The insurance examiner, looking for suspicious claims, will concentrate on the s columns of M in $\mathbb{R}^a$, the **attribute space.** But the sociologist, looking for behavioral profiles, will focus on the a rows of M in $\mathbb{R}^s$, the **sample space.**

(By way of comparison, recall Section 2.5. There we considered the sample space model and only one attribute: the y value of a data point. The feasible values of approximating polynomials $f(x)$ at the x values of data formed what is called a subspace on page 218 and we found the polynomial whose values at the data was nearest the "attibute point" of y values.)

In either the attribute space or the sample space, the investigator uses the the geometry of $\mathbb{R}^n$ to find patterns. In order to more easily express any patterns that turn up (and simplify the geometric interpretation), data is customarily **standardized** in two steps. The process called **mean centering** amounts to a translation of coordinates in the attribute space so that the average value, over all samples, of each attribute is 0.

Example 3.20 (Mean centering, standardization — geometrically)
   Three scatter plots of data with two attributes and six samples are displayed. The top left is a scatter plot of raw data; for the lower left the data has been mean centered and for the right the data has been fully standardized as explained below.

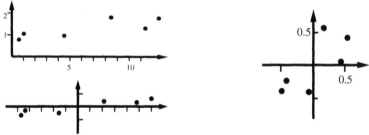

   Notice that the coordinate system origin of the lower left plot is visually centered in the data; that is the purpose of mean centering. Notice also that the data is more rotationally balanced in the right plot; that is the purpose of standardization.   □

Mean centering can be described in matrix terms as follows. Let J be an s-by-s matrix of all 1's. Then MJ/s is a matrix all of whose i-th row entries equal the average value of the i-th attribute and $M - MJ/s = M(I - J/s)$ is mean centered.

The second part of standardization is most easily expressed in the sample space model and is intended to control for the fact that the units used to measure the different attributes may not be comparable (literally apples and oranges). Since such differences have no statistical meaning, the convention is to replace each attribute vector in the sample space with the positive scalar multiple of length 1. (When interpreted in the attribute space, this normalization amounts to requiring the standard deviation of each attribute separately is 1.)

In matrix terms, the process of standardizing a mean centered data matrix M amounts to replacing it with DM, where D is a diagonal matrix whose i-th diagonal value is

$$\sqrt{1/(MM^T)_{ii}}.$$

Data is not standardized unless both $MJ = 0$ and MM^T has all diagonal entries 1.

Example 3.21 (Standardizing data.)

An unscientific study of six Christmas presents was made. The attributes studied were "maximum dimension in feet," "weight in pounds," and "cost in dollars." We wish to **determine if these attributes follow a pattern** and if so, to determine its strength.

gift	1	2	3	4	5	6
ft.	1.58	0.25	1.33	0.96	1.29	0.23
lb.	0.98	1.10	1.83	1.35	1.71	0.98
$	4.60	1.00	8.50	11.30	12.40	0.90

In order to quantify, for example, the extent which, dimension, weight and cost, are statistically related, the data is first "mean centered, " subtracting an appropriate constant from each row so the resulting matrix has rows that sum to zero:

$$M_m = \begin{pmatrix} 0.64 & -.69 & 0.39 & 0.02 & 0.35 & -.71 \\ -.345 & -.225 & 0.505 & 0.025 & 0.385 & -.345 \\ -1.85 & -5.45 & 2.05 & 4.85 & 6.00 & -5.55 \end{pmatrix}.$$

Because the attributes have incomparable units, it is essential that we find a way to treat them equally. The is achieved by multiplying each row by an appropriate positive number so that the resulting matrix has all rows the same "length" (1) in the attribute space $\mathbb{R}^6$.

$$M_s = \begin{pmatrix} 0.4960 & -.53477 & 0.30226 & 0.01550 & 0.27126 & -.55027 \\ -.41456 & -.27037 & 0.60683 & 0.03004 & 0.46263 & -.41456 \\ -.16412 & -.48350 & 0.18186 & 0.43027 & 0.52786 & -.49237 \end{pmatrix}.$$

The columns of the matrix M_s determine a scatter plot of six points in the (three-dimensional) attribute space. They are arranged in a balanced way around the origin. If you ignored the "size" attribute this arrangement would look like the third figure in Example 3.20. In fact, that example stepped through the standardization process for the weight and cost attributes geometrically. □

Assume henceforth that all data has been standardized, so $M = M_s$. Our goal is to decide if the scatter plot (more or less) lies along a single line, or if it (pretty much) falls in a single plane etc. Because the data is standardized, the scatter plot is balanced about the origin, any such line (or plane) will have to go through the origin.

To express the condition more clearly, imagine an arbitrary unit vector **a** in the attribute space ($\mathbb{R}^2$ in the figure and Example 3.20). We must express quantitatively how closely the scatter plot is approximated the line $\mathcal{A}$ with direction **a**. We will then let **a** vary over all unit vectors and look for a winner. This is basically Pearson's 1901 formulation. He called such a line a **principal component** of the data. Quite naturally, he used the square of the distance in the attribute space from a sample point, to the line as his measure of closeness.

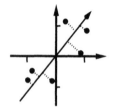

Hotelling in 1933 observed that each sample determined the hypotenuse of a right triangle having one side on $\mathcal{A}$ and that to minimize (the square of) one leg of the triangle is equivalent to maximizing (the square of) the other leg. Thus, he recast the problem as finding a unit vector **a** that maximizes

$$|P_\mathbf{a}(\mathbf{m}_1)|^2 + \cdots + |P_\mathbf{a}(\mathbf{m}_s)|^2 = (\mathbf{a}^T\mathbf{m}_1)^2 + \cdots + (\mathbf{a}^T\mathbf{m}_s)^2,$$

where $\mathbf{m}_j$ is the j-th column of the (normalized data) matrix M and $P_\mathbf{a}$ is the symmetric projector associated with **a** as in Section 2.4. Notice that this formula is in turn, simply the length (squared) of the vector $M^T\mathbf{a}$ in the **sample space**:

(3.1)   Maximize $\mathbf{a}^T M M^T \mathbf{a}$, for unit vectors $\mathbf{a} \in \mathbb{R}^s$.

This formulation leads us to the ***correlation matrix*** $C = MM^T$. The best way to understand this symmetric matrix is to work in the six dimensional sample space. In this space the ROWS of M determine three unit vectors $\mathbf{u}_i$, $i = 1, 2, 3$ and $c_{ij} = C_{ij}$ is the cosine of the angle between $\mathbf{u}_i$ and $\mathbf{u}_j$ (See the remark in the paragraph preceding Example 2.13 on page 79). Statisticians call c_{ij} the ***correlation*** of attributes i and j. If $\mathbf{u}_i$ and $\mathbf{u}_j$ are perpendicular, then $c_{ij} = 0$ and the associated attributes are ***uncorrelated***. Any matrix of the form MM^T can be interpreted in this way as the matrix of dot products of the rows of M. Such a matrix is called a ***Gramm*** matrix.

At this point a number of special properties of symmetric matrices in general, and Gramm matrices in particular come into play. To maintain continuity, we will simply list what is needed and come back to these theorems at the end of the section.

1. Any real symmetric matrix is diagonalizable.

2. The eigenvalues of a Gramm matrix are greater than or equal to zero.

3. A real symmetric matrix has a diagonalizing matrix P such that $P^{-1} = P^T$.

Assume that the correlation matrix has distinct eigenvalues and apply these results to $MM^T = C$ in Hotelling's formulation (3.1). Then there is a diagonal matrix D and a matrix P such that $CP = PD$; equivalently $MM^T = C = PDP^{-1} = PDP^T$, by result 3.

Observe that result 3 also implies that $\mathbf{u} = P^T\mathbf{a}$ is a unit vector if and only if **a** is a unit vector and substitute this equation into (3.1) to obtain:

Maximize $\mathbf{u}^T D\mathbf{u}$, for **u** unit vectors $\mathbf{u} = P^T\mathbf{a} \in \mathbb{R}^s$.

By result 2. and our assumption that C and so also D has distinct non negative eigenvalues $\lambda_1 > \cdots > \lambda_a \geq 0$. The expression to be maximized is simply:

(3.2)   Maximize $\lambda_1 u_1^2 + \cdots + \lambda_a u_a^2$ for $u_1^2 + \cdots + u_a^2 = 1$.

Practice problem 3 hints at how to show that this expression achieves its maximum value when $u_1 = 1$ and the other $u_i = 0$. This implies that the maximum in (3.1) is achieved for **a** an eigenvector associated with the dominant eigenvalue of C.

In summary, Pearson looked for a principal component, that is, a linear combination of attributes or **profile** that best describes the variation of the attributes across the entire population. We have shown that a scatter plot has a principal component, in this sense, whenever the correlation matrix C has a unique dominant eigenvalue with geometric multiplicity one and that the profile is an associated eigenvector. Assuming C to have no eigenvalues with geometric multiplicity greater than one, result 3 above implies eigenvectors of C are mutually perpendicular, so the principal component can be removed and a second uncorrelated profile calculated etc. This process produces exactly the component decomposition of C as defined in Section 3.4.

By Definition, the total **variance** of standardized data is the trace of C. By Problem 11 of Section 3.1, it is the sum of the eigenvalues of C. Since C has non-negative eigenvalues, a profile's strength is measured by the associated C-eigenvalue's size.

Example 3.22 (The principal components of a correlation matrix.)

Consider again the correlation matrix $C = M_s M_s^T$ of the Christmas present data in part b of the preceding example. It has eigenvalues

$$\lambda_1 = 2.274, \lambda_2 = .5348, \lambda_3 = .1912.$$

The principal component $C_1(A) = C_{\lambda_1}$ accounts for $76\%(= \lambda_1/trace(C))$ of the total variance. The second component C_{λ_2} accounts for an additional $20\% = \lambda_2/trace(C)$ of the total variance.

Because correlation matrices are numerical, round off errors may prevent the computation of their eigenvectors by linear system system solution as done in Section 3.1. The alternative eigenvector method, outlined just before Example 3.11 on page 142, is more "stable" and gives useful information for such matrices.

The matrix $(C - \lambda_2)(C - \lambda_3)$ has λ_1-eigenvectors as columns but percentages are used here. Therefore eigenvectors are **normalized** by dividing each entry by the sum of all entries — as with Markov chains (see Example 3.17 on page 160).

$$(C - \lambda_2)(C - \lambda_3) \doteq \begin{pmatrix} 1.03 & 1.12 & 1.20 \\ 1.11 & 1.20 & 1.29 \\ 1.20 & 1.29 & 1.40 \end{pmatrix}$$ Therefore the normalized dominant eigenvector of C is: $\doteq \begin{pmatrix} .308 \\ .332 \\ .359 \end{pmatrix}.$

The strongest gift profile (76%) back at the Christmas tree is determined by the coordinates of thenormalized λ_1- eigenvector. This gift profile, **size** consists of 31% dimension, 33% weight and 36% cost.

Apart from this **size** gift profile, there is a second strong pattern accounting for 83% of the remaining variation (and 20% of the total gift variation). This second gift profile, **packaging** is positively reflected by 52% of dimension but negatively reflected by 39% weight and 9% cost — the relative coordinates of a λ_2-eigenvector.   □

Real symmetric matrices are diagonalizable

Suppose A is ANY real symmetric matrix and write $\mu_A(x) = (x - \lambda)^m g(x)$ with $g(\lambda) \neq 0$. By the minimal polynomial theorem, we must show $m = 1$. By definition of the $\mu_A(x)$,

$$(A - \lambda I)^m g(A) = 0 \text{ but } (A - \lambda I)^{m-1} g(A) \neq 0.$$

Let $\mathbf{v}$ be the first nonzero column of $(A - \lambda I)^{m-1} g(A)$. Then $(A - \lambda I)\mathbf{v}$ is the first column of a zero matrix, and so equals $\mathbf{0}$. In other words, $\mathbf{v}$ is a λ-eigenvector of A, and as in Section 3.1, $(A - \lambda I)\mathbf{v} = \mathbf{0}$. If $m \geq 2$, multiply this equation on the left by $g(A)(A - \lambda I)^{m-2}$:

$$\mathbf{0} = g(A)(A - \lambda I)^{m-2}[(A - \lambda I)\mathbf{v}] = (A - \lambda I)^{m-1} g(A)\mathbf{v}.$$

But look at the first nonzero row of the symmetric matrix $(A - \lambda I)^{m-1} g(A)$; it is $\mathbf{v}^T$, and we have shown that $\mathbf{v}^T \mathbf{v} = \mathbf{0}$, $\mathbf{v} \neq \mathbf{0}$. This is a contradiction and the only way out is that is $m \leq 1$. This completes the proof that C is diagonalizable.

$C= MM^T$ has non negative real eigenvalues

Now suppose $C = MM^T$. Then C is symmetric and therefore diagonalizable. Let $\mathbf{v}$ be an eigenvector of C. Then

$$0 \leq |M^T \mathbf{v}| = (M^T \mathbf{v})^T M^T \mathbf{v} = \mathbf{v}^T (\lambda \mathbf{v}) = \lambda |\mathbf{v}|.$$

This implies that $0 \leq \lambda \in \mathbb{R}$. This shows that C has non negative real eigenvalues.

A diagonalizing matrix P such that $P^T P=1$ exists

Let be t the number of distinct eigenvalues of A (also the degree of $\mu_A(x)$). If λ has geometric multiplicity 1 with associated basic eigenvector $\mathbf{v}$ then set $P_\lambda = \frac{1}{|\mathbf{v}|}\mathbf{v}$, and call this process **scaling**. Note that $P_\lambda^T P_\lambda = 1$

Suppose λ has geometric multiplicity $m > 1$ and let V be a matrix whose columns are the basic eigenvectors, $\mathbf{v}_1 \ldots \mathbf{v}_m$ associated with λ. Observe that V has full rank (every column of $RREF(V)$ is pivotal). Also start with an empty matrix P_λ.

Remove the last column from V calling the resulting matrix W. Scale the removed vector, as above, so that it has length 1. Call the result $\mathbf{w}$ and append it to P_λ as the last column.

Observe that equation $(\mathbf{w}^T W)\mathbf{x} = 0$ is a non trivial homogeneous linear system with just one equation. Form the basic solutions of this linear system into a matrix X and replace V with WX. Suppose $\mathbf{z} = V\mathbf{y}$ is a linear combination of the columns of V (see Example 2.7 on page 69). Then

$$0 = (\mathbf{w}^T W)X\mathbf{y} = \mathbf{w}^T(WX)\mathbf{y} = \mathbf{w}^T V\mathbf{y} = \mathbf{w}^T \mathbf{z}.$$

In other words $\mathbf{z}$ is perpendicular to $\mathbf{w}$. Observe also that the new V has one fewer columns than the old V, but still has full rank.

Repeat the last two paragraphs until V is gone and P_λ has m columns. Observe that if $1 \le j < j \le m$, then j-th column of P_λ is, by construction, a linear combination of the basic solutions of $((P_\lambda)_i)^T\mathbf{x} = 0$. It follows that $P_\lambda^T P_\lambda = I_m$.

Define P to be the blocked matrix (see Example 2.20 page 99) $(P_{\lambda_1}|\ldots|P_{\lambda_t})$ and consider the blocked matrix product $P^T P$. Each block on the main diagonal is an identity matrix by the preceding paragraph. Each off diagonal block is zero by Problem 5 of Section 3.1. This shows $P^T P = I$. Since every column of P is an eigenvector of A, $P^{-1}AP = P^T AP$ is diagonal.

SUMMARY

Approximating a given square matrix A by one of small rank can save time and space, as well as highlight the most significant information it conveys. The optimal such approximation of rank r, with respect to a variety of measures of "matrix closeness," is sum of the components A_λ, starting with dominant eigenvalues and working downward by absolute value.

This kind of **matrix compression** is the basis of **principal component analysis**, an important multivariate statistical technique. Multivariate data with a attributes recorded for s samples can be tabulated in an a-by-s matrix M. There are two higher-dimensional geometric models. The **sample model** has one vector in $\mathbb{R}^s$ for each attribute (rows of M), and the **attribute model** has one vector in $\mathbb{R}^a$ for each sample (columns of M).

Data is **standardized** by first **mean centering**, then "scaling" so each attribute has standard deviation 1. For a standardized data matrix M_s, the **correlation matrix** $C = M_s M_s^T$ has entries that are the cosines of angles between the respective attribute vectors in the sample model. If C has a unique dominant eigenvalue λ of multiplicity 1, then the relative components of an associated eigenvector determines an attibute **profile** that accounts for $(\lambda/s)(100\%)$ and the variance of the data.

Looking further

A typical application of principal component analysis is "ordination" the identification and study of relationships between ecological communities in an ecosystem.

A remarkably clear exposition of the geometric view of multivariate statistics appears in Chapter 2 of Krzanowski.[8] In actual practice, matrix components are computed by a completely different method than we have used. Computational issues are addressed in an appendix of[9]. A more readable treatment that emphasizes applications is contained in[10]. Finally, a fun website with "detailed examples" of many statistical methods is http://lib.stat.cmu.edu/DASL/.

Practice Problems

1. Six fictitious snack foods are compared for calories, cost and nutritional value in the indicated data. Standardize the data matrix and compute the principal component of the associated correlation matrix. Describe the extent to which each attribute contributes to the principal component and the percentage of the data variance for which it accounts.

	a	b	c	d	e	f
calories	0.25	0.09	0.4	0.61	0.4	−0.78
cost	0.62	0.11	0.88	0.01	0.3	0.81
nutrition	−0.05	−0.28	0.04	−0.11	0.1	0.57

2. Six imaginary students compare their "internet surfing time," "course grades" and "general level of Happiness" Standardize the data matrix and compute the principal component of the associated correlation matrix. Describe the extent to which each attribute contributes to the principal component and the percentage of the data variance for which it accounts.

	a	b	c	d	e	f
surfing	0.46	−0.2	0.52	0.31	0.65	0.46
grades	−0.24	0.63	0.73	0.76	−0.68	0.67
happiness	−0.38	0.23	0.14	0.58	−0.92	−0.08

3. Let $a, b, x, y \in \mathbb{R}$. Suppose $a > b \geq 0$ and $x^2 + y^2 = 1$. Show that if $y \neq 0$ then

$$ax^2 + by^2 = b + (a - b)x^2 < a.$$

   Use this to determine the maximum value of the expression in equation (3.2).

---

[8]**Principles of Multivariate Analysis** by W. J. Krzanowski, Oxford University Press, 1988.

[9]**Principal Component Analysis** by I. T. Jolliffe, Springer-Verlag, New York, 1993.

[10]**Applied Factor Analysis in the Natural Sciences** by Richard Reyment and K. G. Jöreskog, Cambridge University Press, 1993.

Chapter 4

Determinants

The *determinant* of the n-by-n matrix A, written $|A|$ or $det(A)$, is a polynomial function of the entries of A. If the matrix A has numerical entries, then $det(A)$ is a number. The invention of determinants is generally attributed to Leibniz (the German philosopher and mathematician), but he failed to publish his work and determinants do not appear in his correspondence until well after 1683, when they were published by the samurai mathematician 関 孝和 Takakazu Seki Kowa.[1] As such, determinants predate Gauß' work by at least 120 years and Cayley's introduction of matrices by more than 150 years! Determinants represent a fundamental mathematical invariant that has been pivotal in the development of modern mathematics.

However, determinants and their role in courses like this one are now, and have long been, controversial.[2] More than 60 years ago, C. C. MacDuffee wrote, "In fact, the importance of the concept of determinants has been and currently is, vastly over-emphasized... perhaps ninety percent of matric theory can be developed without mentioning a determinant (sic.)"[3]

In addition, determinants present an expository challenge. The most elegant definition is not used in computation, and their most important properties don't seem at first to have much to do with either definition or method of computation. Technically correct support of their properties requires both attention to detail and mathematical sophistication. Consequently, determinant presentations can be unmotivated and mysterious.

For these reasons, we overlay a review of earlier work using many of the tools that we have developed as motivation for determinantal results. We begin with the most important properties of 2-by-2 determinants and use a simple recursive definition for the general case. Each general determinantal theorem is then an exercise in recursion. This theoretical development is interwoven with computational tools and applications. Only at the end do we come around to the full picture and see why determinants are the subject of such a long-running pedagogical controversy.

---

[1] http://www-gap.docs.st-and.ac.uk/history/Mathematicians/Seki.html.

[2] "Down with determinants" by S. Axler, Amer. Math. Monthly, 102 (1995), 139–154.

[3] **Vectors and Matrices** by C. C. MacDuffee p. v, Mathmatical Association of America (1943).

4.1   Area and composition of linear functions

Objective 1

Given square matrix A, factor it as a product of elementary matrices and RREF(A). Use a sequence of transform plots to realize the associated linear function f_A as a composition of elementary linear functions.

Gauss-Jordan elimination could have been presented in a different way using matrix multiplication. This multiplicative version of elimination corresponds to a sequence of transform plots. This leads to the important fact that matrix multiplication corresponds to composition of linear functions.

Pick your favorite elementary row operation and let E be the result of applying it to the identity matrix.

$$\begin{pmatrix} 1 & 0 \\ 0 & 1 \end{pmatrix} \xrightarrow{\;(-1)_1\;} \begin{pmatrix} 1 & 0 \\ -1 & 1 \end{pmatrix} = E$$

Now compare A and EA for A of appropriate shape. **Notice that left multiplication by E has the same effect on A as the original row operation**

$$A = \begin{pmatrix} 1 & 2 \\ 1 & 3 \end{pmatrix} \xrightarrow{\;(-1)_1\;} \begin{pmatrix} 1 & 2 \\ 0 & 1 \end{pmatrix} = \begin{pmatrix} 1 & 0 \\ -1 & 1 \end{pmatrix} \begin{pmatrix} 1 & 2 \\ 1 & 3 \end{pmatrix} = EA.$$

For this reason, E is called an **elementary matrix** and the associated linear function f_E is an **elementary linear function**.

Gauss-Jordan elimination consists of applying a sequence of elementary row operations to transform a matrix A into RREF(A). We can rephrase it in matrix multiplication language. For every matrix A, there is a sequence of elementary matrices $E_1 \ldots E_s$ such that

$$A \xrightarrow{E_1} \ldots \xrightarrow{E_s} RREF(A) \quad \text{or equivalently} \quad RREF(A) = E_s \ldots E_1 A.$$

Because each elementary row operation can be reversed by another elementary row operation, each elementary matrix is invertible and has an elementary matrix as its inverse. We sometimes write the preceding equation as

$$A = F_1 F_2 \ldots F_s RREF(A) \quad \text{where } F_i = E_i^{-1}.$$

The Es and Fs are in **reversed order** because $(AB)^{-1} = B^{-1}A^{-1}$ (recall Problem 7 of Section 2.3 and the formula on page 73).

Example 4.1 (Elementary matrices.)

a. The matrix reduction:

$$A = \begin{pmatrix} 1 & 2 \\ 1 & 3 \end{pmatrix} \xrightarrow{\;(-1)_1\;} \begin{pmatrix} 1 & 2 \\ 0 & 1 \end{pmatrix} \xrightarrow{\;(-2)_2\;} \begin{pmatrix} 1 & 0 \\ 0 & 1 \end{pmatrix} = RREF(A)$$

is equivalent to the sequence of matrix multiplications:

$$E_1 A = \begin{pmatrix} 1 & 0 \\ -1 & 1 \end{pmatrix} \begin{pmatrix} 1 & 2 \\ 1 & 3 \end{pmatrix} = \begin{pmatrix} 1 & 2 \\ 0 & 1 \end{pmatrix} ; \; E_2(E_1 A) = \begin{pmatrix} 1 & -2 \\ 0 & 1 \end{pmatrix} \begin{pmatrix} 1 & 2 \\ 0 & 1 \end{pmatrix} = \begin{pmatrix} 1 & 0 \\ 0 & 1 \end{pmatrix}.$$

It leads to the equivalent formulas where $F_i = E_i^{-1}$,

$$E_2 E_1 A = \begin{pmatrix} 1 & -2 \\ 0 & 1 \end{pmatrix} \begin{pmatrix} 1 & 0 \\ -1 & 1 \end{pmatrix} \begin{pmatrix} 1 & 2 \\ 1 & 3 \end{pmatrix} = \begin{pmatrix} 1 & 0 \\ 0 & 1 \end{pmatrix}$$

$$A = F_1 F_2 = \begin{pmatrix} 1 & 0 \\ 1 & 1 \end{pmatrix} \begin{pmatrix} 1 & 2 \\ 0 & 1 \end{pmatrix}.$$

b. Of course elementary matrices work with rectangular matrices too. The process

$$A = \begin{pmatrix} 0 & 0 & 1 & 2 \\ 2 & 4 & -2 & 0 \\ 2 & 4 & -1 & 2 \end{pmatrix} \xrightarrow{RREF} \begin{pmatrix} 1 & 2 & 0 & 2 \\ 0 & 0 & 1 & 2 \\ 0 & 0 & 0 & 0 \end{pmatrix} = RREF(A)$$

is achieved by the sequence row operations:

$$\begin{array}{c} \times \end{array} \; (1/2) \to \begin{pmatrix} 1 & 2 & -1 & 0 \\ 0 & 0 & 1 & 2 \\ 2 & 4 & -1 & 2 \end{pmatrix} \begin{array}{c} (-2) \\ (-1) \end{array} \begin{pmatrix} 1 & 2 & -1 & 0 \\ 0 & 0 & 1 & 2 \\ 0 & 0 & 0 & 0 \end{pmatrix} (1)$$

and corresponds to the matrix equation:

$$RREF(A) = \begin{pmatrix} 1 & 1 & 0 \\ 0 & 1 & 0 \\ 0 & 0 & 1 \end{pmatrix} \begin{pmatrix} 1 & 0 & 0 \\ 0 & 1 & 0 \\ 0 & -1 & 1 \end{pmatrix} \begin{pmatrix} 1 & 0 & 0 \\ 0 & 1 & 0 \\ -2 & 0 & 1 \end{pmatrix} \begin{pmatrix} \frac{1}{2} & 0 & 0 \\ 0 & 1 & 0 \\ 0 & 0 & 1 \end{pmatrix} \begin{pmatrix} 0 & 1 & 0 \\ 1 & 0 & 0 \\ 0 & 0 & 1 \end{pmatrix} A.$$

There is also an important connection between linear functions and matrix multiplication; that is in fact how Cayley first came to define matrix multiplication in 1855. Just as we chained together elementary row operations and changes in coordinate systems, we also chain together or, ***compose***, linear functions.

Suppose linear functions $f_R : \mathbb{R}^m \to \mathbb{R}^n$ and $f_S : \mathbb{R}^n \to \mathbb{R}^t$ have associated matrices R and S. Then the composition $f_S \circ f_R$ is the vector function that results from first applying f_R and then applying f_S:

$$f_S \circ f_R(\mathbf{x}) = f_S(f_R(\mathbf{x})):$$

WARNING: Note that $f_S \circ f_R$ means "first do f_R then do f_S".
We used backward arrows when changing coordinates for just this reason.

We leave it as an exercise to check that the composition of two linear functions is itself a linear function. A second important exercise shows that the matrix associated with the composition $f_S \circ f_R$ is the matrix product SR. The shape compatibility requirement for matrix multiplication is nothing more than the matching of ranges and domains of the associated functions.

Now the factorization of A as product of elementary matrices and RREF(A) is also, by the above association, a factorization of the linear function f_A as a product of elementary linear functions. We can understand this matrix factorization at a geometric level by chaining together several transform plots.

Example 4.2 (Transform plots and elementary linear function composition.)
The matrix factorization in part a of the preceding example can also expressed by
the composition of two linear functions and the chaining together of their associated
transform plots. We have shaded a unit square in the domain and its image in the
range of each part as a way of keeping track of what is happening.

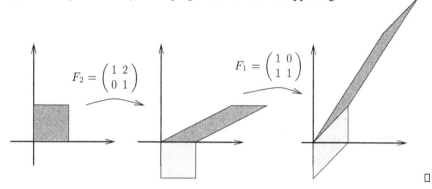

Objective 2

Given a transform plot of a linear function $f_A : \mathbb{R}^2 \to \mathbb{R}^2$ that tracks a figure, use
area and orientation to compute $det(A)$.

If you blast out the inverse of an arbitrary 2-by-2 matrix, rather like part b of Example 2.10 on page 73, you get

$$A^{-1} = \frac{1}{a_{11}a_{22} - a_{12}a_{21}} \begin{pmatrix} a_{22} & -a_{12} \\ -a_{21} & a_{11} \end{pmatrix}.$$

The denominator of that fraction is the **determinant** of A. For any 2-by-2 matrix A,
$det(A)$ (or simply $|A|$) is, by definition,

$$det(A) = |A| = a_{11}a_{22} - a_{12}a_{21}.$$

We will eventually see that there is an analogous formula for the inverse of an arbitrary n-by-n matrix A and, it too, has a fraction with denominator $det(A)$. But these
formulas are not nearly as important as the geometric meaning of determinants.

Determinants really have to do with "measure" (area or volume in two and three
dimensions) and how it is changed by a linear function. The area of the squares
and their images may differ in a more complicated example than Example 4.2. For
instance, in the reduction

$$\begin{pmatrix} 2 & 4 \\ 1 & 3 \end{pmatrix} \xrightarrow{\frac{1}{2}} \begin{pmatrix} 1 & 2 \\ 1 & 3 \end{pmatrix} \xrightarrow{(-1)_1} \begin{pmatrix} 1 & 2 \\ 0 & 1 \end{pmatrix} \xrightarrow{(-2)_2^1} \begin{pmatrix} 1 & 0 \\ 0 & 1 \end{pmatrix}$$

the **standard unit square** S, defined on page 42, would first have its height cut in
half and then proceed as in Example 4.2.

In general, we know that the image $f_A(S)$ of S under f_A is the parallelogram determined by the images of its corners. We have also seen in Example 1.19 on page 44 that each square of the domain graph paper maps to a parallelogram geometrically congruent to $f_A(S)$ and together these parallelograms form a grid. It follows[4] that the linear function f_A associated with the matrix A changes the area of any region $\mathcal{R}$ in the same way as the standard unit square S. **The most important property of 2-by-2 determinants is that they measure this change in area.**

$$AREA(f_A(\mathcal{R})) = |det(A)|AREA(\mathcal{R}) \text{ for any region } \mathcal{R}.$$

But what about the sign of the determinant? Surely the expression $a_{11}a_{22} - a_{12}a_{21}$ is sometimes negative, and area is never negative, so the absolute value in the area equation is essential. But what does a negative determinant mean geometrically?

It is easy to guess the answer (we offer no proof) by picking a few simple matrices that have negative determinant and studying their transform plots. For example, if you work with

$$\begin{pmatrix} 1 & 0 \\ 0 & -1 \end{pmatrix} \text{ or } \begin{pmatrix} 0 & 1 \\ 1 & 0 \end{pmatrix}$$

you quickly discover that the orientation of the axes is reversed. In the "before" part of the transform plot you traveled counterclockwise to go from the positive x-axis to the positive y-axis, but in the "after" part you travel clockwise to go from the positive x'-axis to the positive y'-axis. In practice it may be more direct to work directly with the indicated figure than with the coordinate axes.

Understanding this change of **orientation** in higher dimensions is one of the hardest parts of the geometric meaning of determinants.

Example 4.3 (From transform plot to determinant.)

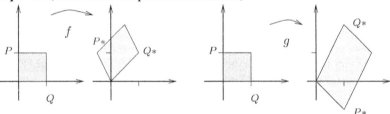

Because of the labeling of points P, Q and their images $P*, Q*$, we can tell that linear function f preserves orientation but g reverses it. Comparing the areas of the images of the unit squares, we conclude that $det(f) = 3/2$ while $det(g) = -3$. This is confirmed by computing the associated matrices as in Example 1.20 on page 45

$$\begin{pmatrix} 1 & 1 \\ \frac{-1}{2} & 1 \end{pmatrix} \text{ and } \begin{pmatrix} 1 & 1 \\ 2 & -1 \end{pmatrix}.$$

Then, for example, $AREA(f(S)) = |det \begin{pmatrix} 1 & 1 \\ \frac{-1}{2} & 1 \end{pmatrix} |AREA(S) = (1 - (-\frac{1}{2}))1.$ □

---

[4]To be rigorous here would require calculus.

A more important question is: **What is the geometric meaning of *det(A) = 0* ?** It can only mean that the unit square maps to a parallelogram of zero area. Again, you will understand it best if you try an example. Consider

$$A = \begin{pmatrix} 1 & 1 \\ 1 & 1 \end{pmatrix}$$

and compute the image of the standard unit square. Do you see that the figure is flattened to a line in the range plot?

Geometric proof that determinants measure change in area

Actually, this amounts to nothing more than working through a general version of Example 4.2 but it gets a little involved and may be skipped on first reading.

Suppose that the coordinates of P in the figure are the first column of A, and the coordinates of Q in the figure are the second column of A. This is equivalent to the oblique parallelogram being the transform plot under f_A of the standard unit square S (see Example 1.20 on page 45). Since P is not on the Y-axis, $a_{11} \neq 0$.

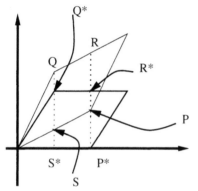

Apply a "*row+" row operation to force a_{21} to zero and a new matrix

$$B = \begin{pmatrix} a_{11} & a_{12} \\ 0 & a_{22} - \frac{a_{21}}{a_{11}}a_{12} \end{pmatrix}.$$

Imagine drawing the transform plot for B. The image under f_B of the standard unit square is $f_B(S)$ looks like the middle figure in Example 4.2.

In comparing $f_A(S)$ and $f_B(S)$, notice that corresponding points have the same first coordinates, so they lie on the same vertical line. (Of course, P corresponds to P^* because $b_{21} = 0$.) In fact, each point in $f_A(S)$ of the figure corresponds to its star (*) partner in $f_B(S)$. Some point S is on the line OP and has the same first coordinate as Q, namely a_{21}. Moreover, the line OP has slope a_{12}/a_{11}, so S has coordinates $(a_{21}\ a_{21}(a_{12}/a_{11}))^T$. Now

$$|\overrightarrow{SQ}| = |a_{22} - a_{21}(a_{12}/a_{11})| = |\overrightarrow{S^*Q^*}|,$$

since the coordinates of Q^* are given by the second column of B. Triangles $\triangle OQS$ and $\triangle OQ^*S^*$ have bases SQ and S^*Q^* of the same length and have the same height, by inspection, so have the same area. Also the parallelograms $PSQR$ and $P^*Q^*S^*R^*$ have the same area, and $\triangle PR(P+Q)$ and $\triangle P^*R^*(P^*+Q^*)$ have the same area. Thus, AREA($f_A(S)$) = AREA($f_B(S)$). Finally,

$$\text{AREA}(f_B(S)) = \begin{vmatrix} x\text{-coordi-} \\ \text{nate of } P^* \end{vmatrix} \times \begin{vmatrix} y\text{-coordi-} \\ \text{nate of } Q^* \end{vmatrix} = |a_{11}||a_{22} - \frac{a_{21}}{a_{11}}a_{12}| = |det(A)|.$$

Notice that this argument does not require that the points P and Q be in the first quadrant as in the figure. Can you craft an argument that covers the excluded case $a_{11} = 0$?

SUMMARY

Elementary matrices correspond to elementary row operations and elimination can be expressed in terms of matrix multiplication,

$$A = F_1 F_2 \ldots F_s RREF(A) \text{ where } A \xrightarrow{E_1} \cdots \xrightarrow{E_s} RREF(A) \text{ and } F_i = E_i^{-1}.$$

The composition of two linear functions is again a linear function. The product AB of two matrices A and B is the matrix of $f_A \circ f_B$.

The ***determinant*** of a 2-by-2 matrix A is $det(A) = a_{11}a_{22} - a_{12}a_{21}$. The absolute value $|det(A)|$ is a scaling factor describing how area changes from domain to range under the linear function f_A. If $det(A)$ is negative, the orientation of the coordinate axes is reversed.

Looking further

Calculus students know that substitution or "elur niahc" (chain rule in reverse)

$$\int f(x)dx = \int f(g(t))\frac{dx}{dt}dt, \text{ where } x = g(t)$$

is a fundamental method of integration.

In the calculus of several variables, this method is even more important, but complicated by the fact that the $\frac{dx}{dt}$ must involve several variables. Because integration has to do with area, it is no surprise that determinants must arise and that $\frac{dx}{dt}$ is replaced by the **Jacobian determinant**.

$$* * * * * * *$$

Practice Problems

1. Write a sequence of elementary matrices that reducesthe matrix to $RREF$. Draw a sequence of transform plots as in Example 4.2 that reflect your reduction.

   a. $A = \begin{pmatrix} 1 & -1 \\ 1 & 1 \end{pmatrix}$ b. $A = \begin{pmatrix} 1 & 1 \\ 1 & 1 \end{pmatrix}$ c. $A = \begin{pmatrix} 0 & 1 \\ 1 & 2 \end{pmatrix}$

2. Write a sequence of elementary matrices that reduces the matrix to $RREF$. Then use the sequence to factor A into a product of elementary matrices followed by $RREF(A)$.

   a. $A = \begin{pmatrix} 1 & 2 & 3 \\ 0 & 1 & 1 \\ 1 & 2 & 3 \end{pmatrix}$ b. $A = \begin{pmatrix} 1 & 1 & 5 \\ 2 & 3 & 12 \\ 0 & 1 & 2 \end{pmatrix}$ c. $A = \begin{pmatrix} 1 & 1 & 5 & 3 \\ 2 & 3 & 12 & 6 \\ 0 & 1 & 2 & 1 \end{pmatrix}$

3. Use the transform plots in Example 3.1 on pages 110 and 111 to visually compute the determinant of each matrix. Verify using the algebraic formula.

   a. $A = \begin{pmatrix} 2 & 0 \\ 0 & 1 \end{pmatrix}$ b. $B = \frac{1}{4} \begin{pmatrix} 5 & 3 \\ 3 & 5 \end{pmatrix}$ c. $C = \begin{pmatrix} 0 & 1 \\ -1 & 2 \end{pmatrix}$

4. Consider the unit circle $C = \{(x, y) : x^2 + y^2 = 1\}$. Use your answer to Problem 3 to, compute the area of C' the image of C under each of the linear functions. In each case sketch C'.

5.

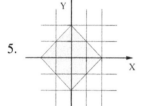

The figure to left has a grid with 1 by 1 squares and is the domain of a transform plot. In each part the figure is the associated range plot for a matrix A.
i) Estimate the determinant geometrically.
ii) Find the matrix A.
iii) Confirm i) using ii).

a.

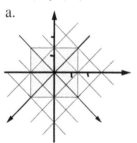

b.

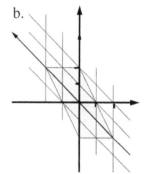

c.

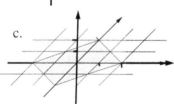

6. Suppose $f_A, f_B : \mathbb{R}^2 \to \mathbb{R}^2$ are two linear functions. Show that the composition $f_A \circ f_B$ defined by $f_A \circ f_B(\mathbf{x}) = f_A(f_B(\mathbf{x}))$ satisfies both the conditions:

$$f_A \circ f_B(\mathbf{x} + \mathbf{y}) = f_A \circ f_B(\mathbf{x}) + f_A \circ f_B(\mathbf{y}) \text{ for all } \mathbf{x}, \mathbf{y} \in \mathbb{R}^2$$
$$f_A \circ f_B(\alpha \mathbf{x}) = \alpha f_A \circ f_B(\mathbf{x}) \qquad\qquad \text{for all } \alpha \in \mathbb{R},\ \mathbf{x} \in \mathbb{R}^2$$

and is therefore a linear function.

7. Show the matrix of the linear function $f_A \circ f_B$ in problem 6 is AB. Use the fact, shown on page 47 at the end Section 1.6, that the columns of M are just the images of the columns of the identity matrix under the associated linear function f_M.

4.2 Computing determinants

Objective 1

Compute the determinant of an n-by-$n(n \leq 4)$ matrix using the cofactor definition.

For any n-by-n matrix we will eventually see that there is a formula for $det(A)$ in terms of the entries in A analogous to the 2-by-2 case. This **permutation expansion** is a sum of $n! (= (1)(2) \cdots (n-1)(n))$ terms. Each term in the sum is a product of n entries from the matrix. In addition, half of these products have a positive and half a negative sign.

Because this formula involves so much addition and subtraction, it is extremely ill-suited to numerical computation using floating point (computer) arithmetic (see the Calculators section on page 192).

A serious challenge to authors is — just how do you get most directly to a useful general determinant definition? We introduced recursion in Section 3.3 in anticipation of this dilemma. As with digraph walks and difficult enumeration problems, it is sensible to give a recurrence relation.

We need two preliminary definitions. Let A be an n-by-n matrix, $n \geq 3$. The i, j **minor**, $m_{ij}(A)$, of A is

$$m_{ij}(A) = \left\{ \begin{array}{l} \text{the } n-1\text{-by-}n-1 \text{ matrix obtained from } A \text{ by} \\ \text{deleting the } i\text{-th row and } j\text{-th column of } A. \end{array} \right\}$$

The i, j **cofactor** of A, denoted by A_{ij} is the number

$$A_{ij} = (-1)^{i+j} det(m_{ij}(A)).$$

WARNING: Do NOT confuse the cofactor A_{ij} with matrix entry a_{ij}.

As mentioned at the beginning of this chapter, there are two commonly used notations $det(A)$ and $|A|$ for the determinant of A. (We generally use the first one to avoid mysterious constructions like "the absolute value of $|A|$." On the other hand sometimes $|A|$ saves typesetting space.)

As also mentioned in the chapter introduction, there are many equivalent formulas for $det(A)$, any one of which could be taken as the starting point. We **arbitrarily** choose a formula that will later be called the first column cofactor expansion, but for now it is **OUR definition** of the **determinant** of the n-by-n matrix A:

$$det(A) = |A| = a_{11}A_{11} + a_{21}A_{21} + \cdots + a_{n1}A_{n1}.$$

(The lower case letters refer to entries in A, and the uppercase letters refer to cofactors of A.) We will eventually have a cofactor expansion for each row and for each column of A.

Example 4.4 (Determinant definition.)
a.   The first column minors of the matrix $A = \begin{pmatrix} 1 & 2 & 3 \\ 8 & 9 & 4 \\ 7 & 6 & 5 \end{pmatrix}$ are

$$m_{11}(A) = \begin{pmatrix} 9 & 4 \\ 6 & 5 \end{pmatrix} \quad m_{21}(A) = \begin{pmatrix} 2 & 3 \\ 6 & 5 \end{pmatrix} \quad m_{31}(A) = \begin{pmatrix} 2 & 3 \\ 9 & 4 \end{pmatrix}.$$

The first column cofactors of A are

$$A_{1,1} = (-1)^{1+1} det(m_{11}(A)) = 21$$
$$A_{2,1} = (-1)^{2+1} det(m_{21}(A)) = 8$$
$$A_{3,1} = (-1)^{3+1} det(m_{31}(A)) = -19.$$

Therefore, by definition,

$$det(A) = a_{11}A_{11} + a_{21}A_{21} + a_{31}A_{31} = 1(21) + 8(8) + 7(-19) = -48.$$

b.   To compute $det(B)$ for the indicated B, we must sum 4 terms, each of which is the product of an entry in the first column of B, with the corresponding cofactor. Since B is upper triangular, there is only one nonzero entry in the first column of B and the sum collapses to

$$B = \begin{pmatrix} 1 & 2 & 3 & 4 \\ 0 & 5 & 6 & 7 \\ 0 & 0 & 8 & 9 \\ 0 & 0 & 0 & 1 \end{pmatrix}.$$

$$det(B) = b_{11}B_{11} + 0 + 0 + 0 = 1(-1)^{1+1} det(m_{11}(B)) = det(m_{11}(B)).$$

But of course, the $1, 1$ minor $m_{11}(B)$ of B is also upper triangular, and the same argument leads to

$$det(B) = 1(5)(m_{11}(m_{11}(B)) = 1(5)det \begin{pmatrix} 8 & 9 \\ 0 & 1 \end{pmatrix} = 1(5)(8)(1) = 40. \qquad \square$$

Suppose that three square matrices A, B, C that match perfectly except for the first column are given. Then they have the same cofactors $A_{i1} = B_{i1} = C_{i1}$ for each value of i. If further, the first column of C is the first column of A plus the first column of B (so $a_{i1} + b_{i1} = c_{i1}$ for each i), then

$$
\begin{aligned}
det(A) + det(B) &= a_{11}A_{11} + \cdots + a_{n1}A_{n1} + b_{11}B_{11} + \cdots + b_{n1}B_{n1} \\
&= a_{11}C_{11} + \cdots + a_{n1}C_{n1} + b_{11}C_{11} + \cdots + b_{n1}C_{n1} \\
&= (a_{11} + b_{11})C_{11} + \cdots + (a_{n1} + b_{n1})C_{n1} \\
&= c_{11}C_{11} + \cdots + c_{n1}C_{n1} = det(C).
\end{aligned}
$$

This shows "first column linearity." This seemingly special situation portends a key determinantal property called **line linearity**.

But line linearity is a logical leap from our definition. As intermediate "stepping stones" we have five basic properties.

Basic properties of determinants

Suppose A, B and C are square matrices of the same size.

1. If B is obtained from A by interchanging two rows, then $det(B) = -det(A)$.

2. If A is upper triangular, then $det(A) = a_{11} \cdots a_{nn}$.

3. Suppose the entries of A, B and C coincide except for the i-th row, and $a_{ij} + b_{ij} = c_{ij}$ for each j. Then $det(A) + det(B) = det(C)$.

4. If B is obtained from A by multiplying a row by r, then $det(A) = \frac{1}{r} det(B)$.

5. If B is obtained from A by adding a multiple of one row to another, then $det(B) = det(A)$.

We show below that these five basic properties of determinants follow from the definition. It is extremely important that you understand exactly what each of these properties says and how to use it. **Verify that each of these properties really holds for any 2-by-2 matrix.**

Basic properties 1, 4 and 5 describe what happens to the determinant of a matrix when you apply an elementary row operation and basic property 2 allows the computation of the determinant of any matrix in REF. Basic property 3, called **row linearity**, is used to establish basic property 5. It, like the special situation just above, is a special case of line linearity.

A simple but useful consequence of basic property 1 is that a matrix A with two identical rows has determinant zero. Indeed, a row swap leaves the matrix unchanged and so its determinant is (the only number) equal to its negative. Therefore, $det(A) = 0$.

To establish the basic properties, we argue by *recursion* or *mathematical induction*. This means that in showing something to be true for n-by-n matrices, we assume it to be true for smaller matrices. Complete, detailed proofs would be very long. Mere outlines are presented.

It is strongly suggested that you use a general 3-by-3 matrix and track through each argument with your example first (as we did with Frobenius' theorem on page 141).

Basic property 1: (Row swap.)
Suppose that the rows being swapped are s and t. Compare the minors $m_{i1}(A)$ and $m_{i1}(B)$, $i \neq s, t$. They are equal in all rows but two, and these two are swapped. Because the minors have smaller size, we apply recursion to conclude that

$$det(m_{i1}(A)) = -det(m_{i1}(B)) \text{ for } i \neq s, t.$$

Now add the formulas for $det(A)$ and $det(B)$. Notice all but four terms cancel!

$$det(A) + det(B) = (-1)^{s+1} det(m_{s1}(A)) + (-1)^{t+1} det(m_{t1}(A))$$
$$+ (-1)^{s+1} det(m_{s1}(B)) + (-1)^{t+1} det(m_{t1}(B)).$$

Suppose for the moment that $t = s + 1$. Then $m_{s1}(A) = m_{t1}(B)$ and $m_{t1}(A) = m_{s1}(B)$ because the deleted row was/got swapped. Using these formulas the four remaining terms cancel, so $det(A) + det(B) = 0$! This proves property 1 when the rows are adjacent.

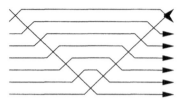 When the swapped rows are not adjacent, we leave it as an exercise to show that you can build up any row swap with an odd number of adjacent row swaps. The figure shows how to swap two rows that are separated by seven rows using thirteen adjacent row swaps.

Basic property 2: (Upper triangular.)
This argument is modeled in Example 4.1 B. Use the definition of det(A) and notice that all terms but the first are zero because $a_{i1} = 0$ for $i \geq 2$. Notice that $m_{11}(A)$ is also upper triangular and apply recursion.

Basic property 3: (Row linearity.)
Use basic property 1 and apply a row swap if necessary to reduce to the case where $i = 1$. Observe that for $k \geq 2$ the minors $m_{k1}(A)$, $m_{k1}(B)$ and $m_{k1}(C)$ satisfy the condition of this basic property but have smaller size. Apply recursion to conclude that
$$det(m_{k1}(A)) + det(m_{k1}(B)) = det(m_{k1}(C)) \text{ for } k \geq 2.$$
Now use the first column cofactor expansion (or definition) for the determinant of all three matrices to compute $det(A) + det(B) - det(C)$. Notice that all but three terms cancel because of the equations just obtained.
$$det(A) + det(B) - det(C) = a_{11}A_{11} + b_{11}B_{11} - c_{11}C_{11}$$
Finally, observe $m_{11}(A) = m_{11}(B) = m_{11}(C)$ since A, B and C match in all but the first row. The remaining terms cancel because $a_{11} + b_{11} = c_{11}$.

Basic property 4: (Row multiply.)
Again use basic property 1 and apply a row swap if necessary to reduce to the case where the affected row is the first. Observe that for $k \geq 2$ the minors $m_{k1}(A)$ and $m_{k1}(B)$ satisfy the condition of this basic property but have smaller size. Apply recursion to conclude that
$$det(m_{k1}(B)) = r det(m_{k1}(A)) \text{ for } k \geq 2.$$
Now use the first column cofactor expansion (our definition) for each of the matrices to compute $det(B) - r det(A)$. Notice that all but two terms cancel because of the equations just obtained
$$det(B) - r det(A) = b_{11}B_{11} + r a_{11}A_{11}.$$
Finally, recall that we started by arranging that, $r a_{11} = b_{11}$ and $A_{11} = B_{11}$ so the remaining terms cancel.

Basic property 5: (Row sum.)

Again use basic property 1 and apply a row swap if necessary to reduce to the case where the multiplied row is the first. Use basic property 1 again and apply a second row swap if necessary to reduce to the case where the other affected row is the second. Observe that for $k \geq 3$ the minors $m_{k1}(A)$ and $m_{k1}(B)$ satisfy the condition of this basic property but have smaller size. Apply recursion to conclude that

$$det(m_{k1}(A)) = det(m_{k1}(B)) \text{ for } k \geq 3.$$

Now use the first column cofactor expansion (our definition) for each of these matrices to compute $det(A) - det(B)$. Notice that all but four terms cancel because of the equations just obtained. Next use the equations $b_{11} = a_{11}$, $b_{21} = ra_{11} + a_{21}$ and $m_{12}(B) = m_{12}(A)$ to eliminate as many b's as possible and regroup:

$$
\begin{aligned}
det(A) - det(B) &= a_{11}A_{11} + a_{21}A_{21} - b_{11}B_{11} - b_{21}B_{21} \\
&= a_{11}A_{11} + a_{21}A_{21} - a_{11}B_{11} - (ra_{11} + a_{21})A_{21} \\
&= a_{11}(det(m_{11}(A)) - det(m_{11}(B)) - r(-det(m_{21}(A))).
\end{aligned}
$$

Now notice that the minors $m_{11}(A)$ and $m_{11}(B)$ match perfectly in all rows but the first, so basic property 3 (together with basic property 4 applied to $m_{21}(A)$) implies that $det(m_{11}(B)) = det(m_{11}(A)) + rdet(m_{21}(A))$. In other words, a miracle occurs and we are done!

Objective 2

Compute the determinant of an n-by-$n(n \leq 4)$ matrix using Gaussian elimination and the block triangular formula.

Basic properties 1, 4 and 5 explain what happens to $det(A)$ when an elementary row operation is applied to A. Because Gaussian elimination uses elementary row operations to reduce an arbitrary matrix to an upper triangular matrix (REF), these properties combine with basic property 2 to give an important alternative computational method. You can compute the determinant of a matrix while REFing the matrix. (You could RREF it, too, but the determinant does not change when you buy the last R.)

Example 4.5 (Determinants by reduction.)

a   Above each row operation a number r is written that is determined by one of the basic determinant properties so that

$$det(\text{before}) = r \, det(\text{after})$$

$$
\begin{array}{cccc}
-1 & 1 & 1 & 1 \\
\begin{pmatrix} 4 & 3 & 1 \\ 0 & 2 & 5 \\ 1 & 0 & -3 \end{pmatrix} &
\begin{pmatrix} 1 & 0 & -3 \\ 0 & 2 & 5 \\ 4 & 3 & 1 \end{pmatrix} \xrightarrow{(-4)} &
\begin{pmatrix} 1 & 0 & -3 \\ 0 & 2 & 5 \\ 0 & 3 & 13 \end{pmatrix} \xrightarrow{(-1)} &
\begin{pmatrix} 1 & 0 & -3 \\ 0 & -1 & -8 \\ 0 & 3 & 13 \end{pmatrix} \xrightarrow{(3)}
\end{array}
$$

$$\begin{array}{ccccccc}
& (-1)(-11) & & & 1 & & & 1 \\
\begin{pmatrix} 1 & 0 & -3 \\ 0 & -1 & -8 \\ 0 & 0 & -11 \end{pmatrix} & \begin{array}{c}(-1)\to \\ (-\frac{1}{11})\to\end{array} & \begin{pmatrix} 1 & 0 & -3 \\ 0 & 1 & 8 \\ 0 & 0 & 1 \end{pmatrix} & \xrightarrow[(3)]{} & \begin{pmatrix} 1 & 0 & 0 \\ 0 & 1 & 8 \\ 0 & 0 & 1 \end{pmatrix} & \xrightarrow[(-8)]{} & \begin{pmatrix} 1 & 0 & 0 \\ 0 & 1 & 0 \\ 0 & 0 & 1 \end{pmatrix}.
\end{array}$$

It is a simple matter to chain it all together to conclude that

$$det \begin{pmatrix} 4 & 3 & 1 \\ 0 & 2 & 5 \\ 1 & 0 & -3 \end{pmatrix} = (-1)(1)(1)(1)(-1)(-11)(1)(1)det \begin{pmatrix} 1 & 0 & 0 \\ 0 & 1 & 0 \\ 0 & 0 & 1 \end{pmatrix} = -11.$$

Notice that the last two row operations that transformed the matrix from REF to RREF didn't change the determinant at all.

b. Usually we don't write all that down. We just keep track of how the determinant changes as we progress. In this tally the "*row+"s are free and the "row swaps" just switch signs.

$$det \begin{pmatrix} 2 & 2 & 5 & 1 \\ 2 & 1 & 1 & 3 \\ 1 & 0 & 1 & 0 \\ 0 & 1 & 3 & 2 \end{pmatrix} = -det \begin{pmatrix} 1 & 0 & 1 & 0 \\ 2 & 1 & 1 & 3 \\ 2 & 2 & 5 & 1 \\ 0 & 1 & 3 & 2 \end{pmatrix} = \cdots = -det \begin{pmatrix} 1 & 0 & 1 & 0 \\ 0 & 1 & -1 & 3 \\ 0 & 0 & 5 & -5 \\ 0 & 0 & 4 & -1 \end{pmatrix}.$$

Now "factor" 5 out of the third row and do one more row operation and use basic property 3

$$= -5det \begin{pmatrix} 1 & 0 & 1 & 0 \\ 0 & 1 & -1 & 3 \\ 0 & 0 & 1 & -1 \\ 0 & 0 & 0 & 3 \end{pmatrix} = (-5)(1)(1)(1)(3) = -15.$$

c. This method even applies when the matrix has entries that are not constants (compare to Example 1.9 on page 21).

$$\begin{vmatrix} 8-x & 9 & 9 \\ 3 & 2-x & 3 \\ -9 & -9 & -10-x \end{vmatrix} = \cdots = \begin{vmatrix} -1-x & 3+3x & 0 \\ 3 & 2-x & 3 \\ 0 & -3-3x & -1-x \end{vmatrix} \begin{array}{l}\frac{1}{x+1}\to \\ \\ \frac{1}{x+1}\to\end{array}$$

$$= (x+1)^2 \begin{vmatrix} -1 & 3 & 0 \\ 3 & 2-x & 3 \\ 0 & -3 & -1 \end{vmatrix} = \cdots = (x+1)^2 \begin{vmatrix} -1 & 3 & 0 \\ 0 & 2+9-9-x & 0 \\ 0 & -3 & -1 \end{vmatrix} \begin{array}{l}(-1)\to \\ \frac{1}{2-x}\to \\ (-1)\to\end{array}$$

$$= (x+1)^2(-1)^2(2-x)det \begin{pmatrix} 1 & -3 & 0 \\ 0 & 1 & 0 \\ 0 & 3 & 1 \end{pmatrix} = (x+1)^2(-1)^2(2-x).$$

$\square$

Block matrices and determinants

As introduced in Section 2.7 with Example 2.20 on page 99, a large matrix M that is viewed as small matrix with "matrix entries" is called a ***block*** or ***partitioned matrix***.

At the end of this section we give the permutation expansion of $det(A)$ and from it obtain the following **block triangular generalization** of basic property 2:

$$det(M) = det(A)\,det(D) \text{ provided } M = \begin{pmatrix} A & B \\ 0 & D \end{pmatrix} \text{ and } A, D \text{ are square.}$$

(Needless to say, C is not required to be square and one of the matrices A or D might itself be block triangular.)

This formula can significantly shorten computation.

Example 4.6 (Determinants and block triangular matrices.)

a. The following matrix is easily seen to be block triangular by drawing the horizontal and vertical lines indicated:

$$M = \begin{pmatrix} 1\,2\,3\,4\,5 \\ 2\,3\,4\,5\,1 \\ 0\,0\,1\,2\,3 \\ 0\,0\,5\,1\,2 \\ 0\,0\,0\,1\,2 \end{pmatrix} = \begin{pmatrix} 1\,2\,|\,3\,4\,5 \\ 2\,3\,|\,4\,5\,1 \\ \overline{0\,0\,|\,1\,2\,3} \\ 0\,0\,|\,5\,1\,2 \\ 0\,0\,|\,0\,1\,2 \end{pmatrix} = \left(\frac{A\,|\,B}{C\,|\,D} \right),$$

where $A = \begin{pmatrix} 1 & 2 \\ 2 & 3 \end{pmatrix}$, $B = \begin{pmatrix} 3 & 4 & 5 \\ 4 & 5 & 1 \end{pmatrix}$, $C = \begin{pmatrix} 0 & 0 \\ 0 & 0 \\ 0 & 0 \end{pmatrix}$ and $D = \begin{pmatrix} 1 & 2 & 3 \\ 5 & 1 & 2 \\ 0 & 1 & 2 \end{pmatrix}$ and

noticing that C is a zero matrix,

$$det(D) = det \begin{pmatrix} 1\,2\,3 \\ 5\,1\,2 \\ 0\,1\,2 \end{pmatrix} \xrightarrow{\text{row ops}} det \begin{pmatrix} 1 & 2 & 3 \\ 0 & -9 & -13 \\ 0 & 1 & 2 \end{pmatrix}.$$

Thus, $det(D) = 1\{(-9)(2) - (-13)(1)\} = -5.$

Therefore, $det(M) = det(A)det(D) = \{(1)(3) - (2)(2)\}det(D) = 5.$

b. (Compare to part b of Example 4.5 on the preceding page.)

$$det \begin{pmatrix} 2\,2\,5\,1 \\ 2\,1\,1\,3 \\ 1\,0\,1\,0 \\ 0\,1\,3\,2 \end{pmatrix} = -det \begin{pmatrix} 1\,|\,0 & 1\,0 \\ 0\,|\,1 & -1\,3 \\ 0\,|\,2 & 3\,1 \\ 0\,|\,1 & 3\,2 \end{pmatrix} = -det \begin{pmatrix} 1 & -1 & 3 \\ 2 & 3 & 1 \\ 1 & 3 & 2 \end{pmatrix}$$

$$= -det \begin{pmatrix} 1 & -1 & 3 \\ 0 & 5 & -5 \\ 0 & 4 & -1 \end{pmatrix} = -\{(5)(-1) - (-5)(4)\} = -15.$$

□

A combination of the cofactor expansion, elimination and block matrix methods is usually more efficient than any one method by itself.

Calculators

The first item in the matrix math menu is DET. As mentioned in the calculator part of Section 3.5, this key evaluates numerical determinants to the accuracy of the calculator (about 14 digits).[5]

The **Hilbert matrix** H_n of size n has entries $(H_n)_{ij} = \frac{1}{i+j-1}$. Hilbert matrices are notoriously difficult to work with numerically. Try computing $det(H_4)$ and $det((H_4)^3)$ on your calculator. The calculator says that the first is nonzero but the second is zero! (Contrary to the second fundamental property of determinants that we establish in the next section!) The calculator BLEW IT!

It had to add and subtract very large numbers. The resulting cancellation completely overwhelmed the 14 significant digits with which the calculator works leaving nothing! The message is that if you really have to know the determinant of a large matrix, use a computer and sophisticated software. It won't be cheap or easy.

The permutation expansion of *det(A)*

This topic can be skipped without serious loss of continuity.

A **permutation** (matrix) is a square matrix that has all 0's but for exactly one 1 in each row and each column. The "list form" of an n-by-n permutation σ is the list of length n whose j-th entry $\sigma(j)$ is the row containing the 1 in the j-th column of σ. The 3-by-3 identity matrix is a permutation and its list form is simply 123. Notice that a permutation σ can be reduced to RREF using only row swaps and therefore $det(\sigma) = \pm 1$.

All told, there are $n!$ permutations of size n. The set of all such permutations is written $\mathcal{S}_n$. The permutation expansion of $det(A)$ is a sum of terms, one term for each $\sigma \in \mathcal{S}_n$ (that "$\in$" sign means "in"— remember?). Because there are so many terms to be summed, we write the permutation expansion in a compact form using "summation notation." Here is the formula:

$$det(A) = \sum_{\sigma \in \mathcal{S}_n} det(\sigma) a_{1\sigma(1)} a_{2\sigma(2)} \cdots a_{n\sigma(n)}.$$

Example 4.7 (The permutation expansion.)

There are six permutations in $\mathcal{S}_3$. Their matrices, lists, determinants and corresponding terms in the permutation expansion of $det(A)$ are:

Matrix:	$\begin{pmatrix} 1&0&0 \\ 0&1&0 \\ 0&0&1 \end{pmatrix}$	$\begin{pmatrix} 0&1&0 \\ 1&0&0 \\ 0&0&1 \end{pmatrix}$	$\begin{pmatrix} 0&1&0 \\ 0&0&1 \\ 1&0&0 \end{pmatrix}$	$\begin{pmatrix} 0&0&1 \\ 0&1&0 \\ 1&0&0 \end{pmatrix}$	$\begin{pmatrix} 0&0&1 \\ 1&0&0 \\ 0&1&0 \end{pmatrix}$	$\begin{pmatrix} 1&0&0 \\ 0&0&1 \\ 0&1&0 \end{pmatrix}$
list:	123	213	312	321	231	132
$det(\sigma)$:	1	-1	1	-1	1	-1
term:	$a_{11}a_{22}a_{33}$	$a_{12}a_{21}a_{33}$	$a_{13}a_{21}a_{32}$	$a_{13}a_{22}a_{31}$	$a_{12}a_{23}a_{31}$	$a_{11}a_{23}a_{32}$

---

[5]TI-83 Graphing Calculator Guidebook, p B10.

The permutation expansion of the general 3-by-3 matrix is therefore:

$$det \begin{pmatrix} a_{11} & a_{12} & a_{13} \\ a_{21} & a_{22} & a_{23} \\ a_{31} & a_{32} & a_{33} \end{pmatrix} = \begin{matrix} a_{11}a_{22}a_{33} & + a_{13}a_{21}a_{32} & + a_{12}a_{23}a_{31} \\ -a_{12}a_{21}a_{33} & - a_{13}a_{22}a_{31} & - a_{11}a_{23}a_{32} \end{matrix}.$$

There is a simple way to remember this expansion, called the "basket weave" formula.

Since the permutation expansion of the determinant of a four by four matrix involves $4! = 24$ terms (and a four by four basket weave would have eight terms), a four by four basket weave would miss $16/24$ of the terms!
Please don't even think about it!!
□

Compared to writing it down, the permutation expansion of $det(A)$ is relatively easy to obtain. Simply apply basic property 3, row linearity, as much as possible recursively. It goes like this: use row linearity with respect to the last row of A repeatedly until you have

$$det(A) = \begin{vmatrix} a_{11} & a_{12} & \cdots & a_{1n} \\ \vdots & \vdots & \vdots & \vdots \\ a_{m1} & a_{m2} & \cdots & a_{mn} \\ a_{n1} & 0 & \cdots & 0 \end{vmatrix} + \begin{vmatrix} a_{11} & a_{12} & \cdots & a_{1n} \\ \vdots & \vdots & \vdots & \vdots \\ a_{m1} & a_{m2} & \cdots & a_{mn} \\ 0 & a_{n2} & \cdots & 0 \end{vmatrix} + \cdots + \begin{vmatrix} a_{11} & a_{12} & \cdots & a_{1n} \\ \vdots & \vdots & \vdots & \vdots \\ a_{m1} & a_{m2} & \cdots & a_{mn} \\ 0 & 0 & \cdots & a_{nn} \end{vmatrix}.$$

Next use row multiply (basic property 4) to obtain zeros and ones in the last row:

$$a_{n1} \begin{vmatrix} a_{11} & a_{12} & \cdots & a_{1n} \\ \vdots & \vdots & \vdots & \vdots \\ a_{m1} & a_{m2} & \cdots & a_{mn} \\ 1 & 0 & \cdots & 0 \end{vmatrix} + a_{n2} \begin{vmatrix} a_{11} & a_{12} & \cdots & a_{1n} \\ \vdots & \vdots & \vdots & \vdots \\ a_{m1} & a_{m2} & \cdots & a_{mn} \\ 0 & 1 & \cdots & 0 \end{vmatrix} + \cdots + a_{nn} \begin{vmatrix} a_{11} & a_{12} & \cdots & a_{1n} \\ \vdots & \vdots & \vdots & \vdots \\ a_{m1} & a_{m2} & \cdots & a_{mn} \\ 0 & 0 & \cdots & 1 \end{vmatrix}.$$

(Here m is used for $n - 1$ to simplify typesetting.)

Now repeat this process for the second to last row. Each one of these terms spawns n new terms. But in each case exactly one of those new terms has the last two rows equal. The "simple but useful consequence of basic property 1" implies that all of these matrices have determinant zero.

Repeat this process for each row in turn. For each new row, the number of newly spawned terms with two rows equal increases by one. In the end this gives us the $n!$ terms of the permutation expansion of $det(A)$.

Now let's sketch how the permutation expansion leads to the block triangular matrix determinant formula. Suppose we are given a block triangular matrix

$$M = \begin{pmatrix} A & B \\ 0 & D \end{pmatrix} \text{ where } A \text{ is } a\text{-by-}a \text{ and } D \text{ is } d \text{ by } d, a + d = n.$$

First, imagine the product $P = det(A)det(D)$ where each determinant is expressed in its permutation expansion. In order to avoid confusion, label the columns of A with numbers 1 to a and the columns of D with numbers $a + 1$ to n. When P is expanded fully there will be $a!d!$ terms, one for each of the n-by-n permutations having only numbers $\leq a$ in the first a terms of its list.

The key is to observe that $\sigma \in S_n$ satisfies this condition, if and only if the last d terms in its list are all $\geq a + 1$. (The first a entries have already been specified, and no term appears twice in a permutation list.)

Now consider the permutation expansion of a block upper-triangular matrix M. Any of the $n!$ terms in the sum that happens to come from a permutation σ that has a number greater than a appearing in the first a terms of its list will vanish because the product $m_{1\sigma(1)} m_{2\sigma(2)} \cdots m_{n\sigma(n)}$ will have a zero term coming from the (2,1) block of zeros. Therefore, the only permutations that need be considered have numbers $1 \ldots a$ in their first a terms. This accounts for all of the terms $1 \ldots a$, and as above, the last d terms in its list are all $\geq a + 1$.

We have shown that the permutation expansion of M simplifies to a sum with exactly the same terms as appear in the product of the permutation expansion of $det(A)$ and $det(D)$, and the block triangular deteminant formula follows.

SUMMARY

Let A be an n-by-n matrix. The i, j **minor**, $m_{ij}(A)$, of A is the $n-1$-by-$n-1$ matrix obtained from A by deleting the i-th row and j-th column of A. The i, j **cofactor** of A, A_{ij}, is $A_{ij} = (-1)^{i+j} det(m_{ij}(A))$. The **determinant** of an n-by-n of A is defined recursively by the **cofactor expansion** with respect to the first column:

$$det(A) = |A| := a_{11}A_{11} + a_{21}A_{21} + \cdots + a_{1n}A_{1n}.$$

Basic properties of determinant include that the determinant of an upper triangular matrix is the product of its diagonal entries, how determinants change under row operations and **row linearity**.

We use row reduction to compute determinants by keeping track of how each row operation changes the determinant. In computing determinants by reduction, we can also use use the **block triangular** matrix determinant formula.

We also obtain the permutation expansion

$$det(A) = \sum_{\sigma \in \mathcal{S}_n} det(\sigma) a_{1\sigma(1)} a_{2\sigma(2)} \cdots a_{n\sigma(n)}.$$

Practice Problems

1. Evaluate each of the following determinants using the definition (first column cofactor expansion recursively).

a. $det \begin{pmatrix} 2 & -40 & 17 \\ 0 & 1 & 11 \\ 0 & 0 & 3 \end{pmatrix}$ b. $det \begin{pmatrix} 1 & 2 & 3 \\ 3 & 7 & 6 \\ 1 & 2 & 3 \end{pmatrix}$ c. $det \begin{pmatrix} 3 & -1 & 2 \\ 6 & -2 & 4 \\ 1 & 7 & 3 \end{pmatrix}$

2. Evaluate the determinants of each of the following matrices by reducing the matrix to an upper triangular matrix.

a. $\begin{pmatrix} 2 & 3 & 4 \\ 0 & 0 & -3 \\ -1 & 2 & 7 \end{pmatrix}$ b. $\begin{pmatrix} 2 & 1 & 1 \\ 4 & 2 & 3 \\ 1 & 3 & 0 \end{pmatrix}$ c. $\begin{pmatrix} 6 & 6 & 6 \\ -3 & 3 & 1 \\ 3 & 9 & 2 \end{pmatrix}$

d. $\begin{pmatrix} 3 & 1 & 4 \\ 1 & 7 & 3 \\ 5 & -12 & 5 \end{pmatrix}$ e. $\begin{pmatrix} -1 & 2 & 1 & 2 \\ 1 & 2 & 4 & 1 \\ 2 & 0 & -1 & 3 \\ 3 & 2 & -1 & 0 \end{pmatrix}$ f. $\begin{pmatrix} 4 & 6 & 8 & -6 \\ 0 & -3 & 0 & -1 \\ 3 & 3 & -4 & -2 \\ -2 & 3 & 4 & 2 \end{pmatrix}$

3. Evaluate the determinants of each of the following matrices by reducing the matrix to block triangular form.

a. $\begin{pmatrix} 2 & 3 & 4 \\ 0 & 0 & -3 \\ -1 & 2 & 7 \end{pmatrix}$ b. $\begin{pmatrix} 1 & 5 & 8 & -3 \\ 6 & 7 & 77 & 8 \\ 0 & 0 & -2 & 3 \\ 0 & 0 & 9 & 5 \end{pmatrix}$ c. $\begin{pmatrix} 0 & 0 & 2 & 2 & 2 \\ 3 & -3 & 3 & 3 & 3 \\ -4 & 4 & 4 & 4 & 4 \\ 4 & 3 & 1 & 9 & 2 \\ 1 & 2 & 1 & 3 & 1 \end{pmatrix}$

4. List all four by four permutations that lead to a nonzero term in the permutation expansion of the matrix in part b of Problem 3.

5. Give a recursive proof that any rowSwap is achieved with an odd number of swaps of adjacent rows.

6. Give a proof of the block diagonal determinant formula, using only the first column cofactor expansion (our definition).

7. Show that the first column cofactor expansion can be obtained from the permutation expansion, thereby showing that these two possible starting points are equivalent.

4.3   Fundamental properties of determinants

Objective 1

Compute the determinant of an n-by-n ($n \leq 5$) matrix using arbitrary cofactor expansions, matrix reduction or block matrices as required.

We are now in a position to develop a full toolbox for determinant evaluation. The general tools apply equally well to any row or column of a matrix. To simplify exposition we use *line* to mean "row or column." The remaining tools are based on two **fundamental** determinant formulas:

THEOREM Suppose A, B are n-by-n matrices. Then
   1. $det(A^T) = det(A)$
   2. $det(AB) = det(A)det(B)$.

The **first fundamental formula** implies that anything we have proven about determinants for rows holds also for columns. This means that we can use "elementary column operations" or work with (block) lower triangular matrices. It further means that basic property 3, "row linearity," extends to "line linearity."

It also means that there is a "first row cofactor expansion, " and when combined with row swaps, a cofactor expansion with respect to any row. Finally, by a second application of the first fundamental formula, there is a cofactor expansion with respect to any line of A. All of these cofactor expansions combine to give a new (but definitely not improved) formula for A^{-1}. To do this we introduce the *classical adjoint,* $A^{@}$, of A (also written $adj(A)$)

$$(A^{@})_{ij} := A_{ji}.$$

WARNING: Did you notice that those sneaky subscripts switched places?

Consider the product

$$A^{@}A = \begin{pmatrix} A_{11} & A_{21} & \cdots & A_{n1} \\ A_{12} & A_{22} & \cdots & A_{n2} \\ \vdots & \vdots & \vdots & \vdots \\ A_{1n} & A_{2n} & \cdots & A_{nn} \end{pmatrix} \begin{pmatrix} a_{11} & a_{12} & \cdots & a_{1n} \\ a_{21} & a_{22} & \cdots & a_{2n} \\ \vdots & \vdots & \vdots & \vdots \\ a_{n1} & a_{n2} & \cdots & a_{nn} \end{pmatrix}.$$

Notice that the $(1, 1)$ entry of this product is exactly the first column cofactor expansion of $det(A)$. In fact the (jj) entry is the j-th column cofactor expansion of $det(A)$.

Consider $(A^{@}A)_{1,2}$ and compare it to the first column cofactor expansion of the matrix that matches A, except that its first column repeats the second column of A

$$(A^{@}A)_{1,2} = A_{11}a_{12} + A_{21}a_{22} + \cdots + A_{n1}a_{n2} = det \begin{pmatrix} a_{12} & a_{12} & \cdots & a_{2n} \\ a_{22} & a_{22} & \cdots & a_{2n} \\ \vdots & \vdots & \vdots & \vdots \\ a_{n2} & a_{n2} & \cdots & a_{nn} \end{pmatrix}.$$

Therefore $(A^{@}A)_{1,2} = 0$ by the simple but useful consequence of basic property 1 **for columns**. In fact, ALL off-diagonal entries of $A^{@}A$ are zero for essentially the same reason. The promised formula for A^{-1} follows:

$$A^{@}A = det(A)I, \quad \text{so when} \quad det(A) \neq 0 \quad A^{-1} = \frac{1}{det(A)} adj(A).$$

This formula is really great for really small matrices. It was, after all, our very first determinant formula:

$$\begin{pmatrix} a_{11} & a_{12} \\ a_{21} & a_{22} \end{pmatrix}^{-1} = \frac{1}{a_{11}a_{22} - a_{12}a_{21}} \begin{pmatrix} a_{22} & -a_{12} \\ -a_{21} & a_{11} \end{pmatrix}.$$

The **second fundamental formula** leads to many uses of determinants rather than evaluation tools. It points directly to why determinants are important within mathematics. It is also an elegant, unified summary of earlier work. Basic properties 1, 4 and 5 are summarized in the simple formula:

$$det(F)det(C) = det(FC)$$

for any elementary matrix F. As emphasized in Section 4.1, the geometric meaning of determinants is an "area scaling factor" for linear functions. This fundamental formula implies that the composition of two linear functions has "area scaling factor" equal to the product of the "area scaling factors" of the two parts — which makes a great deal of sense if you think about it.

The second fundamental formula easily implies that similar matrices have the same determinant. Indeed if $A = PBP^{-1}$, then

$$det(A) = det(PBP^{-1}) = det(P)det(BP^{-1})$$
$$= det(BP^{-1})det(P) = det(BP^{-1}P) = det(B).$$

(The point here is that the middle equality is switching the order of multiplication of numbers, NOT matrices!)

We now turn to the justification of the two fundamental formulas. They use section 4.1 and are easier than the 5 basic properties "proofs." Nonetheless, these proofs may be skipped on a first reading.

Both arguments depend on the use of elementary matrices and the factorization of A as a product of elementary matrices

$$A = F_1 F_2 \cdots F_s RREF(A).$$

(This use of Fs for elementary matrices follows Section 4.1 and distinguishes these elementary matrices from those arising in the reduction process.) By repeated application of the formula $(AB)^T = B^T A^T$, this formula can be transformed into the equally impressive:

$$A^T = (F_1 F_2 \cdots F_s RREF(A))^T = \cdots = RREF(A)^T F_s^T \cdots F_2^T F_1^T.$$

Suppose B is a square matrix, possibly even an identity matrix. Repeatedly apply the elementary matrix determinant formula as follows:

$$det(AB) = det(F_1 \, F_2 \cdots F_s RREF(A)B)$$
$$= det(F_1) \, det(F_2 \cdots F_s RREF(A)B) = \cdots$$
$$det(F_1) \, det(F_2) \cdots det(F_s) \, det(RREF(A)B).$$

If $RREF(A) \neq I$, its last row is zero. Direct computation reveals that the last row of $RREF(A)B$ is also zero! Of course this implies that $det(RREF(A)B) = 0$. Thus we have shown

"**If** $RREF(A) \neq I$, **then** $det(AB) = 0$, **for any** B."

All of this is common to the proofs of the separate formulas, but we now must treat them separately.

Consider the **first fundamental formula**: $det(A) = det(A^T)$. In this case we use the idea of the above calculation twice with $B = I$, once for A and once for A^T. Recall the formula $(A^T)^{-1} = (A^{-1})^T$ from problem 8, Section 2.3 — sure thing!

It implies that $RREF(A) \neq I$ if and only $RREF(A^T) \neq I$, hence $det(A) = 0$ if and only if $det(A^T) = 0$.

The point is that we need only consider the case when $RREF(A) = I = RREF(A^T)$. Observe that the transpose of an elementary matrix is also elementary and has the same determinant; so as above

$$det(A^T) = det(I^T F_s^T \cdots F_2^T F_1^T) = det(I^T) \, det(F_s^T) \, det(F_{s-1}^T \cdots F_2^T F_1^T) = \cdots$$
$$= det(I^T) \, det(F_s^T) \cdots det(F_1^T) = det(F_1) \cdots det(F_s) det(I)$$
$$= det(F_1) \, (det(F_2 \cdots F_s I)) = det(F_1 \cdots F_s) = det(A).$$

It is critical to realize that the rearrangement of terms in the middle of the second row is of numbers — **not matrices**.

This completes the proof of the first fundamental property.

The **second fundamental formula** $det(AB) = det(A)det(B)$ is easier. As shown above, if $RREF(A) \neq I$ then $det(AB) = 0$ for any B, $B = I$ included. Thus,

if $RREF(A) \neq I$ then $det(A) = 0$ and $det(AB) = 0$ so $det(A)det(B) = det(AB)$.

There remains the possibility that $RREF(A) = I$, in which case

$$det(AB) = \cdots = det(F_1) \, det(F_2) \cdots det(F_s) \, det(RREF(A)B)$$
$$= det(F_1) \, det(F_2) \cdots det(F_s) \, det(I) \, det(B) =$$
$$= (det(F_1) \, det(F_2) \cdots det(F_s) \, det(I)) \, det(B) = \cdots = det(A)det(B).$$

This completes the proof of the second fundamental property.

Objective 2

Use properties of determinants to verify formulas without computation.

As already repeatedly mentioned, determinants are not an effective computational tool. Nonetheless, they are quite important and have a long history as a formal mathematical tool. Generally speaking, these applications follow from the basic properties and fundamental formulas that we have developed, NOT from computation.

In this objective we illustrate a few of the more celebrated such uses of determinants. Please understand that this is not a numerical objective. Rather, **it presents determinants in their strongest role as a theoretical tool.**

Vandermonde determinants

In Section 2.3 we used Lagrange interpolants to show that a square Vandermonde matrix with distinct entries in its second column is invertible:

$$V(x_0 \ldots x_n) = \begin{pmatrix} 1 & x_0 & x_0^2 & \cdots & x_0^n \\ 1 & x_1 & x_1^2 & \cdots & x_1^n \\ \cdots & \cdots & \cdots \\ 1 & x_n & x_n^2 & \cdots & x_n^n \end{pmatrix}.$$

This also follows from determinants. All that is needed is $det(V(x_0, \ldots, x_n)) \neq 0$. But we do better, and actually obtain a RECURSIVE formula for this determinant:

$$det(V(x_0 \ldots x_n)) = det(V(x_0 \ldots x_{n-1})) \left[(x_n - x_0) \ldots (x_n - x_{n-1}) \right], n \geq 1.$$

The first case is correct because $det(V(x_0, x_1)) = det \begin{pmatrix} 1 & x_0 \\ 1 & x_1 \end{pmatrix} = x_1 - x_0.$

Consider the function
$$f(t) = det(V(x_0 \ldots x_{n-1}t)).$$
This means that we are thinking of first n rows as fixed numbers BUT the last row has as its entries the successive powers of the variable t.

The last row cofactor expansion hands us a polynomial in t. Each involved cofactor involves only $x_0 \ldots x_{n-1}$, and every occurrence of t comes from a power of t in the last row. Notice especially that the last term in this expansion is exactly $t^n det(V(x_0 \ldots x_{n-1}))$, and that none of the other terms involves this high a power of t. Thus, $f(t)$ has degree n and the leading coefficient $f_n = det(V(x_0 \ldots x_{n-1}))$.

Next, notice that
$$f(x_0) = f(x_1) = \ldots f(x_{n-1}) = 0,$$
because when $t = x_i$, the matrix $V(x_0 \ldots x_i \ldots x_{n-1}x_n)$ has two identical rows!

Amazingly, this is all that is needed! It completely determines $f(t)$ because $f(t)$ has degree n. Conclude that

$$f(t) = f_n \left[(t - x_0) \cdots (t - x_{n-1}) \right]$$
$$= det(V(x_0, \ldots, x_{n-1})) \left[(t - x_0) \ldots (t - x_{n-1}) \right].$$

The result follows by setting $t = x_n$.

It follows (by recursion) that $det(V(x_0, \ldots, x_n))$ is the product of all terms

$$(x_i - x_j) \text{ where } 0 \le i < j \le n.$$

None of these terms is zero unless $x_i = x_j$, for some $i \ne j$.

Implicit determinantal formulas

Sometimes it is possible to give a simple-looking formula for a complicated problem using determinants. These formulas can be quite elegant and can often be proven without direct computation. They can be also very helpful in analysis of the possible degenerate cases of the original problem. Here, and in the exercises, are a few examples.

Example 4.8 (Determinant formulas without computation.)

a. Suppose two points $P(p_x, p_y) \ne Q(q_x, q_y)$ in $\mathbb{R}^2$ are given. Show that the points (x, y) on the line through P, Q satisfy $f(x, y) = 0$, where

$$f(x, y) = det \begin{pmatrix} 1 & p_x & p_y \\ 1 & q_x & q_y \\ 1 & x & y \end{pmatrix}.$$

This problem has exactly the same elements that appeared in the Vandermonde determinant calculation.

   i. **Describe the nature of the function f.**
   By line linearity (wrt, the last row), the formula for $f(x, y) = \cdots$ is linear in x, y. Therefore, the equation $f(x, y) = 0$ is a STRAIGHT LINE.

   ii. **Use properties of determinants to locate zeros of f.**
   If we substitute the coordinates for P or Q for x, y then the determinant has two identical rows and therefore $f(p_x, p_y) = f(q_x, q_y) = 0$.

   iii. **Show that i and ii completely determine f.**
   Two (distinct) points uniquely determine a straight line. Therefore, the equation of the line through P and Q is $f(x, y) = 0$.

   Notice that the original geometric problem is degenerate exactly when $P = Q$. How does the determinantal formula behave in this situation?

b. Show that the circle through the points $P(-2, 3)$, $Q(2, 3)$, $R(2, -2)$ is given by the formula

$$det \begin{pmatrix} 1 & 1 & 1 & 1 \\ x & -2 & 2 & -2 \\ y & 3 & 3 & -2 \\ x^2 + y^2 & 13 & 13 & 8 \end{pmatrix} = 0.$$

Let $f(x, y)$ be the determinant of the indicated matrix and notice, by the first column cofactor expansion, $f(x, y) = A + Bx + Cy + D(x^2 + y^2)$ for some numbers A, B, C, D.

Describe the nature of the function f.

Notice that the standard form for a circle (of radius r and center (x_0, y_0)), $(x - x_0)^2 + (y - y_0)^2 = r^2$, can be written as $a + bx + cy + d(x^2 + y^2) = 0$ for some numbers a, b, c, d depending on x_0, y_0, r. (I know you can work out the formulas explicitly — but it would be very unappreciated to do so!)

The point is that the formula $f(x, y) = 0$ and the formula for a circle are algebraically equivalent.

Use properties of determinants to locate zeros of f.

Notice that $f(-2, 3) = f(2, 3) = f(2, 2) = 0$ because for each of these substitutions, the resulting matrix has two equal columns.

Show that these conditions completely determine $f(x,y)$.

If you think about it, this can mean only one thing. There is but one circle through three (noncollinear) points. I know it isn't fair, but this is really elementary geometry.

This amounts to showing that there is a unique point equidistant from P, Q and R. This follows, geometrically, by showing that the perpendicular bisectors of PQ and QR are not parallel, and therefore must meet.

Notice that this argument has exactly the same structure as part a of this example.

- First you show the formula, however unexpected, has a fighting chance.
- Then you show that it actually works in a bunch of special cases.
- And finally you show there is only one formula that passes all the tests you have concocted.

If you look closely, and use the transpose of part a, we have done better than the usual formula because our version actually gives the right answer even when the circle "degenerates" and the given points are collinear.

c. Suppose $A \neq 0$ is a 2-by-2 matrix. Show that

$$|det(A)| \leq \sqrt{a_{11}^2 + a_{21}^2}\sqrt{a_{12}^2 + a_{22}^2},$$

and describe the case of equality.

Although this can easily be proven algebraically, it is much more insightful to think geometrically. We saw in Section 4.1 that $|det(A)|$ is the area of the image of the standard unit square under the linear function $f_A(S)$. Also

$$\sqrt{a_{11}^2 + a_{21}^2}\sqrt{a_{12}^2 + a_{22}^2}$$

is the product of lengths of two adjacent sides of the parallelogram $f_A(S)$. Thus, the claimed inequality amounts to saying that the area of a parallelogram is less than or equal to the product of lengths of two adjacent sides. Clearly, equality occurs exactly when it is a rectangle and the sides are perpendicular. This can be expressed as $A^T A$ is diagonal, and would not come quickly to mind from the totally algebraic proof. □

Objective 3

Given a 2-by-2 integer matrix, compute its Smith normal form and interpret the associated lattice graphically.

Finally, we consider a purely mathematical question that suits determinants perfectly. **In this objective we require all matrices and vectors to have integer entries.**

Let A be an n-by-n matrix. Way back when we first used transform plots to understand linear systems in Section 1.6, we saw that the image $im(f_A)$ of the linear function f_A consists of all linear combinations of the columns of A and is the set of all $\mathbf{b}$ such that $A\mathbf{x} = \mathbf{b}$ for some $\mathbf{x}$.

In this situation f_A is an **integer linear function** and $im_{\mathbb{Z}}(f_A)$, or for short $\mathbf{A}$, is called a **lattice**. The subscript $\mathbb{Z}$ is a reminder that $\mathbf{x}$ is required to have integer entries. When we tracked the grid on graph paper with transform plots in parts a and c of Example 1.19 on page 44 we described an integer linear function and displayed its image lattice by means of a grid.

WARNING: Technically, we are talking about three different object types: matrices, grids and lattices with the font based notation A, $\mathcal{A}$ and $\mathbf{A}$ for related objects.

The grid $\mathcal{A}$ associated with the matrix A is the arrangement of lines that is the image of the normal graph paper under the associated linear function f_A — just as in Section 1.6. To go from the grid $\mathcal{A}$ to the lattice $\mathbf{A}$, simply forget about the lines and only remember their points of intersection.

Not surprisingly, the same lattice can arise from many different matrices and their various associated grids. For example the **normal** grid $\mathcal{I}$ coming from an ordinary piece of graph paper has associated matrix the identity matrix I_2. The associated **normal lattice** $\mathbf{I}$ is also realized by any other (integer) matrix Q with associated grid $\mathcal{Q}$ for which the starting parallelogram has area 1. Of course you know this condition is equivalent to $|det(Q)| = 1$.

There is a geometric way to see this too. First notice that the starting parallelogram (the image of the standard unit square under the linear function f_A as on page 42) for Q has four points of the lattice $\mathbf{I}$ as corners, but no interior points of the starting parallelogram in $\mathbf{I}$. Then imagine "tiling" the entire plane with parallelograms congruent to the starting one. Since each normal lattice point winds up at a corner of four such tiles, the grid $\mathcal{Q}$ realizes the normal lattice.

In general, essentially the same argument shows that two matrices X and Y represent the same lattice $\mathbb{X} = \mathbb{Y}$ whenever there is a matrix Q, with $|det(Q)| = 1$, such that $X = YQ$. When this happens we say that the grids **X** and **Y** are *equivalent*. You can see two different equivalent grids if you compare the first two figures in (either part) of Example 4.9.

Because we are interested in lattices, not grids, we are willing to work with either the matrix Y or the matrix X = YQ, $|det(Q)| = 1$, rather like how we moved between directed line segments representing the same geometric vector in Section 1.6.

But there is more fun!

Two lattices, **A** and **B**, are said to be *equivalent* if there is an integer matrix P such that $|det(P)| = 1$ such that $\mathbb{B} = P\mathbb{A}$. This means that each of the infinitely many points in (the lattice) **A** is mapped by the linear function f_P to one of the points in (the lattice) **B**. The requirement $|det(P)| = 1$ is there to insure that P^{-1} has integer entries and so $f_{P^{-1}}$ has a matrix with integer entries.

Okay, so we start with an integer matrix A. It determines a grid $\mathcal{A}$ and a lattice **A**. We obtain an equivalent grid (and the same lattice) if we replace A by AQ for any integer matrix Q for which $|det(Q)| = 1$. But we could also change to a different, but equivalent, lattice by replacing this matrix with PAQ where P is any integer matrix with $|det(P)| = 1$. So, at the matrix level, where we really do operate, we see that two integer matrices A and B represent equivalent lattices if and only if

$$B = PAQ \text{ for integer matrices } P, Q \text{ such that } |det(Q)| = |det(P)| = 1.$$

It is time to introduce some more language. Say two integer matrices A and B are *LR-adjacent* if there is an elementary matrix such that

$$A = EB \text{ or } A = BE; \; E \text{ elementary matrix }, |det(E)| = 1.$$

Further, say A and M are *LR-equivalent* if there is a chain of LR-adjacent matrices starting with A and ending with M.

Suppose A and B are LR-equivalent. Then, and only then, is $B = PAQ$ where P and Q are each products of elementary matrices each of which has determinant ± 1. LR-equivalence is the matrix realization of lattice equivalence.

YOU have become experts in a game called RREF. Now we have a new game called SMITH or "LR-equivalence" to play. Here are its rules:

> Start with an integer matrix A, apply area preserving elementary row AND column operations to obtain the simplest possible matrix B that is LR-equivalent to A.

The standard "simplest possible" matrix is unique and is called the *Smith normal form* of A (achieved by the MAPLE command "ismith(A)").

The Smith normal form of A is obtained by a strategy similar to Gaussian elimination. First find the greatest common divisor d of the matrix entries $\{a_{ij}\}$. Then use area preserving row and column operations to put d in the $(1,1)$ position. Then use area preserving row and column operations to kill everything else in the first row or column. Finally, reduce to the smaller working matrix by ignoring the first row and column and repeat.

Example 4.9 (Lattices and Smith normal form.)

a. The figure depicts the step-by-step reduction of the matrix G to Smith normal form. The basic rectangle associated with the grid is shaded, and a few elements from the normal grid not in G are represented by dots.

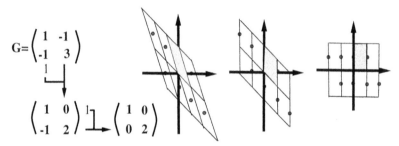

The first step is a column operation, so the grid changes but the lattice does not. The second step is a row operation and so it amounts to a transform plot of the linear function f_E where E is an elementary matrix associated with the row operation.

b. Here is an example where you have to work to get started because the greatest common divisor of the matrix entries is not a matrix entry.

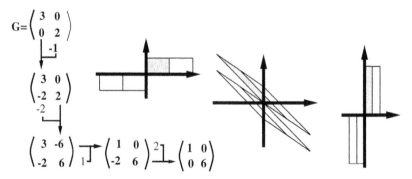

We have only included the middle matrix as an intermediate figure. The first two steps do not change the actual lattice, only the grid representing the lattice, while the last two steps determining a different but equivalent lattice.   □

Unlike the reduced row echelon form, there is a way to determine the Smith normal form of a matrix A without row operations. That's the good news. The bad news is that this other method involves the determinant of every possible square submatrix.

THEOREM Suppose A is an integer matrix. Then there is a unique matrix S (the Smith normal form of A) that is LR-equivalent to A such that $s_{ij} = 0$ if $i \neq j$ and s_{ii} is an non-negative integer that divides s_{jj} when $i \leq j$.

Moreover, the product $s_{11} \cdots s_{ii}$ is the greatest common divisor of the determinants of the i-by-i submatrices of A. Two lattices are equivalent if and only if any matrices representing them have the same Smith normal form.

Proof of this remarkable theorem is quite beyond our means.

SUMMARY

A *line* of a matrix is a row or column. Determinants satisfy the fundamental identities

$$det(AB) = det(A)det(B) \qquad det(A^T) = det(A).$$

It follows that determinants are **line linear** and that there is a cofactor expansion with respect to lines. The *classical adjoint* of a matrix has (i, j) entry equal to the cofactor $A_{j,i}$ and satisfies

$$A^{@}A = det(A)I.$$

Determinants are powerful theoretical tools. The Vandermonde determinant is the most important implicit determinantal formula. Such formulas can be proven without explicit computation.

If the matrix A has integer entries then $\mathbf{A} = im_{\mathbb{Z}}(f_A)$ is *lattice*. The lattice $\mathbf{A}$ is also *represented* by any matrix B that is obtained from A by elementary area preserving column operations. At the matrix level, this means that $A = BQ$, $|det(Q)| = 1$.

Two lattices $\mathbf{A}$, $\mathbf{B}$ are equivalent if there is an invertible integer linear function f_P mapping one to the other. At the matrix level, this means $A = PB$, $|det(P)| = 1$. Two matrices are *LR-equivalent* if and only if they represent equivalent lattices. Every integer matrix is LR-equivalent to a unique matrix in Smith normal form.

Looking further

Lattices are basic in algebraic number theory. This beautiful, classical subject has recently enjoyed a renaissance. It includes the world's best tools for factoring large integers. This is just what is needed to decrypt the widely used RSA public-key crypto system.[6]

---

[6]**A Course in Number Theory and Cryptography** by Neal Koblitz, Springer-Verlag, New York, (1994).

Practice Problems

1. Find the determinant of the following matrices by any manual method.

a. $\begin{pmatrix} 1 & 8 & -7 \\ 1 & 10 & -9 \\ 2 & 18 & -13 \end{pmatrix}$
b. $\begin{pmatrix} 1 & 5 & 8 & -3 \\ 6 & 7 & 77 & 8 \\ 0 & 0 & -2 & 3 \\ 0 & 0 & 9 & 5 \end{pmatrix}$
c. $\begin{pmatrix} -2 & -5 & 7 & -9 \\ 4 & 18 & -15 & 20 \\ 4 & 18 & -12 & 27 \\ -4 & -10 & 17 & 9 \end{pmatrix}$

d. $\begin{pmatrix} 3 & 1 & 0 \\ -2 & 0 & 1 \\ 1 & 3 & -1 \end{pmatrix}$
e. $\begin{pmatrix} 3 & 1 & -2 & 2 \\ 2 & 0 & 1 & 4 \\ 0 & 1 & 3 & 5 \\ -1 & 2 & 0 & -3 \end{pmatrix}$
f. $\begin{pmatrix} 2 & -2 & 0 & 0 & -3 \\ 3 & 0 & 3 & 2 & 1 \\ 0 & 1 & -2 & 0 & 2 \\ -1 & 2 & 0 & 3 & 0 \\ 0 & 4 & 1 & 0 & 0 \end{pmatrix}$

2. Evaluate the following determinants by inspection. Explain.

a. $det \begin{pmatrix} 1 & 2 & 3 & 4 \\ 0 & 5 & 0 & 0 \\ 0 & 6 & 7 & 9 \\ 0 & 7 & 0 & 1 \end{pmatrix}$
b. $det \begin{pmatrix} 1 & 2 & 3 & 4 \\ 5 & 6 & 7 & 0 \\ 8 & 9 & 0 & 0 \\ 1 & 0 & 0 & 0 \end{pmatrix}$
c. $det \begin{pmatrix} 1 & 0 & 0 & 2 \\ 0 & 3 & 4 & 0 \\ 0 & 5 & 6 & 0 \\ 7 & 0 & 0 & 8 \end{pmatrix}$

3. Give a recursive proof for the fundamental determinant properties arguing by recursion on the number s of elementary matrices required in the factorization of A on page 197.

4. Suppose three distinct points $P(p_x, p_y), Q(q_x, q_y), R(r_x, r_y)$ in $\mathbb{R}^2$ are given. Describe the points (x, y) for which $f(x, y) = 0$, where

$$f(x, y) = det \begin{pmatrix} 1 & p_x & p_x^2 & p_y \\ 1 & q_x & q_x^2 & q_y \\ 1 & r_x & r_x^2 & r_y \\ 1 & x & x^2 & y \end{pmatrix}.$$

Compare to Section 1.7.

5. Suppose three points $P(p_x, p_y), Q(q_x, q_y), R(r_x, r_y)$ in $\mathbb{R}^2$ are given and have distinct x-coordinates. Show that all the points (x, y) on the circle through P, Q, R satisfy $f(x, y) = 0$, where

$$f(x, y) = det \begin{pmatrix} 1 & p_x & p_y & p_x^2 + p_y^2 \\ 1 & q_x & q_y & q_x^2 + q_y^2 \\ 1 & r_x & r_y & r_x^2 + r_y^2 \\ 1 & x & y & x^2 + y^2 \end{pmatrix}.$$

6. Use elementary matrices to show Problem 12 of Section 1.2.

7. For each of the following matrices, find its Smith normal form and indicate steps taken.

a. $\begin{pmatrix} 2 & -1 \\ 1 & 2 \end{pmatrix}$
b. $\begin{pmatrix} 2 & 3 \\ -3 & 2 \end{pmatrix}$
c. $\begin{pmatrix} 1 & 2 \\ 3 & 4 \end{pmatrix}$

8. Suppose the matrix A has Smith normal form S. Show that $det(A) = det(S)$.

4.4   Further applications

Objective 1

Find the characteristic polynomial/equation of an n-by-n matrix $(n \leq 4)$.

It is finally time to present the most common appearances of determinants in contemporary matrix algebra books. These are alternative methods for important problem types that we have already discussed and really only make connections with the mainstream literature.

The common element is the adjoint formula $A^@ A = det(A)I$ formula. It implies:

$$det(A) = 0 \text{ if and only if } A\mathbf{x} = \mathbf{0} \text{ for some } \mathbf{x} \neq \mathbf{0}.$$

You already know that the problem of finding eigenvalues and eigenvectors of a matrix is too difficult to attack directly. Instead, an intermediate step — a polynomial whose zeros are the eigenvalues — is introduced. For most of Chapter 3, this polynomial was the minimal polynomial. Recall that λ is an eigenvalue of A if

$$A\mathbf{x} = \lambda\mathbf{x}, \text{ for some } \mathbf{x} \neq \mathbf{0}.$$

The idea is to use determinants to eliminate the eigenvector $\mathbf{x}$ from this equation. Rewrite it and regroup:

$$\mathbf{0} = \lambda\mathbf{x} - A\mathbf{x} = \lambda I\mathbf{x} - A\mathbf{x} = (\lambda I - A)\mathbf{x},$$

where I is the identity matrix. Thus, the problem of finding the eigenvalues and eigenvectors of A comes down to solving:

$$(\lambda I - A)\mathbf{x} = \mathbf{0} \text{ for some } \mathbf{x} \neq \mathbf{0}.$$

By the above fundamental property of determinants, this is equivalent to

$$det(\lambda I - A) = 0.$$

The *characteristic polynomial* of the matrix A is the polynomial

$$p_A(x) = det(xI - A).$$

The *characteristic equation* of A is $det(xI - A) = 0$. The eigenvalues of A are the zeros of the characteristic polynomial. At the end of this objective, we sketch a proof of the celebrated **Cayley Hamilton theorem** that $p_A(x)$ always has the minimal polynomial $\mu_A(X)$ as a factor at the end of this section.

Notice that the characteristic polynomial is always of degree n and has the same zeros as the minimal polynomial, but it may have some zeros to higher multiplicity. For example, the n-by-n identity matrix has minimal polynomial $x - 1$ while its characteristic polynomial is $(x - 1)^n$.

Example 4.10 (The characteristic polynomial.)

a. By part c of Example 4.5 on page 190) the matrix $A = \begin{pmatrix} 8 & 9 & 9 \\ 3 & 2 & 3 \\ -9 & -9 & -10 \end{pmatrix}$ has characteristic polynomial

$$p_A(x) = det \begin{pmatrix} x-8 & -9 & -9 \\ -3 & x-2 & -3 \\ 9 & 9 & x+10 \end{pmatrix} = \cdots = x^3 - 2x - 2.$$

b. Let $A = \begin{pmatrix} 2 & 1 & 2 \\ 1 & 2 & 2 \\ 1 & 1 & 3 \end{pmatrix}$. Then $p_A(x) = det \begin{pmatrix} x-2 & -1 & -2 \\ -1 & x-2 & -2 \\ -1 & -1 & x-3 \end{pmatrix}$

$$= -det\left(\begin{pmatrix} 0 & (x-2)^2-1 & 2-2x \\ -1 & x-2 & -2 \\ 0 & 1-x & x-1 \end{pmatrix} \right) = -det \begin{pmatrix} (3-x)(1-x) & 2(1-x) \\ x-1 & 1-x \end{pmatrix},$$

by the cofactor expansion with respect to the first column.

Thus, $p_A(x) = (x-1)((3-4x+x^2)-(2-2x)) = (x-1)^2(x-5)$.

c. Let $A = \begin{pmatrix} 2 & 5 & 3 & -6 \\ -2 & 4 & 0 & 2 \\ -4 & -10 & -6 & 12 \\ 1 & 7 & 3 & -5 \end{pmatrix}$. Then $p_A(x) = det(xI - A) =$

$$-det \begin{pmatrix} 2-x & 5 & 3 & -6 \\ -2 & 4-x & 0 & 2 \\ -4 & -10 & -6-x & 12 \\ 1 & 7 & 3 & -5-x \end{pmatrix} = -det \begin{pmatrix} 0 & 7x-9 & 3x-3 & 4-3x-x^2 \\ 0 & 18-x & 6 & -8-2x \\ 0 & 18 & 6-x & -8-4x \\ 1 & 7 & 3 & -5-x \end{pmatrix},$$

by elementary row operations. Now the cofactor expansion with respect to the first column, followed by subtracting the third row from the second, gives:

$$det \begin{pmatrix} 7x-9 & 3x-3 & 4-3x-x^2 \\ -x & x & 2x \\ 18 & 6-x & -8-4x \end{pmatrix} = xdet \begin{pmatrix} 0 & 10x-12 & -14+11x-x^2 \\ -1 & 1 & 2 \\ 0 & 24-x & 28-4x \end{pmatrix}$$

by elementary row operations. Finally, by the cofactor expansion with respect to the first column and the 2-by-2 formula, this is

$$p_A(x) = -x((10x-12)(28-4x) + (24-x)(x^2-11x+14)) = x(x^3 + 5x^2 - 50x).$$

It follows that A has eigenvalues $x = -10$, $x = 0$ and $x = 5$.   □

Recall from Section 3.1 that the geometric multiplicity of an A-eigenvalue λ is the number of basic eigenvectors associated with λ. The ***algebraic multiplicity*** of λ is, by definition, the multiplicity of $(x - \lambda)$ as a factor of $p_A(x)$.

It is not hard to show that the matrix A is diagonalizable if and only if each eigenvalue has the same algebraic multiplicity as geometric multiplicity (see Problem 4).

Although it is much more difficult to apply this criterion than to compute the minimal polynomial, it is all that is offered students of most texts.

The Cayley Hamilton theorem[7]

This theorem asserts that, for any square matrix A, the characteristic polynomial $p_A(x)$ is divisible by the minimal polynomial $\mu_A(x)$.

This theorem implies the shortcut at the bottom of page 122. Indeed, let $f(x)$ be the product of the polynomials output by a looping minimal polynomial algorithm and suppose $deg(f) \geq n$. We know that $f(x)$ is a factor of $\mu_A(x)$ which in turn is a factor of $p_A(x)$ by the Cayley Hamilton theorem. But $p_A(x)$ has degree n, so in fact $deg(f) = deg(p_A(x))$. Now two monic polynomials of the same degree cannot have one as a divisor of the other unless they coincide. Therefore, $f(x) = \mu_A(x) = p_A(x)$, as claimed.

Our proof requires a preliminary result. Suppose $B = P^{-1}AP$ is similar to A. By Problem 7 of Section 3.2, we know that $\mu_A(x) = \mu_B(x)$. The characteristic polynomial of A and B also coincide. Indeed,

$$p_B(x) = det(xI - B) = det(xI - P^{-1}AP) = det(P^{-1}(xI - A)P)$$
$$= det(P^{-1})p_A(x)det(P) = p_A(x)det(P)^{-1}det(P) = p_A(x).$$

By Frobenius' theorem, there is an upper triangular matrix T that is similar to A.

Putting all of this together, it is sufficient to prove the Cayley Hamilton theorem for upper triangular matrices. We do this by recursion on the size of the matrix.

The initial case is trivial.

Suppose T is upper triangular with $t = T_{11}$ and $U = m_{11}(T)$ the $(1,1)$ minor matrix. Then U is upper triangular of smaller size, so $p_U(U) = 0$, by recursion.

On the other hand, $p_T(x) = (x - t)p_U(x)$, since T is upper triangular. Now, of course, $p_T(T) = (T - tI)(p_U(T)) = XY$, for some matrices X and Y. But the matrix X is zero in the first column, and Y is zero in all but the first row. It follows that $p_A(x)$ annihilates T. Since $\mu_T(x)$ is the least degree monic polynomial annihilating T, the result follows (see Problem 5 of Section 3.2).

Objective 2

Use Cramer's rule to solve the linear system $A\mathbf{x} = \mathbf{b}$ when appropriate.

In many physical applications, a linear model is used to predict the values of certain physical expressions in a specific situation. If the linear model does not have a unique solution, then it is regarded as being incomplete. For example, any linear system

---

[7]Cayley and Hamilton only treated the cases $n \leq 4$. Frobenius gave the first complete proof of this result in 1878.

modeling an electrical circuit that does not have a unique solution is incomplete, because the laws of physics demand that the circuit behave in just one way at a time! When dealing with small linear systems for which it is known in advance that there is a unique solution, some people use determinants and ***Cramer's rule***.

Here is our last proof. And with it we come full circle, "solving" the problem we started with — how to solve $A\mathbf{x} = \mathbf{b}$?

Let A_i be the matrix obtained from A by replacing the i-th column by the vector $\mathbf{b}$, and let X_i be the matrix obtained from the identity matrix by replacing the i-th column with $\mathbf{x}$. Then $|X_i| = x_i$ by the i-th row cofactor expansion. Now consider the matrix product:

$$AX_i = \begin{pmatrix} a_{11} \cdots a_{1n} \\ a_{21} \cdots a_{2n} \\ \vdots \\ a_{n1} \cdots a_{nn} \end{pmatrix} \overset{|\leftarrow i \rightarrow|}{\begin{pmatrix} 1 \cdots x_1 \cdots 0 \\ 0 \cdots x_2 \cdots 0 \\ \vdots & \vdots & \vdots \\ 0 \cdots x_n \cdots 1 \end{pmatrix}} = \overset{|\leftarrow i \rightarrow|}{\begin{pmatrix} a_{11} \cdots b_1 \cdots a_{1n} \\ a_{21} \cdots b_2 \cdots a_{2n} \\ \vdots & \vdots & \vdots \\ a_{n1} \cdots b_n \cdots a_{nn} \end{pmatrix}} = A_i.$$

Use the fundamental multiplicative determinantal property and the formula $|X_i| = x_i$ to obtain:

$$|A_i| = |AX_i| = |A|\,|X_i| = |A|x_i.$$

When $|A| \neq 0$, this implies ***Cramer's rule:***

$$x_i = |A_i|/|A|.$$

Now that is the formula for just the i-th coordinate. If you want the whole vector (and who wouldn't?) you will need a total of $n + 1$ determinants.

Example 4.11 (Cramer's rule.)

The linear system
$$\begin{aligned} x_1 \quad\quad + 2x_3 &= 6 \\ -3x_1 + 4x_2 + 6x_3 &= 30 \\ -x_1 - 2x_2 + 3x_3 &= 8 \end{aligned}$$

can be solved by evaluating the four 3-by-3 determinants:

$$|A| = \det \begin{pmatrix} 1 & 0 & 2 \\ -3 & 4 & 6 \\ -1 & -2 & 3 \end{pmatrix} = 44, \quad |A_1| = \det \begin{pmatrix} 6 & 0 & 2 \\ 30 & 4 & 6 \\ 8 & -2 & 3 \end{pmatrix} = -40,$$

$$|A_2| = \det \begin{pmatrix} 1 & 6 & 2 \\ -3 & 30 & 6 \\ -1 & 8 & 3 \end{pmatrix} = 72 \text{ and } |A_3| = \det \begin{pmatrix} 1 & 0 & 6 \\ -3 & 4 & 30 \\ -1 & -2 & 8 \end{pmatrix} = 152.$$

Then Cramer's rule tells us that: $x_1 = \frac{-40}{44}$, $x_2 = \frac{72}{44}$ and $x_3 = \frac{152}{44}$. $\square$

SUMMARY

The **characteristic polynomial** of A is $det(xI - A)$ and has the same zeros as the minimal polynomial of A, namely the eigenvalues of A.

Cramer's rule gives a simple-looking formula for the solution to a nonhomogeneous linear system in the case that the coefficient matrix is invertible. If the coefficient matrix is not invertible, Cramer's rule does not apply.

---

* * * * * * *

Practice Problems

1. Find the characteristic polynomial for each of the following matrices.

$$
\text{a. } \begin{pmatrix} 0 & 0 & 2 & 0 \\ 1 & 0 & 1 & 0 \\ 0 & 1 & -2 & 0 \\ 0 & 0 & 0 & 1 \end{pmatrix}
\quad
\text{b. } \begin{pmatrix} 10 & -9 & 0 & 0 \\ 4 & -2 & 0 & 0 \\ 0 & 0 & -2 & -7 \\ 0 & 0 & 1 & 2 \end{pmatrix}
\quad
\text{c. } \begin{pmatrix} 4 & 2 & 12 & 0 \\ 0 & -1 & 0 & 1 \\ 1 & 1 & 5 & 1 \\ 1 & 1 & 4 & 1 \end{pmatrix}
$$

2. For each of the following matrices, find the characteristic polynomial, the eigenvalues and the basic eigenvectors.

$$
\text{a. } \begin{pmatrix} 1 & 5 & -4 \\ -2 & -10 & 8 \\ -2 & -7 & 7 \end{pmatrix}
\quad
\text{b. } \begin{pmatrix} 4 & -3 & 0 \\ 4 & -1 & -2 \\ 1 & -3 & 3 \end{pmatrix}
\quad
\text{c. } \begin{pmatrix} -2 & -2 & -1 & 3 \\ -1 & -3 & 0 & 2 \\ -3 & -6 & -1 & 6 \\ -1 & -2 & 0 & 1 \end{pmatrix}
$$

3. Use Cramer's rule to solve the following linear systems.

$$
\text{a. } \begin{aligned} 2x_1 + 2x_2 + x_3 &= 7 \\ x_1 + 2x_2 - x_3 &= 0 \\ -x_1 + x_2 + 3x_3 &= 1 \end{aligned}
\qquad
\text{b. } \begin{aligned} 2x_1 + x_2 - x_3 &= 4 \\ x_1 \phantom{{}+x_2} + x_3 &= 2 \\ -x_2 + 5x_3 &= 1 \end{aligned}
$$

4. Use Frobenius' theorem to show that a matrix is diagonalizable if and only if each eigenvalue's algebraic multiplicity equals its geometric multiplicity.

A pedagogical postscript

We used Gauss-Jordan elimination to obtain the general solution of an arbitrary linear system. Cramer's rule is a simply stated formula that solves any linear system with a square coefficient matrix that is known to have a unique solution.

We first computed eigenvalues by building up annihilator polynomials, then the minimal polynomial. Eigenvalues can also be computed using determinants by finding the zeros of the characteristic polynomial. Computation of the characteristic polynomial requires the matrix entries to involve a totally new indeterminant x.

The eigenvalues of a matrix also have a natural similarity invariant meaning. Determinants have a natural geometric meaning that is LR-equivalence invariant. By increasing the complexity of the matrix entries and moving to the weaker LR-equivalence invariant setting, the characteristic polynomial conceptually moves the "eigenvalue problem" much farther from its original statement than does the minimal polynomial, to say nothing about how difficult determinants are conceptually or computationally.

It is time to address the relative merits of these vastly different methods. One can compare these methods in a number of different ways. Perhaps the most basic is the generality of the problem they solve. (For example, Gauss-Jordan deals with ANY linear system, while Cramer's rule is useless unless the linear system has a square coefficient matrix and a unique solution.) Another important comparison might be based on which approach leads most easily to important additional information. (For example, the minimal polynomial leads to a diagonalization criterion and to components, while the characteristic polynomial conveys no such information.) Other ways to compare these alternative methods include considering which is most efficiently presented or most computationally efficient.

Because this is after all a text, I believe none of these criteria is the most important. I believe the most important criterion is: "which method conveys the deepest, most long lasting understanding?"

I invite you to consider some pedagogical issues. I am pleased to be able to cite the influential work of G. Polya.[1] In a chapter titled "On learning, teaching and learning teaching," Polya refers to the philosopher Spinoza's "Treatise on the Improvement of the Mind" in which Spinoza distinguishes four different levels of knowledge.

Mechanical Knowledge
A student has learned a rule by heart, accepting it on authority, without proof, but is able to use the rule and apply it correctly.

Inductive Knowledge
The student has tried the rule on simple cases and convinced him/herself that it works correctly in all cases tested.

Rational Knowledge
The student has understood a demonstration or proof of the rule.

Intuitive Knowledge
The student conceives the rule clearly and distinctly, and is so convinced of it that she/he cannot doubt its truth.

Polya goes on to observe, "As dedicated teachers we may find ways to anchor a new fact in the experience of the student, connect it with formerly learned facts, cement its knowledge by application. We can only hope that the student's well-anchored, well-connected, well-cemented, well-organized knowledge will finally become intuitive."

Now I posit that to be pedagogically sound, we must build on and reinforce earlier understanding repeatedly. The underlying justifications should be accessible from several different perspectives and supported with numerous relevant applications.

Now you be the judge. Consider our solution of linear systems in Section 1.4 and Cramer's rule. Also compare the Lagrange interpolant formula (appearing in Section 2.3) for the inverse of a Vandermonde matrix to the Vandermonde determinant proof on page 195 and compare our polynomial interpolation in Section 1.6 with Problem 4 Section 4.3.

Which is more anchored in your background experience? Which is more connected to other learned facts? Which lends more insight and understanding?

Consider finally the minimal polynomial algorithm and the characteristic polynomial computation. Which connects best with other learned material? Which has supporting application (separate from the solved problem — eigenvalues)? Which melts the problem away, revealing the solution as you work? Which leads most easily to deeper understanding? And most importantly, are the differences important enough to choose one presentation over the other?

---

[1] **Mathematical Discovery: On understanding, learning, and teaching problem solving** by George Polya, J Wiley & Sons, New York 1965.

Looking backward

Maybe there is also time to revisit another of Polya's favorite topics,[2] the discussion of the use of language and notation that started the first two chapters. Perhaps you do not sufficiently appreciate what every advertising agent lives on, namely, that language shapes your thoughts and attitudes. Language can be both powerful and ambiguous — every pun exploits this fact.

Notation is the language of mathematics; it can be powerful and is also potentially ambiguous. We have used matrix notation to accentuate the essential aspects of our problems and to build a new arithmetic. On the other hand, the determinant has a very elegant notation that gives no hint of the difficult computation it represents.

Like language, notation can be used both to enlighten and obscure, to motivate and to intimidate. Simple terms like "imaginary unit," "real number," "free variable" and "defective matrix," no less than "politically correct," "military intelligence" and "moral majority," set tones and project images that must be accepted with some care.

---

[2]**How to Solve It** by G. Polya, Princeton University Press (1945), Princeton NJ, pp. 134 – 141.

Appendix: The abstract setting

Mathematically speaking, this book has actually been about something other than vectors and matrices. It is about a class of more abstract objects called linear spaces. Because this is not an abstract course, we have talked all around the subject, giving many examples and aspects of linear spaces, never quite telling the whole story.

This appendix briefly outlines the abstract setting and thereby gives the proper context for the simple themes that we have repeatedly presented. Almost anything you want to say about or do with linear spaces comes down to a matrix computation, so we already know a lot about working with linear spaces.

A linear space **L** comes with its own number system $\mathbb{F}$ of "scalars." Although there are many other interesting and useful possibilities, we will assume the number system $\mathbb{F}$ of scalars is one of $\mathbb{Q}, \mathbb{R}, \mathbb{C}$ and sometimes $\mathbb{Z}$.

For **L** to be an $\mathbb{F}$-linear space, there must be defined a "scalar multiplication" and a "vector addition," and these operations must satisfy certain axioms.

Abstract Linear Space

The set **L** is an $\mathbb{F}$-linear space if both vector addition and scalar multiplication are defined. This means that for any $a, b \in$ **L**, there is a method provided to compute their sum $a + b \in$ **L**, and also for any $\alpha \in \mathbb{F}$ and any $b \in$ **L**, there is a method provided to compute their scalar product $\alpha b \in$ **L**.

Moreover, these operations satisfy the following axioms:

axiom 1. There is an element $0 \in$ **L** such that $a + 0 = a$ for all $a \in$ **L**.

axiom 2. For each $a \in$ **L**, there is a unique $-a \in$ **L** such that $-a + a = 0$.

axiom 3. $(a + b) + c = a + (c + b)$ for all $a, b, c \in$ **L**.

axiom 4. $(\alpha + \beta)(a + b) = \alpha a + \beta a + \alpha b + \beta b$ for all $\alpha, \beta \in \mathbb{F}$ and all $a, b \in$ **L**.

axiom 5. $(\alpha \beta) a = \alpha(\beta a)$ for all $\alpha, \beta \in \mathbb{F}$ and all $a \in$ **L**.

axiom 6. If $\alpha a = 0$ then $\alpha = 0$ or $a = 0$. Also $1a = a$ for all $a \in$ **L**.

WARNING: The exact form of these axioms is not standard and the first part of axiom 6 is unnecessary for $\mathbb{F} \neq \mathbb{Z}$. These fine points are addressed in the notes.

The first linear space we considered was $\mathbb{R}^n$ in Section 1.5. But not every linear space we have seen consists of vectors. The first example was the set of all polynomials with coefficients in $\mathbb{F}$ that appeared in Section 1.7.

We have also seen linear spaces whose elements are matrices. The set of ALL m by n matrices with coefficients in $\mathbb{F}$ was recognized to be a linear space in Section 2.2 when we discussed properties of matrix arithmetic.

In each case the set comes with an addition and a scalar multiplication under which it is "closed" (in the sense of Section 2.1). These operations also satisfy all six axioms. It must be understood, however, that in many instances the linear spaces studied have additional structure beyond being a linear space, and often we studied several linear spaces at once. For example, both polynomials and square matrices have their own multiplication as well. We also used each of polynomials and matrices to study the other. That is okay; in fact, kind of interesting.

Because this is an **axiomatic presentation**, the terms "scalar multiplication" and "vector addition" are **not necessarily the familiar coordinate-wise operations**, but rather a part of the **definition** of a linear space. That sentence bears repeating. It means that the fundamental operations might be undiscovered at this time. Still we are able to discuss "relative truths" — in Steinmetz's words (see page 61).

These operations could easily have some other meaning that you or I have never even seen before. That is the responsibility of the person presenting **L** to be a linear space. Officially, we have **no need to know**. Nonetheless, we will use exactly the same notation as in the text; the only difference is that to check some assertion, we now have only the **axioms** to work with, while in Section 1.5 we checked formulas coordinate by coordinate. Should we falter and rely on our experience or intuition, sooner or later posterity will discover our blunder — and not treat us kindly for it.

Subgeometries and Subspaces

Given an $\mathbb{F}$-linear space **L**, there are two essentially different ways to obtain a new linear space — change the number system or change the "vector set."

Suppose $\mathbb{E}$ is possibly a **smaller** number system than $\mathbb{F}$. An $\mathbb{E}$-*subgeometry* of **L** is an $\mathbb{E}$-linear space that is a subset of **L** with the same operations as **L**. For example, $\mathbb{C}^n$ contains $\mathbb{R}^n$ contains $\mathbb{Q}^n$ contains $\mathbb{Z}^n$, and each is a subgeometry of its predecessors.

Subgeometries reflect a change in "rules" and not necessarily in size. The matrix model of $\mathbb{C}$ that appears in Section 2.2 changes only the number system. In it $\mathbb{C}$ is regarded as an $\mathbb{R}$-subgeometry of itself. This has the effect of "forgetting" complex multiplication. Finally, $\mathbb{C}$-scalar multiplication is interpreted as an $\mathbb{R}$-linear function. This kind of "forgetting" of structure is actually more useful than it first seems.

Notice that when checking if a subset U is an $\mathbb{E}$-subgeometry, the meaning of addition and scalar multiplication is inherited from the original linear space. Therefore, none of the listed axioms need be checked. The important thing is to check the sum of any two elements of U is again in U (closure under addition) and that any $\mathbb{E}$-scalar multiple of an element of U is again in U (closure under $\mathbb{E}$-scalar multiplication).

The $\mathbb{F}$-subgeometries of an $\mathbb{F}$-linear space are the most important. They are called subspaces. Thus, a nonempty subset U of **L** is a *subspace* of the $\mathbb{F}$-linear space **L** provided it is closed under addition and $\mathbb{F}$-scalar multiplication. Any subspace of **L** is automatically an $\mathbb{F}$-linear space in its own right.

Suppose A is an m by n matrix having entries in $\mathbb{F}$. There are two subspaces with which we worked intensively.

The F-**null space**, $Null_\mathbb{F}(A)$, is the set of $\mathbf{x} \in \mathbb{F}^n$ such that $A\mathbf{x} = \mathbf{0}$. $Null_\mathbb{F}(A)$ appeared in Section 1.5 as the set of solutions to the homogeneous linear system with coefficient matrix A. We usually talked as if $\mathbb{F}$ was $\mathbb{R}$, but actually worked with $\mathbb{Q}$ until forced to work with $\mathbb{C}$ because of complex eigenvalues. (You don't really think it was a coincidence that all those examples worked out to have simply expressed coordinates, do you?)

The F-**column space** of the matrix A, written $Col_\mathbb{F}(A)$, is the set of all $\mathbf{b} \in \mathbb{F}^m$ such that $\mathbf{b} = A\mathbf{x}$ for some $\mathbf{x} \in \mathbb{F}^n$. $Col_\mathbb{F}(A)$ appeared in Section 1.6 as the set all vectors $\mathbf{b}$ for which the nonhomogeneous system $A\mathbf{x} = \mathbf{b}$ is consistent and was discussed algebraically in part b of Example 2.7. Here again we usually took $\mathbb{F}$ to be $\mathbb{Q}$ while talking as if it was $\mathbb{R}$. But we also considered $\mathbb{Z}$ when considering lattices in Section 4.3.

We also worked with subspaces of linear spaces other than $\mathbb{F}^n$. Here are just a couple of examples. The set of all polynomials that annihilate a fixed square matrix A appeared in Section 3.2 and is a subspace of $\mathbb{F}[x]$. Suppose A is a fixed n by n matrix; much of Chapter 3 focused on the set, called $\mathbb{F}[A]$, of matrices having the form $f(A)$ for some $f(x) \in \mathbb{F}[x]$. $\mathbb{F}[A]$ is a subspace of the linear space of all n by n matrices.

There is a lot to be said about what changes when the number system varies (for example what happens to those complex eigenvector of a real matrix if we stuck to $\mathbb{R}$) but that is not the purpose of this appendix. The reason for presenting different number systems here is really to display the variety of examples before us, and to be able to point out when special properties of the number system come into play.

For the rest of this appendix, we fix $\mathbb{F}$ and drop explicit reference to it except when necessary. Thus, the principal subgeometries considered will be subspaces.

Dimension
The theory of linear spaces applies to all of the above examples. It keys on the fact that the concepts of linear combination and linear independence, as introduced in Section 1.5, actually make perfect sense in any $\mathbb{F}$-linear space.

The zero vector $\mathbf{0}$ is in every subspace ($\mathbf{0} = 0\mathbf{x}$, for any $\mathbf{x}$). Because we have infinitely many scalars, axiom 6 implies that every subspace containing a nonzero element has infinitely many elements. Technically, the set containing only $\mathbf{0}$ also qualifies as a subspace. It is called the **zero subspace**.

A subspace of **L** can be built from ANY subset S of **L**. The **span** of S, written $\langle S \rangle$, is the set of of all linear combinations of elements of S. It is a subspace because a linear combination of linear combinations is itself a linear combination. If **U** is a subspace of **L** and $\mathbf{U} = \langle S \rangle$ then S **spans U**.

To illustrate "span," let A be an m by n matrix. Then the columns of A, viewed as a set of n vectors in $\mathbb{F}^m$, span $Col(A)$. Also, the set $\mathcal{B}$ of basic solutions to the homogeneous linear system $A\mathbf{x} = \mathbf{0}$ spans the null space of A.

A subset of **L** that both spans **L** and is linearly independent is called a ***basis***. Bases (plural of basis) play a central role in the theory of linear spaces.

The set of columns of the identity matrix I_n form the ***natural*** basis, $\mathcal{N}_n$, of $\mathbb{F}^n$. The set of monomials $1, x, \ldots, x^n$ is a basis for $\mathbb{F}_n[x]$, because every polynomial is a linear combination of monomials (span), and the only linear combination of monomials that is the zero function is the trivial one (linear independence).

Because the set $\mathcal{B}$ of basic solutions to $A\mathbf{x} = \mathbf{0}$ has this property, as shown in problem 6 of Section 1.5, it is a basis for $Null(M)$. At last you know my little secret. The name "basic solutions" was really a pun!

The big theorem about bases of linear spaces is that any two bases have the same size. This common size is the ***dimension*** of **L**. The key step in the proof of this theorem appears as a theorem on page 33 (but the proof is much harder if $\mathbb{F}$ is $\mathbb{Z}$).

Note that $\mathbb{R}^n$ has dimension n by this definition but $\mathbb{F}_n[x]$ has dimension $n+1$. Also the dimension of $Null(A)$ is the number of nonpivotal columns of $RREF(A)$ and we will see that the dimension of $Col(A)$ is the rank of A.

A line or plane in $\mathbb{R}^3$ is a subspace if and only if it contains the origin ($\mathbf{0}$ is in every subspace!), in which case it has dimension 1 or 2, respectively. The other lines and planes are sometimes called ***affine*** subspaces, and as you know they represent the solution sets to nonhomogeneous linear systems with three variables.

Homogeneous Coordinates

The introduction of homogeneous coordinates in Section 2.6 could be viewed as embedding $\mathbb{R}^2$ in $\mathbb{R}^3$ as the plane $z = 1$ (recall the balloon was NOT in the water, and the ships were on the surface of the flat ocean). This embedding leads to a correspondence of affine subspaces of $\mathbb{R}^2$ with certain genuine subspaces of $\mathbb{R}^3$ (determined by the observer's line of sight).

We could have done something analogous (adding a dimension/variable to simplify possibilities) with linear systems by inventing a new variable, say c. Multiply each equation's constant term by c and subtract the result from both sides. This turns the nonhomogeneous system into a homogeneous system with the extra variable c (listed last). Next, RREF the fattened coefficient matrix and solve the homogeneous system. Notice that c is a free variable whenever the original system is consistent. Finally, set $c = 1$ to return to the treatment of nonhomogeneous systems appearing in Chapter 1.

All of this seems like a silly way of avoiding nonhomogeneous linear systems — until you see the connection with homogeneous coordinates. Then it gives both insight and unification far beyond the their common "homogeneous" name.

The dimension theory of linear spaces is easier when $\mathbb{F}$ is not $\mathbb{Z}$. Then any spanning set has a subset that is a basis, and any linearly independent set can be extended to form a basis. However, neither of these important results holds for $\mathbb{Z}$-linear spaces, as illustrated in the notes.

Also, it is a definite possibility that a linear space has infinite dimension. For example, the set of differentiable real-valued functions is an $\mathbb{R}$-linear space of infinite dimension that is of great importance. We work only with finite-dimensional linear spaces, because matrices are not sufficient to deal with those of infinite dimension.

Subspaces associated with a linear system

Suppose that A is an m by n matrix of rank r. Let's interpret the solution of the homogeneous linear system $A\mathbf{x} = \mathbf{0}$ in terms of subspaces and bases. But please understand that there is no "magic bullet" in the abstract version. It will NOT help you solve any linear system; in fact it is more likely to reveal some "new but NOT improved" way to do what you already know is basically easy.

The objective here is instead to build on your knowledge of linear systems to help your understanding of the abstract setting. Let's begin with interpreting the construction of basic solutions when A is in RREF and then move on to what happens when we move from A to $RREF(A)$.

Suppose firstly that a matrix is already in RREF, and to emphasize this fact let's call the matrix $RREF$ rather than A. We start with two sets of natural basis vectors. Let $\mathcal{PC}$ be the vectors in the natural basis, $\mathcal{N}_m$, of $\mathbb{F}^m$ that appear as pivotal columns in $RREF$. In addition, let $\mathcal{NC}$ be the vectors in the natural basis, $\mathcal{N}_n$, of $\mathbb{F}^n$ that are **zero** in the positions of the leading coefficients of the rows $RREF$.

Observe that the column space $Col(RREF)$ is spanned by $\mathcal{PC}$.

As you know, the null space $Null(RREF)$ consists of the set of all solutions to the homogeneous system $RREF\mathbf{x} = \mathbf{0}$. Further, $Null(RREF)$ has dimension $n - r$, which is exactly the size of $\mathcal{NC}$. BUT a vector in $\mathcal{NC}$ is not in $Null(A)$ unless the associated column of A is all zeros (see page 34).

Suppose the j-th column of $RREF$ is nonpivotal and associated with $\mathbf{n} \in \mathcal{NC}$. Recall from Section 1.5 that the associated basic solution, $\mathbf{x_n}$, has its j-th coordinate equal to 1 and all other nonpivotal column coordinates zero. Also the r linear equations associated with the nonzero ROWS of $RREF$ are used to determine the remaining coordinates of $\mathbf{x_n}$.

Each of these rows has as its transpose a basis vector of the r-dimensional subspace $Col(RREF^T)$ of $\mathbf{F}^n$. The subspace $\langle Col(RREF^T), \mathbf{n} \rangle$ has dimension $r+1$ and the basic solution $\mathbf{x_n}$ is the intersection of this subspace and $Null(A)$. The dimension theory of linear spaces insures that this intersection is always non-zero, but that is anything but NEWS.

In order to complete the picture, we must understand what happens to $Col(A)$ and to $Null(A)$ during the reduction process. Perhaps the major point in Chapter 1 was that the solutions to $A\mathbf{x} = \mathbf{0}$ are precisely the solutions to $RREF(A)\mathbf{x} = \mathbf{0}$. In other words, A and $RREF(A)$ have the SAME null space. Its dimension is (still), $n - r$, and is the number of nonpivotal columns of $RREF(A)$.

Although $Col(A)$ and $Col(RREF(A))$ may be quite different subspaces, the **same** column positions of A and of $RREF(A)$ determine vectors comprising a basis. To see this we show that matrix reduction is not just an algorithm to solve linear systems! It is also a "left-leaning" algorithm to a pick basis of $Col(A)$.

Recall, from Section 4.1, that there are invertible matrices P and $Q = P^{-1}$ (products of elementary matrices) such that

$$PA = RREF(A) \text{ and } QRREF(A) = A.$$

Of particular interest is the second equation $QRREF(A) = A$. Think of $RREF(A)$ as a bunch of columns, each of which specifies a certain linear combination of the columns of Q that equals the associated column of A(as in Example 2.7 on page 69). With this interpretation it is clear that the first r (the rank of A) columns of Q are precisely the columns of A that become pivotal in $RREF(A)$ and that they form a basis $\mathcal{P}\mathcal{C}$ of $Col(A)$.

More than this, the process of building $RREF(A)$ from the left to the right is revealed to be testing whether the j-th column A_j of A is linearly dependent on the columns to its left, and if not, accepting A_j as a part of a set of linearly independent columns of A. Then, in either case, A_j is a linear combination of the partial basis and the coefficients of this linear combination are entered in the j-th column of $RREF(A)$. Thus, **the dimension of the column space of A is the rank of A.**

Coordinates

In a certain sense, $\mathbb{F}^n$ is the **only** n-dimensional $\mathbb{F}$-linear space, $n < \infty$. Our problem is that we have relied too heavily on the natural basis to understand it. We now show how any basis of an $\mathbb{F}$-linear space leads to the concept coordinates and the major objective — finding a way to treat all bases equally.

Suppose $\mathbf{L}$ is an n-dimensional $\mathbb{F}$-linear space. This means that every basis of $\mathbf{L}$ has has n elements. Let $\mathcal{B} = \{\mathbf{b}_1, \dots, \mathbf{b}_n\}$ be a basis for $\mathbf{L}$. This, in turn, means that $\mathcal{B}$ is linearly independent and $\mathcal{B}$ spans $\mathbf{L}$. Because $\mathcal{B}$ spans $\mathbf{L}$, each $\mathbf{v} \in \mathbf{L}$ is expressible as a linear combination of elements of $\mathcal{B}$, say

$$\mathbf{v} = \alpha_1 \mathbf{b}_1 + \cdots + \alpha_n \mathbf{b}_n; \alpha_i \in \mathbb{F}.$$

If there were to be a second expression of $\mathbf{v}$ as a linear combination of elements of $\mathcal{B}$, and some coefficients did not match this expression perfectly, then the difference of the two equations for $\mathbf{v}$ would express $\mathbf{0} = \mathbf{v} - \mathbf{v}$ as a nontrivial linear combination of the elements of $\mathcal{B}$, contradicting the fact that $\mathcal{B}$ is linearly independent.

This shows that each $\mathbf{v} \in \mathbf{L}$ is expressible UNIQUELY as a linear combination of elements of any basis. Use this fact to define a very useful function with the unusual name $(\ )_\mathcal{B}$, called "coordinates with respect to $\mathcal{B}$" from $\mathbf{L}$ to $\mathbb{F}^n$ by

$$(\mathbf{v})_\mathcal{B} = (\alpha, \dots, \alpha_n)^T, \text{ whenever } \mathbf{v} = \alpha_1 \mathbf{b}_1 + \cdots + \alpha_n \mathbf{b}_n.$$

Using only the axioms for a linear space, one can show that if you have two vectors in $\mathbf{L}$ and add their coordinate vectors (in $\mathbb{F}^n$), you get the same result as if you first summed the vectors in **L** and then took the coordinates of that sum. Similarly, it doesn't matter where you do scalar multiplication, before or after taking coordinates, you get the same result.

NOTE: Remember, addition in **L** is completely formal and may not have a direct connection with coordinates, like geometric vector addition in Section 1.6 or even worse! The point is that, in the above paragraph, addition is used with several (possibly) different meanings in the same sentence. This is a point quite similar to the parsing problem mentioned on page 144. Really, it is a consequence of our use of powerful notation and why mathematicians are **very slow readers**!

In any case, $(\ )_\mathcal{B}$ gives a one-to-one correspondence between **L** and $\mathbb{F}^n$ that transports the algebraic structure of **L** into the algebraic structure of $\mathbb{F}^n$. Such a map is called an *isomorphism* of linear spaces. The sense in which $\mathbb{F}^n$ is the only n-dimensional $\mathbb{F}$-linear space is that any other n-dimensional $\mathbb{F}$-linear space is isomorphic with $\mathbb{F}^n$.

But this isomorphism is a mixed blessing.[1] In order to get it, we had to work with the basis $\mathcal{B}$ of **L**, and there are infinitely many bases. The choice of basis is critical. A random basis may or may not be well-suited to the task at hand.

Consider, for example, the original problem of solving $A\mathbf{x} = \mathbf{0}$. Any basis of $Null(A)$ would do to obtain all solutions. But we chose the basic solutions because they arise most naturally from the RREF.

Consider also polynomials of degree at most n. In Section 2.3, we saw that when a dataset of size $n + 1$ is specified and the task is polynomial interpolation, the fundamental Lagrange interpolants actually make a much more convenient basis than monomials. You might even say that the fundamental Lagrange interpolants are **the natural bases** for functions on the dataset because their **values** are zeros and ones just like the **coordinates** of the natural basis of $\mathbb{F}^n$.

Finally, consider the linear function f_A associated with a diagonalizable matrix A. In Chapter 3 we used the diagonalizing matrix P to reduce all sorts of matrix questions about A to numerical questions about eigenvalues. We will show that this process was in fact changing to a new basis, namely the columns of P, that is very well suited to the study of A.

**Careful basis choice can transform a hard problem to an easy one.
Do you like easy problems?**

Transition matrices
Suppose $\mathcal{B} = \{\mathbf{b}_1, \ldots, \mathbf{b}_n\}$ and $\mathcal{C} = \{\mathbf{c}_1, \ldots, \mathbf{c}_n\}$ are bases. We now study how $(\mathbf{v})_\mathcal{B}$ and $(\mathbf{v})_\mathcal{C}$ are related, for $\mathbf{v} \in \mathbf{L}$. Really this is just an abstract version of Section 2.6. In the end we see that, even when working with an abstract linear space, **matrices** cannot be avoided if you really want to be a "basis egalitarian."

---

[1] "Mathematical education is still suffering from the enthusiasm which the discovery of this isomorphism has aroused" – from page 13 of **Geometric Algebra** by E. Artin, Interscience, New York, 1957.

Just as in Section 2.6, the key is to express the elements of one basis as linear combinations of the other. Define $t_{ij} \in \mathbb{F}$, $i, j = 1 \ldots n$ by

$$\mathbf{b}_j = t_{1j}\mathbf{c}_1 + \cdots + t_{ij}\mathbf{c}_i + \cdots + t_{nj}\mathbf{c}_n.$$

These t_{ij} do exist because $\mathcal{C}$ spans $\mathbf{L}$, and $\mathbf{b}_i$ is uniquely expressible as a linear combination of the elements $\mathcal{C}$ because it is a linearly independent set.

Now $(\mathbf{v})_\mathcal{B} = (\beta_1 \ldots \beta_n)^T$ means that $\mathbf{v} = \beta_1\mathbf{b}_1 + \cdots + \beta_n\mathbf{b}_n$, and upon substitution and regrouping, this is expressed as a linear combination of the $\mathbf{c}$'s as follows:

$$
\begin{aligned}
\mathbf{v} = \ & \beta_1(t_{11}\mathbf{c}_1 + \cdots + t_{i1}\mathbf{c}_i + \cdots + t_{n1}\mathbf{c}_n) + \ldots \\
& + \beta_j(t_{1j}\mathbf{c}_1 + \cdots + t_{ij}\mathbf{c}_i + \cdots + t_{nj}\mathbf{c}_n) + \ldots \\
& + \beta_n(t_{1n}\mathbf{c}_1 + \cdots + t_{nj}\mathbf{c}_j + \cdots + t_{nn}\mathbf{c}_n) \\
= \ & (t_{11}\beta_1 + \cdots + t_{1j}\beta_j + \cdots + t_{1n}\beta_n)\mathbf{c}_1 + \ldots \\
& + (t_{i1}\beta_1 + \cdots + t_{ij}\beta_j + \cdots + t_{nj}\beta_n)\mathbf{c}_i + \ldots \\
& + (t_{1n}\beta_n + \cdots + t_{nj}\beta_j + \cdots + t_{nn}\beta_n)\mathbf{c}_n.
\end{aligned}
$$

This is the (unique) expression of $\mathbf{v}$ as a linear combination of the elements of $\mathcal{C}$. Now imagine the last equation when all the βs are 0 but for one 1. Lo and behold:

$$(\mathbf{v})_C = T_{C \leftarrow B}\,(\mathbf{v})_B, \qquad \text{where the matrix } T_{C \leftarrow B} \text{ has } ij \text{ entry} \\ t_{ij} = \text{the } i\text{-coordinate of } (\mathbf{b}_j)_C.$$

Also as in Section 2.6, we can chain together several basis changes by simply multiplying the associated transition matrices.

Notice that we are not using perpendicular unit vectors in the abstract setting. In fact, we don't even have a meaning for length or angle for vectors in an abstract linear space (what is the angle between two polynomials?).

There is an important abstract version of "length," and therefore "angle," for linear spaces; it involves more axioms and another postulated operation. Such a structure is called a ***normed*** linear space. They have some bases that are really nice (e.g., perpendicular unit vectors). The point is that normed linear spaces are linear spaces with additional "structure."

Linear functions
In Section 1.6 we used transform plots to interpret matrix times vector multiplication as functions from $\mathbb{F}^n$ to $\mathbb{F}^m$. You will not be surprised that there is an abstract version of this, too.

Suppose $\mathbf{L}$ and $\mathbf{M}$ are abstract $\mathbb{F}$-linear spaces. A ***linear function*** $f : \mathbf{L} \to \mathbf{M}$ is just a function that respects both linear space structures. Formally, this amounts to

$$f(\mathbf{x} + \mathbf{y}) = f(\mathbf{x}) + f(\mathbf{y}) \text{ and } f(\alpha\mathbf{x}) = \alpha f(\mathbf{x}) \text{ for } \mathbf{x}, \mathbf{y} \in \mathbf{L}, \alpha \in \mathbb{F}.$$

Recall that the ***domain*** of f is, by definition, $\mathbf{L}$, and its ***codomain*** is $\mathbf{M}$.

We have seen a variety of linear functions. Of course, the functions mapping $\mathbf{x} \in \mathbb{R}^n$ to $A\mathbf{x} \in \mathbb{R}^m$, for fixed matrix A, studied with transform plots in Sections 1.6, 3.1 and 4.1 are linear functions.

We used Vandermonde matrices to obtain a linear function from the coefficients of polynomials in $\mathbb{F}_n[x]$ to their values at specified data points $x_0, \ldots, x_n$ in Sections 1.7 and 2.6. This function was called the "Fourier transform" in Section 2.7.

The remainder function rem_g (in the first objective of) Section 2.7 is actually a linear function from polynomials to polynomials.

Last but not least, the "coordinate" isomorphism $(\ )_\mathcal{B}$:$\mathbf{L} \rightarrow \mathbb{F}^n$ associated with a basis $\mathcal{B}$ of the n-dimensional $\mathbb{F}$-linear space $\mathbf{L}$ (introduced in the preceding part of this appendix) is a particularly useful linear function.

Associated with any linear function f there are two important subspaces. The **kernel** of f, $ker(f)$, is the set of all $\mathbf{x}$ in the domain of f such that $f(\mathbf{x}) = \mathbf{0}$; the **image** of f, $im(f)$, is the set of all $\mathbf{y}$ in the codomain of f such that $\mathbf{y} = f(\mathbf{x})$ for some $\mathbf{x} \in \mathbf{L}$. Any linear function whose kernel is the zero subspace (of its domain) and whose image equals its codomain is an **isomorphism** of linear spaces.

The linear function f_A mapping $\mathbf{x} \in \mathbb{F}^n$ to $A\mathbf{x} \in \mathbb{F}^m$, for a fixed matrix A, has $ker(f_A)$ equal to $Null(A)$ and $im(f_A)$ equal to $Col(A)$.

For an example, take $g(x)$ be a polynomial of degree d as in Section 2.7. Then the kernel of rem_g is the set of all multiples of $g(x)$ in $\mathbb{F}_n[x]$, and its image is all possible remainders, namely $\mathbb{F}_d[x]$ — even if we chose to express the polynomials in terms of fundamental Lagrange interpolants on some weird dataset.

NOTE: This may seem as if we are naming the same thing twice, but we are not. The difference is that the linear function f_A, and subspaces $ker(f_A)$, $im(f_A)$ are "basis free." Later we will be able express these basis free concepts with respect to arbitrary bases, say $\mathcal{B}$ of $\mathbb{F}^n$ and $\mathcal{C}$ of $\mathbb{F}^m$. When so expressed, f_A still has a matrix (called $_C(f_A)_\mathcal{B}$ below) but it may not look at all like A.
(This note is, to some extent, anticipating coming attractions. Don't worry if it doesn't make complete sense now. It is just trying to say that we become "basis egalitarian" by first being "basis free.")

To associate a matrix to a linear function $f: \mathbf{L} \rightarrow \mathbf{M}$ between abstract linear spaces requires bases in $\mathbf{L}$ and $\mathbf{M}$.

As already emphasized, our purpose is to **find bases that make hard problems easy**. But first we must develop machinery to move between bases easily. Suppose $f: \mathbf{L} \rightarrow \mathbf{M}$ is a linear function between finite-dimensional abstract $\mathbb{F}$-linear spaces. Suppose $\mathcal{A} = \{\mathbf{a}_1 \ldots \mathbf{a}_n\}$ is a basis of $\mathbf{L}$ and that $\mathcal{C} = \{\mathbf{c}_1 \ldots \mathbf{c}_m\}$ is a basis of $\mathbf{M}$. The matrix **realization** of f with respect to $\mathcal{A}$ and $\mathcal{C}$, $_C(f)_\mathcal{A}$, is defined by:

$$(f(\mathbf{x}))_C = \ _C(f)_\mathcal{A} \ (\mathbf{x})_\mathcal{A}.$$

In "plain English," the coordinates, $(f(\mathbf{x}))_C$, of $f(\mathbf{x})$ with respect C are obtained by multiplying the matrix realization of f with respect to $\mathcal{A}$ and C, $_C(f)_\mathcal{A}$, by the coordinates, $(\mathbf{x})_\mathcal{A}$, of $\mathbf{x}$ with respect to $\mathcal{A}$.

Just as in Section 1.6 on page 47, the j-th column of $_C(f)_\mathcal{A}$ consists of the coordinates of the image of the j-th domain basis vector when expressed in terms of the codomain basis C. But this time the notation is more impressive because we are keeping track of so much more.

The $\mathcal{A}$-coordinates, $(\mathbf{a}_j)_\mathcal{A}$, of the j-th element of $\mathcal{A}$ is a vector (in $\mathbb{F}^{dim(\mathbf{L})}$) of all 0's except for a 1 in its j-th position. The j-th column of the matrix realization of f, with respect to $\mathcal{A}$ and C, is the C-coordinates of $f(\mathbf{a}_j)$, written $(f(\mathbf{a}_j))_C$ (and in $\mathbb{F}^{dim(\mathbf{M})}$). Using all our notation this reads as: $_C(f)_\mathcal{A}$ has i, j-entry f_{ij} where

$$(f(\mathbf{a}_j))_C = f_{1j}\mathbf{c}_1 + \cdots + f_{mj}\mathbf{c}_m.$$

Finally, we are in a position to deal with more than one basis. Suppose $\mathcal{B}$ is a second basis of $\mathbf{L}$ and that $\mathcal{D}$ is a second basis for $\mathbf{M}$. Then we have transition matrices $T_{\mathcal{A}\leftarrow\mathcal{B}}$ and $T_{\mathcal{D}\leftarrow C}$ that allow us to switch between realizations of f. The matrix realization of f with respect of $\mathcal{B}$ and $\mathcal{D}$ is given by

$$_\mathcal{D}(f)_\mathcal{B} = T_{\mathcal{D}\leftarrow C}\ _C(f)_\mathcal{A}\ T_{\mathcal{A}\leftarrow\mathcal{B}}.$$

All of this is summarized in the diagram. The matrix equation asserts that you can get from the top left to bottom left by following either one arrow straight down or three arrows passing through the top right and bottom right along the way.

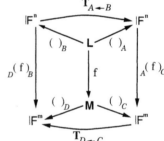

We are finally in a position to describe some common elements of three fundamental problems we have faced:

Linear systems
Since RREF may involve division, the number system $\mathbb{F} = \mathbb{Z}$, the integers, must be excluded.

Start with an m by n matrix A of rank r and let $f_A : \mathbb{F}^n \to \mathbb{F}^m$ be the associated linear function. Then $A = \ _{\mathcal{N}_n}(f_A)_{\mathcal{N}_m}$ where, as earlier, $\mathcal{N}_s$ is the natural basis of $\mathbb{F}^s$, $s = m, n$. Our objective now is to ask what would happen to the matrix realization of f_A if we chose highly compatible bases of $\mathbb{F}^n$ and $\mathbb{F}^m$.

In an earlier part of this appendix, we discussed the subspace interpretation of the basic solutions of the homogeneous linear system $A\mathbf{x} = \mathbf{0}$ and recognized them as a basis for the null space of the matrix A. All there is to add is that they are also a basis for $ker(f_A)$, the kernel of the linear function f_A. It is good enough to take the basis $\mathcal{B}$ consisting of the vectors in $\mathcal{N}_n$ associated with the leading coefficients

$RREF(A)$ (and called $\mathcal{NC}$ on page 221) followed by the basic solutions to $A\mathbf{x} = \mathbf{0}$ in their natural order.

What about a basis $\mathcal{D}$ for the codomain? It seems clear we should start with the columns of A that become pivotal in $RREF(A)$ (called $\mathcal{PC}$ on page 222) because they form a basis for the column space of A. That leaves $m - r$ elements of $\mathcal{D}$. The fact is that these vectors could be anything as long as $\mathcal{D}$ is a basis! This is a reflection of the large amount of freedom in precisely what row operations are used to reduce a matrix.

Now, at last, we have the matrix realization of f_A with respect to the basis $\mathcal{B}$ of $\mathbb{F}^n$ and the basis $\mathcal{D}$ of $\mathbb{F}^m$. It has the block matrix form

$$_\mathcal{B}(f_A)_\mathcal{D} = \left(\begin{array}{c|c} I_r & \\ \hline & \end{array} \right).$$

Now that is my kind of linear system coefficient matrix!

Diagonalizing a matrix
Suppose A is a diagonalizable square matrix with diagonalizing matrix P. Consider the associated linear function f_A, BUT regard P as a transition matrix.

The columns of P are eigenvectors of A and form a basis $\mathcal{E}$ of the domain of f_A, which has of course its natural basis $\mathcal{N}$. As above, the j-th column of $T_{\mathcal{N} \leftarrow \mathcal{E}}$ consists of the $\mathcal{N}$-coordinates of the j-th element of $\mathcal{E}$. Therefore, $P = T_{\mathcal{N} \leftarrow \mathcal{E}}$. Now

$$_\mathcal{N}(f_A)_\mathcal{N} \, T_{\mathcal{N} \leftarrow \mathcal{E}} = AP = PD = T_{\mathcal{N} \leftarrow \mathcal{E}} D.$$

It follows that $D = {}_\mathcal{E}(f_A)_\mathcal{E}$. This says that D is the matrix realization of the linear function f_A with respect to $\mathcal{E}$ and $\mathcal{E}$.

It also explains similarity of matrices. The linear function interpretation of matrix similarity is that the linear function has the same domain and codomain ($\mathbf{L} = \mathbf{M}$), so you are forced to change the bases in both simultaneously.

It even puts the similarity problem posed at the very beginning of Chapter 3 in its original setting. Namely, suppose a linear function is given with domain equal to its codomain. Find a simultaneous basis of the domain and codomain with respect to which the linear function is most easily understood.

Please notice that this setting does not require matrices. The historical fact is that Cauchy addressed this problem before Cayley invented them!

Smith normal form
In Section 4.3 we briefly discussed isometric (area preserving) row and column operations with integer matrices. To obtain the Smith normal form S of A, we found isometric matrices (matrices with determinant ± 1) P and Q such that $PAQ = S$. Although we only worked with 2 by 2 matrices, the methods applied to m by n matrices.

Again we interpret A as defining a linear function, but this time view both P and Q as transition matrices. The difference from similarity is that we regard f_A as having **distinct** linear spaces for its domain and codomain.

Start with natural bases $\mathcal{N}_n$ and $\mathcal{N}_m$ of the domain and codomain, respectively and

$$A = {}_{\mathcal{N}_m}(f_A)_{\mathcal{N}_n}.$$

Because Q is isometric, its inverse has integer entries and therefore the columns of Q form a $\mathbb{Z}$-basis, $\mathcal{Q}$ for $\mathbb{Z}^n$. As above the j-th column of $T_{\mathcal{N}_n \leftarrow \mathcal{Q}}$ consists of the $\mathcal{N}_n$-coordinates of the j-th element of $\mathcal{Q}$. Thus, $Q = T_{\mathcal{N}_n \leftarrow \mathcal{Q}}$.

In a similar way $P = T_{\mathcal{P} \leftarrow \mathcal{N}_m}$ where $\mathcal{P}$ is a $\mathbb{Z}$-basis of the codomain $\mathbb{Z}^m$. We are led to the grand equation

$$T_{\mathcal{P} \leftarrow \mathcal{N}_m} \, {}_{\mathcal{N}_m}(f_A)_{\mathcal{N}_n} \, T_{\mathcal{N}_n \leftarrow \mathcal{Q}} = PAQ = S = {}_{\mathcal{P}}(f_A)_{\mathcal{Q}}.$$

If you return to Section 4.3 and reinterpret the different grids associated to a lattice as a change of basis in the domain of a function, and reinterpret moving to an equivalent lattice as a change of basis in the codomain of the same function, then you see the fundamental nature of the section.

In fact, we have addressed a problem as basic as that addressed by Cauchy and quite analogous to the abstract version of solving linear systems.

Suppose a $\mathbb{Z}$-linear function with domain different from its codomain is given. Find a basis of the domain and a basis of the codomain with respect to which the linear function is most easily understood.

With the Smith normal form, we have also addressed the case excluded at the bottom of page 226. If you have a homogeneous system of linear equations $A\mathbf{x} = \mathbf{b}$ with A having **integer** entries, then there are matrices P, Q invertible over the integers such that the system can be expressed

$$SNF(A)Q^{-1}\mathbf{x} = PAQQ^{-1}\mathbf{x} = P\mathbf{b}.$$

Thus, the original system is equivalent to a series of equations, the i-th of which is

$$d_i(\text{some linear combination of the } xs) = \text{some fixed integer,}$$

and d_i is the i-th diagonal term in the Smith normal for $SNF(A)$ of A.

Notes

1. Set $a = 0$ in axiom 3 to obtain $b + c = c + b$ and commutativity, from which follows associativity $a + (b + c) = (a + b) + c$. Conversely show that commutativity and associativity imply axiom 3.

2. Let $\mathbb{F} = \mathbb{Q}, \mathbb{R}$ or $\mathbb{C}$. Suppose $\alpha \neq 0$ and $\alpha a = 0$ as in axiom 6. Use $\beta = \alpha^{-1}$ in axiom 5 to show that the first part of axiom 6 is redundant in these cases.

3. Let **L** be "clock arithmetic" $\{0 \ldots 11\}$ where $a + b$ is the remainder when the integer sum is divided by 12 (for example $7 + 8 = 3$). Define scalar multiplication similarly. Show that **L** is almost a $\mathbb{Z}$-linear space satisfying all axioms except the first part of axiom 6.

4. Consider the integer matrix $M = \begin{pmatrix} 1 & 2 & 3 \\ 2 & 0 & 0 \end{pmatrix}$. Show that $\begin{pmatrix} 0 \\ 2 \end{pmatrix}$ is in $Col(M)$ but not an integer linear combination of any two of its columns. This shows that there is no integer basis of the $col(M)$ obtainable by deleting some column. Show that there is no third column with integer coordinates that could be added to M^T so the result has determinant 1. This shows that the columns of M^T cannot be augmented to obtain a basis of $\mathbb{Z}^3$.

5. The trace function, Problem 10 of Section 3.1, is a linear function from the linear space of square matrices to $\mathbb{F}$ but the determinant is NOT.

6. Compare the dimension of the $\mathbb{F}$-linear space $\mathbb{F}[A]$ and the degree of $\mu_A(x)$.

Selected Practice Problem Answers

1.1

1. yes   3. no   5. (i) and (iii)   7. (ii)   9. (i)   11. none

1.2

1. $\begin{pmatrix} 25 & 27 & 76 & | & 43 \\ 36 & 51 & 79 & | & 20 \\ 15 & 13 & -9 & | & 76 \end{pmatrix}$
3. $\begin{pmatrix} 62.781 & 0 & -4 & -14.27 & 0 & | & 0 \\ 0 & -67.66 & 0 & -19.1 & 15.67 & | & 0 \\ -37.11 & 0 & -11.13 & 0 & -14.5 & | & -47.62 \end{pmatrix}$ (or switch columns 2 and 5)

5.
$$\begin{aligned}
-23.54x_1 - 56.02x_2 &= -54.82 \\
42.55x_2 + 31.02x_3 &= -0.02 \\
-15.52x_1 + 62.1x_2 - 90.13x_3 &= 0 \\
-82.9x_1 - 5x_3 &= 78.51
\end{aligned}$$

7.
$$\begin{aligned}
67.3x_1 + 25.9x_3 + 6.3x_4 &= 9 \\
56.2x_3 + 11.9x_4 &= -44.6 \\
-23.5x_1 - 42.6x_2 + 28.2x_4 &= -76.2 \\
45.2x_3 + 61x_4 &= 18.9
\end{aligned}$$

9. $\begin{pmatrix} 1 & 8/7 & 9/7 \\ 0 & -3 & -6 \\ 0 & 6/7 & 12/7 \end{pmatrix}$

11. $R1 \leftrightarrow R2; R3 - 2R2 \rightarrow R3$

13. The first row operation changes the first line to a horizontal line that intersects the second line at the same point. The third doesn't change either line, it changes the background graph paper as in part a of Example 1.19.

1.3

1. a. neither
   c. RREF
   e. neither

3. $\begin{pmatrix} 1 & 0 & 0 & | & 0 \\ 0 & 1 & 0 & | & 0 \\ 0 & 0 & 1 & | & 0 \\ 0 & 0 & 0 & | & 1 \end{pmatrix}$; pivotal: 1, 2, 3, 4.

5. $\begin{pmatrix} 1 & 3 & 2 & 0 & 0 & | & -3 \\ 0 & 0 & 0 & 1 & 0 & | & -12 \\ 0 & 0 & 0 & 0 & 1 & | & 4 \end{pmatrix}$; pivotal: 1, 4, 5; nonpivotal: 2, 3, 6.

7. $\begin{pmatrix} 1 & 0 & 0 & | & 0 \\ 0 & 1 & 0 & | & 0 \\ 0 & 0 & 1 & | & 0 \\ 0 & 0 & 0 & | & 1 \end{pmatrix}$; pivotal: 1, 2, 3, 4.

9. $\begin{pmatrix} 5 & 3 & -2 & | & 10 & 2 \\ 14 & 1 & -1 & | & 2 & 4 \\ 1 & -1 & 0 & | & 5 & 6 \end{pmatrix} \rightarrow \begin{pmatrix} 1 & 0 & 0 & | & -1/2 & 0 \\ 0 & 1 & 0 & | & -11/2 & -6 \\ 0 & 0 & 1 & | & -\frac{29}{2} & -10 \end{pmatrix}$

11. Suppose A and B are both in RREF. Then the homogeneous linear system with coefficient matrix A has exactly the same solutions as the homogeneous linear system with coefficient matrix B. Use this to show $A = B$.

1.4

1. none   2. d   3. none

5. a. $\begin{aligned} x_1 &= 2 \\ x_2 &= 3 - 2x_3 \\ x_4 &= -1 \end{aligned}$, $\begin{pmatrix} 2 \\ 3 - 2x_3 \\ x_3 \\ -1 \end{pmatrix}$

   c. $\begin{aligned} x_1 &= 2 - 2x_3 \\ x_2 &= 1 + 3x_3 \\ x_4 &= 3 \end{aligned}$, $\begin{pmatrix} 2 - 2x_3 \\ 1 + 3x_3 \\ x_3 \\ 3 \end{pmatrix}$

7. $\begin{pmatrix} \frac{-1}{3} - 2w \\ w \\ \frac{3}{2} \\ 0 \\ \frac{-1}{3} \end{pmatrix}$ 9. $\begin{pmatrix} -1 \\ 2 \\ -1 \\ 3 \end{pmatrix}$ 11. no solution   13. Use Problem 12 of Section 1.2

1.5

1. a. $\begin{pmatrix} -1 \\ 14 \\ -10 \end{pmatrix}$ $\begin{aligned} x_1 +2x_2 &= -1 \\ 4x_1 -3x_2 +x_3 &= 14 \\ 6x_1 \quad\quad 7x_3 &= -10 \end{aligned}$ c. not defined   e. not defined

3. a. $x_2 \begin{pmatrix} -3 \\ 1 \\ 0 \\ 0 \\ 0 \end{pmatrix} + x_3 \begin{pmatrix} 8 \\ 0 \\ 1 \\ 0 \\ 0 \end{pmatrix} + x_5 \begin{pmatrix} -8 \\ 0 \\ 0 \\ 1 \\ 1 \end{pmatrix} + \begin{pmatrix} 2 \\ 0 \\ 0 \\ 6 \\ 0 \end{pmatrix}$ Standard list: $\begin{pmatrix} -3 \\ 1 \\ 0 \\ 0 \\ 0 \end{pmatrix}, \begin{pmatrix} 8 \\ 0 \\ 1 \\ 0 \\ 0 \end{pmatrix}, \begin{pmatrix} -8 \\ 0 \\ 0 \\ 1 \\ 1 \end{pmatrix}$

c. $x_2 \begin{pmatrix} -5 \\ 1 \\ 0 \\ 0 \\ 0 \end{pmatrix} + x_5 \begin{pmatrix} -8 \\ 0 \\ -6 \\ 3 \\ 1 \end{pmatrix} + \begin{pmatrix} 2 \\ 0 \\ -1 \\ 7 \\ 0 \end{pmatrix}$ Standard list: $\begin{pmatrix} -5 \\ 1 \\ 0 \\ 0 \\ 0 \end{pmatrix}, \begin{pmatrix} -8 \\ 0 \\ -6 \\ 3 \\ 1 \end{pmatrix}$

5. (i) is basic
(iii) is distinguished

6. Suppose a linear combination of basic solutions equals **0**. Use the free variable's coordinates to show each scalar is zero.

8. If some rows are removed, what happens to the general solution of a linear system when some of the equations are omitted? Can you show that the number of free variables cannot decrease? For the columns, how are the ranks related if the original matrix was already in RREF? Imagine the entire reduction process with all row operationss listed and scratch out the affected columns.

1.6

1. a. i) = ii), all distances equal 3   3. a. $\frac{\pi}{4}$ b. $\theta \approx .4636$ c. $\theta \approx .2257$

4.

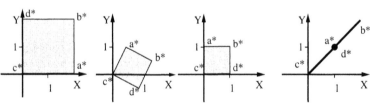

5. a. $\begin{pmatrix} 0 & 1 \\ -1 & 0 \end{pmatrix}$ c. $\begin{pmatrix} 1 & -1 \\ 2 & 2 \end{pmatrix}$ 6. $\mathbf{x} = \pm \frac{1}{\sqrt{6}}(1\ -2\ 1)^T$.

7. Can you show intersecting lines are mapped to intersecting lines? Why is this sufficient?

1.7

1. $y = x^2$ 3. $y = -4x + x^3$ 5. $y = -2 - \frac{11}{2}x + \frac{11}{2}x^2 + f_3(2 - x - 2x^2 + x^3)$

2.1

1. a. $(11 + i)$ c. $(3 + i)$ e. $\frac{1}{5} + \frac{3}{5}i$

2. a. 4   c. $\doteq 5(\cos(.6435) + i\sin(.6435))$
    e. $\doteq 2.24(\cos(2.678) + i\sin(2.678))$ (Angles in radians.)

3. a.$x = \pm i$ 4. See the bottom of page 59

5. $z = x + iy; (x - 1)^2 + y^2 = 1$ 8.   Use Problem 7

2.2

1. a. $\begin{pmatrix} 29 & -36 \\ -40 & -27 \end{pmatrix}$ c. $\begin{pmatrix} -31 & 14 & -43 \\ 17 & -22 & 50 \\ 52 & 24 & -27 \end{pmatrix}$ e. $\begin{pmatrix} 65 & -7 & 3 & -41 \\ -39 & 3 & -18 & 51 \\ -34 & 2 & -24 & 58 \\ -36 & 4 & 0 & 20 \end{pmatrix}$

2. a. $\begin{pmatrix} 2 & 1 \\ 6 & -7 \end{pmatrix}$ c. $(5 \ -6)$ e. $\begin{pmatrix} 8 \\ 15 \\ -24 \end{pmatrix}$

3. a. $\begin{pmatrix} -1 & 0 \\ 15 & 4 \end{pmatrix}$ c. 90   e. $\begin{pmatrix} 8 & 8 \\ 2 & 2 \end{pmatrix}$

5. a. $A = \begin{pmatrix} 1 & 0 & 1 & 0 & 0 \\ 1 & 0 & 1 & 0 & 1 \\ 0 & 1 & 0 & 1 & 1 \\ 1 & 1 & 0 & 0 & 0 \\ 1 & 0 & 0 & 1 & 0 \end{pmatrix}$ c. sum all entries in $(A + A^2 + A^3) = 106$

6. a. $\begin{pmatrix} 32 & 40 & 15 \\ -58 & 109 & 37 \\ 91 & -9 & 93 \end{pmatrix}$ 8. $\begin{pmatrix} 1 & -1 \\ 1 & 1 \end{pmatrix}^2 = \ldots$

2.3

1. $\begin{pmatrix} 34 & 16 & -42 \\ 55 & 18 & -37 \\ 2 & -27 & 0 \end{pmatrix}$ 3. a. $\begin{pmatrix} 32 \\ 108 \end{pmatrix}$ b. $\begin{pmatrix} -111 \\ -204 \\ -123 \end{pmatrix}$ 4.   a. $x^2 - 6x + 5$

5 a. $\begin{pmatrix} -2 & \frac{3}{2} \\ 1 & \frac{-1}{2} \end{pmatrix}$ d. Does not exist   e. $\begin{pmatrix} 1 & 0 & -1 & -2 \\ 0 & 1 & -3 & -4 \\ 0 & 0 & 1 & 0 \\ 0 & 0 & 0 & 1 \end{pmatrix}$ 6. a. $V^{-1} = \begin{pmatrix} 1 & 0 & 0 \\ \frac{-3}{2} & 1 & \frac{-1}{2} \\ \frac{1}{2} & \frac{-1}{2} & \frac{1}{2} \end{pmatrix}$

7.   The equation $Y = IY = (XM)Y = X(MY) = XI = X$ shows $X = Y$, so inverses are unique when they exist. Now $(AB)(B^{-1}A^{-1}) = ((AB)B^{-1})A^{-1} = (A(BB^{-1}))A^{-1} = (AI)A^{-1} = AA^{-1} = I$. Similarly $(B^{-1}A^{-1})(AB) = I$, so $B^{-1}A^{-1}$ satisfies all the conditions that determine $(AB)^{-1}$ uniquely.

9.   See page 139   11.   See top of page 142

2.4

1. a. $\frac{1}{2}\begin{pmatrix} 1 & 1 \\ 1 & 1 \end{pmatrix}$ $\frac{1}{2}\begin{pmatrix} 1 & -1 \\ -1 & 1 \end{pmatrix}$ c. $I_2,$ $\begin{pmatrix} 0 & 0 \\ 0 & 0 \end{pmatrix}$

   e. $P_A = \frac{1}{10}\begin{pmatrix} 6 & 4 & 2 & 0 & -2 \\ 4 & 3 & 2 & 1 & 0 \\ 2 & 2 & 2 & 2 & 2 \\ 0 & 1 & 2 & 3 & 4 \\ -2 & 0 & 2 & 4 & 6 \end{pmatrix}$, $P_{A^\perp} = \frac{1}{10}\begin{pmatrix} 4 & -4 & -2 & 0 & 2 \\ -4 & 7 & -2 & -1 & 0 \\ -2 & -2 & 8 & -2 & -2 \\ 0 & -1 & -2 & 7 & -4 \\ 2 & 0 & -2 & -4 & 4 \end{pmatrix}$

2. a. $\frac{1}{5}\begin{pmatrix} 42 \\ 56 \end{pmatrix}$, $\frac{1}{5}\begin{pmatrix} 8 \\ -6 \end{pmatrix}$ c. $\begin{pmatrix} 0 \\ 0 \\ 1 \end{pmatrix}$, $\begin{pmatrix} 1 \\ 1 \\ 0 \end{pmatrix}$ e. $\frac{1}{6}\begin{pmatrix} 2 \\ -1 \\ 1 \end{pmatrix}$, $\frac{1}{6}\begin{pmatrix} 4 \\ 7 \\ -1 \end{pmatrix}$

4. When the boat, sail and wind are as in the figure, the forward force **F** experienced by the boat is $\sin(\phi)$ times the force **f** experienced by the sail. The force experienced by the sail, perpendicular to the sail, has magnitude $|\mathbf{f}| = |\mathbf{w}|sin(\theta - \phi)$.

2.5

1. a. $f(x) = -\frac{11}{5} - \frac{11}{5}x$ c. $f(x) = \frac{1}{5} - \frac{11}{20}x$

2. a. $f(x) = \frac{16}{5} + \frac{25}{12}x - \frac{1}{12}x^3$, c. $f(x) = -\frac{30}{53} - \frac{83}{106}x + \frac{1}{53}x^2 + \frac{9}{106}x^3$

4. Replace A with WA where W is a diagonal "weight matrix" having entry $w_{ii} > 0$ proportional to the confidence in the i-th observation.

2.6

1. $\begin{pmatrix} 1 & 0 & 2 \\ 0 & 1 & 3 \\ 0 & 0 & 1 \end{pmatrix}$ (4,2) 3. $\begin{pmatrix} \sqrt{3}/2 & -1/2 & -(\sqrt{3}+1)/2 \\ 1/2 & \sqrt{3}/2 & (-1+\sqrt{3})/2 \\ 0 & 0 & 1 \end{pmatrix}$ $(-(3\sqrt{3}+4)/2, -3/2+2\sqrt{3})$

4. a. $\begin{pmatrix} 4/5 & -3/5 & 1/5 \\ 3/5 & 4/5 & -18/5 \\ 0 & 0 & 1 \end{pmatrix}$ c. $\frac{1}{5}\begin{pmatrix} -4 & -3 & -1 \\ 3 & -4 & 7 \\ 0 & 0 & 5 \end{pmatrix}$

5. A translation doesn't even change the vector represented by $\overrightarrow{PQ}$ (see page 38) so it has no effect on its length. Let T_R be a rotation transition matrix. Then the length of $\mathbf{v} = T_r(\overrightarrow{PQ})$ is the square root of $\mathbf{v}^T\mathbf{v}$ (see page 39). Work it out.

2.7

1. a. (1 2 4 8 16) c. $\begin{pmatrix} 1 & 0 & 0 & -1 & 1 \\ 0 & 1 & 0 & -1 & 0 \\ 0 & 0 & 1 & 0 & 0 \end{pmatrix}$

2.
$$\left(\begin{array}{cc} 1 & 1 \\ 1 & -1 \\ & 1 \; w^2 \\ & 1 \; w^4 \\ & \quad 1 \; w \\ & \quad 1 \; w^5 \end{array} \right)$$
$$\left(\begin{array}{cccccc} 1 & 0 & 1 & 0 & 1 & 0 \\ 0 & 1 & 0 & 1 & 0 & 1 \\ 1 & 0 & -1 & 1 & 0 & -1 \\ 0 & 1 & -1 & 0 & 1 & -1 \\ 1 & 0 & -1 & -1 & 0 & 1 \\ 0 & 1 & 1 & 0 & -1 & -1 \end{array} \right)$$

4. a. $(x^2 + x + 1) - x(x + 1) = 1$

4. c. $\frac{1}{42} \left((4x + 5)(2x^3 - 7x^2 + 7x - 2) - (4x - 11)(2x^3 + x^2 + x - 1) \right) = x - \frac{1}{2}$

6. b. As in Example 2.19, the associated matrix can be computed one column at a time.

3.1

1. a. **r** and **s** are eigenvectors each having $\lambda = 6$ as eigenvalues.

   c. $\mathbf{w}_2$ is an eigenvector having eigenvalue $\lambda = 3$.
   $\mathbf{w}_3$ and $\mathbf{w}_4$ are eigenvectors each having eigenvalue $\lambda = 0$.

2. a. λ_2 and λ_3

3. a. $\begin{pmatrix} 1 \\ 1 \end{pmatrix}, \begin{pmatrix} 2 \\ 1 \end{pmatrix}$

   c. $\begin{pmatrix} \frac{1}{2} \\ \frac{-1}{2} \\ 1 \end{pmatrix}, \begin{pmatrix} \frac{-1}{2} \\ \frac{1}{2} \\ 1 \end{pmatrix}, \begin{pmatrix} 1 \\ 1 \\ 0 \end{pmatrix}$

5. Since A is symmetric, $A = A^T$ and so $\mu(\mathbf{v}^T \cdot \mathbf{w}) = \mathbf{v}^T \cdot (\mu\mathbf{w}) = \mathbf{v}^T \cdot (A\mathbf{w}) = \mathbf{v}^T A^T \mathbf{w} = (\mathbf{w}^T A\mathbf{v})^T = \ldots$

6. a. $P = \begin{pmatrix} 2 & -1 \\ 1 & 1 \end{pmatrix}; \quad D = \begin{pmatrix} 1 & 0 \\ 0 & 4 \end{pmatrix}$

7. $f(\lambda)$, see pages 143 and 144.

9. Let $\mathbf{z} = a\mathbf{u} + b\mathbf{w}$ and $a, b \neq 0$ and suppose $\mathbf{z}$ is an A eigenvector with eigenvalue ν. Then $\nu\mathbf{z} = A(\mathbf{z}) = A(a\mathbf{u}+b\mathbf{w}) = \lambda a\mathbf{u}+\mu b\mathbf{w} = \lambda(a\mathbf{u}+b\mathbf{w})+(\mu-\lambda)b\mathbf{w} = \lambda\mathbf{z} + (\mu - \lambda)b\mathbf{w}$. It follows that $(\nu - \lambda)\mathbf{z} = (\mu - \lambda)b\mathbf{w}$. If this is the zero vector then $\nu = \lambda = \mu$. Otherwise $\mathbf{z}$ is a scalar multiple of $\mathbf{w}$ and so they must have the same eigenvalue $\nu = \mu$. A symmetrical argument forces $\nu = \lambda$.

3.2

1. a. $f(x) = x - 4 \quad f(x) = x^2 - 4x$
   c. $f(x) = x^2 - 4x + 7$
   $f(x) = x^3 - 5x^2 + 11x - 7$

   b. $f(x) = x^2 + 1 \qquad f(x) = x - i$
   e. $f(x) = x^2 + 6x$
   $f(x) = x^3 + 2x^2 - 24x$

2. a. $f_1(x) = x - 4 \quad f_2(x) = x$
   c. $f_1(x) = x^2 - 4x + 7$
   $f_2(x) = x - 1$

   b. $f_1(x) = x^2 + 1$
   e. $f_1(x) = x^2 + 6x$
   $f_2(x) = x - 4$

3. a. $y(x) = x^2 - 5x + 4$ c. $y(x) = x^2 - 5x + 6$ e. $y(x) = x^3 - x^2 - 4x + 4$

5. Divide $f(x)$ by $\mu_A(x)$ and let $r(x)$ be the remainder. Show that $r(A) = 0$, so the definition of $\mu_A(x)$ forces $r(x) = 0$.

7. See Example 3.12

3.3

1.
$$\begin{pmatrix} 0 & 1 & 0 & 0 & 0 \\ 0 & 0 & 1 & 1 & 0 \\ 0 & 0 & 0 & 1 & 0 \\ 0 & 0 & 0 & 0 & 1 \\ 1 & 0 & 1 & 0 & 0 \end{pmatrix}$$
$w_k = w_{k-3} + w_{k-4} + w_{k-5}$
$w_i = 0, 0, 3, 4, 5, 3, 7;$ for i from to 7

3. a.
$$\begin{pmatrix} 0 & 1 & 0 & 0 \\ 0 & 0 & 1 & 1 \\ 1 & 1 & 0 & 0 \\ 0 & 0 & 1 & 0 \end{pmatrix},$$
$b_k = b_{k-2}$
$+2b_{k-3} + b_{k-4}$
$b_2 = 4, b_3 = 6,$
$b_4 = 10, b_5 = 16$

b.
$$\begin{pmatrix} 1 & 1 & 0 & 0 \\ 0 & 0 & 1 & 0 \\ 1 & 1 & 0 & 0 \\ 0 & 0 & 0 & 1 \end{pmatrix},$$
$b_k = 2b_{k-1} - b_{k-3}$
$b_2 = 4, b_3 = 6,$
$b_4 = 9, b_5 = 14$

4. Suppose the formula holds for $n-1$. Then use recursion and the equality

$$\left(\frac{n(n+1)}{2}\right)^2 - \left(\frac{(n-1)n}{2}\right)^2 = \left(\frac{n(n+1)}{2} + \frac{(n-1)n}{2}\right)\left(\frac{n(n+1)}{2} - \frac{(n-1)n}{2}\right) = n^3$$

3.4

1. a. $y = x^2 + x - 6$;
$\lambda_1 = -3, \quad \lambda_2 = 2;$
$P = \begin{pmatrix} \frac{-1}{4} & 1 \\ 1 & 1 \end{pmatrix}; D = \begin{pmatrix} -3 & 0 \\ 0 & 2 \end{pmatrix}$

c. $y(x) = x^3 - 4x^2 - 4x + 16$
$\lambda_1 = -2, \quad \lambda_2 = 2, \quad \lambda_3 = 4;$
$P = \begin{pmatrix} \frac{-1}{3} & \frac{-3}{2} & 1 \\ \frac{-1}{3} & \frac{-1}{2} & 1 \\ 1 & 1 & 0 \end{pmatrix}; D = \begin{pmatrix} -2 & 0 & 0 \\ 0 & 2 & 0 \\ 0 & 0 & 4 \end{pmatrix}$

e. $y(x) = x^2 - 2x - 3$
$\lambda_1 = -1$ (mutiplicity 2) $\quad \lambda_2 = 3;$
$P = \begin{pmatrix} -1 & 1 & 1/5 \\ 1 & 0 & 2/5 \\ 0 & 1 & 1 \end{pmatrix}; D = \begin{pmatrix} -1 & 0 & 0 \\ 0 & -1 & 0 \\ 0 & 0 & 3 \end{pmatrix}$

2. b. $A_{-8} = \begin{pmatrix} -2 & 2 & 4 \\ 2 & -2 & -4 \\ 2 & -2 & -4 \end{pmatrix}, A_{-4} = \begin{pmatrix} -1 & 1 & -2 \\ 1 & -1 & 2 \\ -1 & 1 & -2 \end{pmatrix}, A_0 = \begin{pmatrix} 0 & 0 & 0 \\ 0 & 0 & 0 \\ 0 & 0 & 0 \end{pmatrix}$

4. $P^{-1}AP = D$ is symmetric, so $D = D^T = (P^{-1}AP)^T = P^T A^T (P^{-1})^T$.

6. Every row of $M_j R$ is a scalar multiple of its i-th row, so its $RREF$ has only one non-zero row, hence its rank is 1. M and $M_j R/m_{i,j}$ have exactly the same j-th column so $M - M_j R/m_{i,j}$ has j-th column zero. Imagine the RREF process for M and apply it to $M - M_j R/m_{i,j}$. The only difference between the results is that the j-th column is pivotal for one and zero for the other.

3.5

1. a. $\mu_A(x) = x^2 - 4x = (x - 4)x \quad a = 4$

2. b. $P = \begin{pmatrix} 1 & 2 \\ 2 & 5 \end{pmatrix} \quad A^n = \begin{pmatrix} 5(-3)^n - 4 & -2(-3)^n + 2 \\ 10(-3)^n - 10 & -4(-3)^n + 5 \end{pmatrix}$

3. a. $0.5, 1 \quad a = 1$
c. $\mathbb{C}$ abs value $\doteq 0.343$
real $\doteq 0.847, a \doteq .847$
e. $\mathbb{C}$ abs value 2
real -1, $\quad a = 2$

4. a. ± 1 oscillate   5.   a. $\lambda = 3$ $\mathbf{v} = \doteq (2.121\ 2.121)^T$

  c. $\doteq 1.048, \doteq -.373 \pm .208i$, diverge

6. a. $A_\lambda = \begin{pmatrix} 8 & -4 \\ 8 & -4 \end{pmatrix}$ $A_2 = \begin{pmatrix} 4 & -4 \\ 8 & -8 \end{pmatrix}$

3.6

1. $\mu(x) = x^2 - 1.5x + 0.5$ $a = 1$,   Steady state, $(25\%, 50\%, 25\%)^T$

3. $\mu(x) = x^2 - 1.3x + .6$ $a = 1$ Steady state $(50\%, 25\%, 25\%)^T$

5. $\mu(x) = x^2 - 0.8125x - .1875$ $a = 1$ Steady state

6. $\mu(x) = x^3 - 1.55x - 0.3$ three real zeros $a \doteq 1.33$ Diverges

8. $\mu(x) = x^3 - 0.2x^2 - 0.43x - 0.1$ one real zero $a \doteq 0.847$ Decays

3.7

1. $Ms = \begin{pmatrix} 0.08 & -0.065 & 0.21 & 0.40 & 0.21 & -0.854 \\ 0.20 & -0.42 & 0.52 & -0.54 & -0.19 & 0.432 \\ -0.147 & -0.50 & -0.00774 & -0.24 & 0.085 & 0.81 \end{pmatrix}$

Eigenvalues of C:
2.255070923
.5356991686
.209229

$C = \begin{pmatrix} 1 & -0.475 & -0.75 \\ -0.475 & 1 & 0.64 \\ -0.75 & 0.64 & 1 \end{pmatrix}$ $C_\lambda = \begin{pmatrix} 0.75 & -0.69 & -0.80 \\ -0.69 & 0.645 & 0.75 \\ -0.8 & 0.75 & 0.86 \end{pmatrix}$ $\begin{pmatrix} 0.33 \\ -0.31 \\ -0.36 \end{pmatrix}$

2. $Ms = \begin{pmatrix} 0.14 & -0.85 & 0.23 & -0.085 & 0.42 & 0.14 \\ -0.40 & 0.23 & 0.30 & 0.325 & -0.72 & 0.26 \\ -0.26 & 0.26 & 0.18 & 0.55 & -0.72 & -0.0071 \end{pmatrix}$

Eigenvalues of C:
2.333281080
.5933454162
.07337350312

$C = \begin{pmatrix} 1 & -0.48 & -0.57 \\ -0.48 & 1 & 0.92 \\ -0.57 & 0.92 & 1 \end{pmatrix}$ $C_\lambda = \begin{pmatrix} 0.55 & -0.69 & -0.71 \\ -0.69 & 0.866 & 0.89 \\ -0.71 & 0.89 & 0.92 \end{pmatrix}$ $\begin{pmatrix} 0.282 \\ -0.353 \\ -0.364 \end{pmatrix}$

4.1

1. a. There are other possible solutions $A = E_1^{-1} E_2^{-1} E_3^{-1} I_2$

$E_1 = \begin{pmatrix} 1 & 0 \\ -1 & 1 \end{pmatrix}$, $E_2 = \begin{pmatrix} 1 & 0 \\ 0 & 1/2 \end{pmatrix}$, $E_3 = \begin{pmatrix} 1 & 1 \\ 0 & 1 \end{pmatrix}$

2. b. There are other possible solutions $A = E_1^{-1} E_2^{-1} E_3^{-1} E_4^{-1} \begin{pmatrix} 1 & 0 & 3 & 0 \\ 0 & 1 & 2 & 0 \\ 0 & 0 & 0 & 1 \end{pmatrix}$

$E_1 = \begin{pmatrix} 1 & 0 & 0 \\ -2 & 1 & 0 \\ 0 & 0 & 1 \end{pmatrix}$, $E_2 = \begin{pmatrix} 1 & -1 & 0 \\ 0 & 1 & 0 \\ 0 & 0 & 1 \end{pmatrix}$, $E_3 = \begin{pmatrix} 1 & 0 & 0 \\ 0 & 1 & 0 \\ 0 & -1 & 1 \end{pmatrix}$, $E_4 = \begin{pmatrix} 1 & 0 & -3 \\ 0 & 1 & 0 \\ 0 & 0 & 1 \end{pmatrix}$,

3. a. det(A)=2   4.   a. 2

5. a. $\begin{pmatrix} 1 & -1 \\ -1 & -1 \end{pmatrix}$ det(A)= -2

4.2 _____

1. a. 6   c. 0   2. a. 21   c. -180 e. -128
3. a. 21   c. 0   4. 1234   2134   2143   2134

4.3 _____

1. a. 6   c. -960 e. 44   2. a. 35

3. By the fourth row cofactor expansion, $f(x,y) = A + Bx + Cx^2 + Dy$ for some numbers A, B, C, D depending on P, Q, R. The coefficient D is a Vandermonde determinant with distinct values of x and so $D \neq 0$. Therefore the curve $f(x,y) = 0$ is the same as $y = \frac{-1}{D}\left(A + Bx + Cx^2\right)$, a parabola or a straight line if $C = 0$.

   This curve passes through the given data points because when (x,y) equals a data point, the matrix has two rows identical, and its determinant is zero.

   By Section 1.7, three distinct data points uniquely determines such a polynomial.

5. An elementary row, respectively column operation on M is achieved by multiplying M on the left, respectively right, by the associated elementary matrix. Thus, the problem translates to $(E_1 M)E_2 = E_1(M E_2)$, the associative law of matrix multiplication.

6. a. $\begin{pmatrix} 5 & 0 \\ 0 & 1 \end{pmatrix} = \begin{pmatrix} 1 & 0 \\ -2 & 1 \end{pmatrix}\begin{pmatrix} 0 & 1 \\ 1 & 0 \end{pmatrix}\begin{pmatrix} 2 & -1 \\ 1 & 2 \end{pmatrix}\begin{pmatrix} 1 & -2 \\ 0 & 1 \end{pmatrix}\begin{pmatrix} 1 & 0 \\ 0 & -1 \end{pmatrix}$

4.4 _____

1. a. $(x + 2)(x + 1)(x - 1)^2$ c. $x^4 - 9x^3 + 2x^2 + 12x - 6$

2. a. $p(x) = x^3 + 2x^2 - 15x$ c. $p(x) = x^4 + 5x^3$

3. a. $\left(\frac{45}{13} \quad \frac{-11}{13} \quad \frac{23}{13} \right)^T$

Index